Catalytic Technology for Selective Hydrogenation of Benzene to Cyclohexene

Zhongyi Liu · Shouchang Liu · Zhongjun Li

Catalytic Technology for Selective Hydrogenation of Benzene to Cyclohexene

Science Press
Beijing

Springer

Zhongyi Liu
Zhengzhou University
Zhengzhou, China

Shouchang Liu
Zhengzhou University
Zhengzhou, China

Zhongjun Li
Zhengzhou University
Zhengzhou, China

Translated by
Zhongyi Liu
Zhengzhou, China

Wang Erqiang
Zhengzhou, China

ISBN 978-981-15-6410-9 ISBN 978-981-15-6411-6 (eBook)
https://doi.org/10.1007/978-981-15-6411-6

Jointly published with Science Press
The print edition is not for sale in China (Mainland). Customers from China (Mainland) please order the print book from: Science Press.

This Springer imprint is published by the registered company Springer Nature Singapore Pte Ltd.
The registered company address is: 152 Beach Road, #21-01/04 Gateway East, Singapore 189721, Singapore

Preface

Since 1990, China has carried out research on benzene selective hydrogenation catalyst and related basic theories and applications, and has resolved the key technology problems for benzene selective hydrogenation to cyclohexene in China. After 2010, China became the second country in the world to realize the industrialization of the catalytic technology for benzene selective hydrogenation to cyclohexene and held the proprietary intellectual property rights.

This book introduces the catalytic technologies of benzene selective hydrogenation to cyclohexene and its downstream products, the developing history and present status, and the innovative work conducted in China. Selective hydrogenation of benzene to cyclohexene is a complex heterogeneous catalytic system. It involves catalyst, catalytic process, key equipment, and a complete set of production facilities from laboratory to industrialization. Summarizing the research work systematically in the field of study and exploring the heterogeneous catalytic mechanism and the scientific essence of high yield and selectivity of cyclohexene are not only academically valuable, but also can promote technical progress of this research area.

This book reveals the objective laws, reaction mechanism, and scientific essence based on the experimental data, analysis, and characterization results and references, which will guide industrial production. The improved catalyst can enhance the selectivity and yield of cyclohexene with low resource consumption and environmental pollution. It is also valuable for peer specialists and technical workers.

The authors' division for this book is as follows: Liu Zhongyi is the Editor-in-Chief for Chaps. 1, 2, 7, and 10; Liu Shouchang is the Associate Editor for Chaps. 3–6; Li Zhongjun is for Chaps. 8 and 9. Liu Zhongyi and Liu Shouchang are also responsible for the ordination and review of the whole book.

This book is supported by the fund from the Ministry of Science and Technology and the National Natural Science Foundation. We also acknowledge the contribution of graduate students, enterprises, and related technical workers.

There may exist some mistakes due to our limited level and please criticize and point out the defects. We will be deeply grateful!

Zhengzhou, China
October 2016

Zhongyi Liu

Brief Introduction

Selective hydrogenation of benzene to prepare cyclohexene and its downstream products such as cyclohexanone, adipic acid, caprolactam, nylon 6, nylon 66, and bulk chemicals and high-value fine chemicals such as medicine, pesticides, essence, and spices are resource saving and environment friendly, and play an important role in national economic development. The main contents of this book involve the thermodynamics of benzene selective hydrogenation, heterogeneous catalytic kinetics, catalytic mechanism and scientific essence, catalyst of benzene selective hydrogenation, modulation on activity and selectivity of the catalyst, studies on deactivation and regeneration of the catalyst as well as the catalytic process, key equipment, and complete set of production facilities.

This book gives a lot of experimental data, analysis, and characterization maps and important references. It can provide the reference for the related professionals of advanced college and scientific research institutes, senior undergraduate, master's and doctoral graduate students as well as for the personnel of related enterprises where they engage in management or industrial production.

Contents

Chapter 1
An Overview of the Catalytic Selective Hydrogenation Technologies of Benzene into Cyclohexene

Selective hydrogenation of benzene from the fossil and coal companies into cyclohexene and its downstream products including bulk chemicals of cyclohexanone, adipic acid, hexanolactam, nylon 6, nylon 66, and high value fine chemicals of medicine, pesticide, and perfumes are of great significance for the economic development. Compared with the traditional process of benzene, hydrogenation into cyclohexane is safe, resource saving, environmentally friendly, and has giant economic and social benefits for the selective hydrogenation of benzene into cyclohexene.

It has been a dream for a long time in history that selective hydrogenation of benzene into cyclohexene was considered very hard, especially in the field of industry. Many countries such as America, England, German, and Japan conducted universal researches from 1960 to 1990s. And Japan firstly realized the industrialization in 1989, and transfer the technology to China, in 1995 and 2005, respectively. However, they monopolized the catalysts preparation method.

After 1990s, Netherland, German, Italy, India, Sweden, Span, Brazil, and China all continued the research work. In 2010, China, industrialized the whole technologies and broke out the abroad monopoly, becoming the second country in industrializing this technology and hold the proprietary intellectual property rights.

This chapter introduces selective hydrogenation of benzene into cyclohexene and its downstream products, foreign developing history and status in selective hydrogenation technologies of benzene, domestic research and progress on selective hydrogenation technologies of benzene, main technology index on selective hydrogenation of benzene over the world.

Z. Liu et al., *Catalytic Technology for Selective Hydrogenation of Benzene to Cyclohexene*,
https://doi.org/10.1007/978-981-15-6411-6_1

1.1 Selective Hydrogenation of Benzene into Cyclohexene and Its Downstream Products

The technology of benzene selective hydrogenation into cyclohexene and its downstream products could be abbreviated as benzene selective hydrogenation technologies, including catalysts, catalytic processes of benzene selective hydrogenation into cyclohexene and its downstream products, key equipment, and devices. The stock of benzene from fossil and coal companies was catalytic selective hydrogenated into cyclohexene and its downstream products depending on the process, key equipment, and devices. Selective hydrogenation of benzene, namely partial hydrogenation or incomplete hydrogenation, is to hydrogenate the two double bonds in the ring to produce cyclohexene. The downstream chemicals of cyclohexene are involved in a large product chain because of the active double bond. Cyclohexene could be hydrated into cyclohexanol, which could be dehydrated into cyclohexanone, and oxidized into adipic acid directly. The hexanolactam obtained through the ammoximation of cyclohexanone, and rearrangement could be depolymerized into nylon 6. The adipic acid could also be used to prepare nylon 66 and polyesters through the co-polymerization with hexamethylenediamine and dihydric alcohols, respectively. Cyclohexene has been universally used in the field of high value fine chemicals production, such as medicine, pesticide, and perfume. Such compounds mentioned above have a wide market prospect.

From Fig. 1.1, the selective hydrogenation of benzene into cyclohexene is mainly to produce adipic acid (**1**), cyclohexanone (**2**), hexanolactam (**3**), nylon 6 (**4**), nylon 66 (**5**), and such staple chemicals.

It could be seen from Fig. 1.2: Cyclohexene could also be oxidized into cyclohexene oxide (**1**), which is the intermediate of medicines, pesticides, and polymers (**2**). Cyclohexene could be used to prepare the monomer of nylon 1212 cyclododecadiene (**3**) through metathesis, followed by oxidation. Phenylcyclohexane (**5**) could be formed from cyclohexene and benzene, and oxidized into phenylcyclohexane hydroperoxide (**6**), which could also be decomposed into phenol (**7**) and cyclohexanone (**8**). In addition, cyclohexene could be converted into cyclohexadiene (**10**)

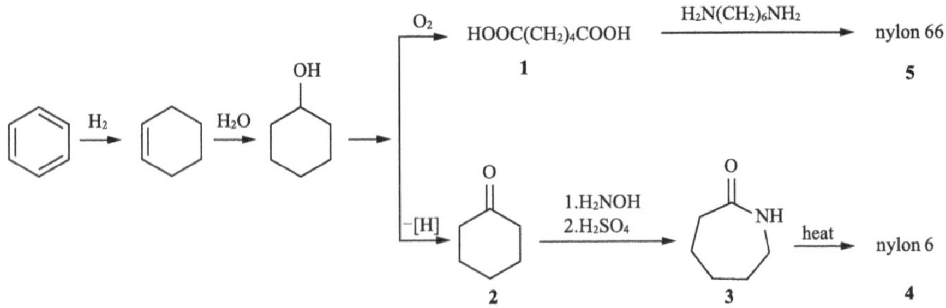

Fig. 1.1 Applied fields of benzene selective hydrogenation into cyclohexene and its downstream products in China

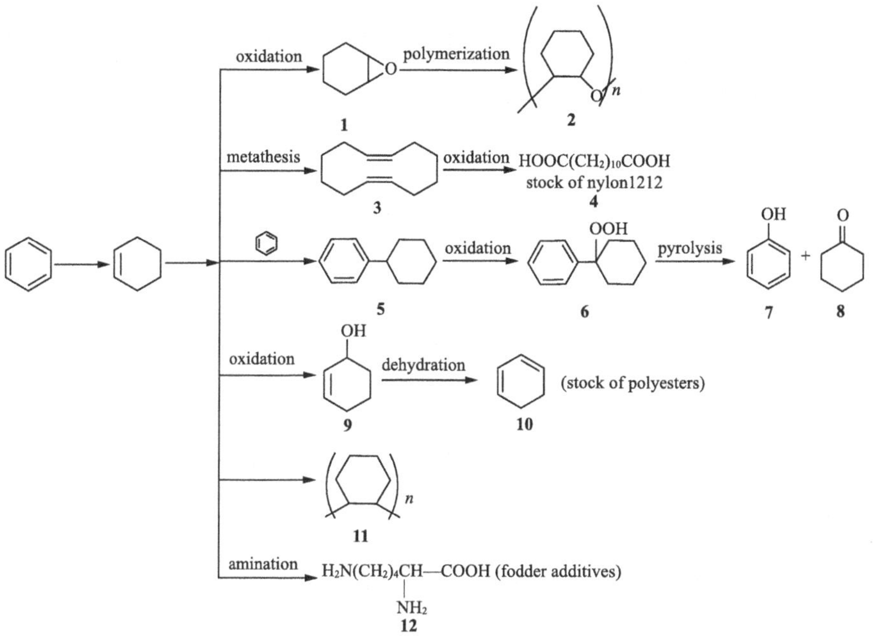

Fig. 1.2 Applied fields of cyclohexene and its downstream products in abroad [1]

through oxidation to cyclohexylene (**9**), followed by dehydration. Otherwise, cyclohexene could also be used to prepare α-amino type additive of L-lysine (**12**), perfume of civetone, rubber anti-aging agent of diphenol cyclohexane and fruit antiseptic [2]. These products development are almost vacancy in China, anyone exploited and launched successfully would be the technology and product support of Chinese middle and small-sized enterprises. It would be the new economic growth point.

Approaches to cyclohexene contained cyclohexanol dehydration, halocyclohexane dehalogenation, and Birch reduction in previous time. However, these methods had a complicated process, high energy consumption, low yield, and high cost, so the cyclohexene was limited in production and narrowed in fabricating high value and fine chemicals. It was reported that the Chinese annual demand of cyclohexene in 2010, was more than 700,000 tons, while the production was less than 30,000 tons, which limited the development and utilization of downstream stocks [3]. The successful development of benzene selective hydrogenation into cyclohexene broke out the technical bottleneck and build the base for the downstream products.

Selective hydrogenation of benzene into cyclohexene could be dated back to 100 years ago. In 1901, it was well known that benzene could be selectively hydrogenated into cyclohexene, but it was easier into cyclohexane due to the relative stability of benzene. Thus, it had still been a dream for the selective hydrogenation of benzene into cyclohexene [4]. Up to 1972, this process was made giant progress by Drinkard from Dupont company. The yield of cyclohexene could up to 30%

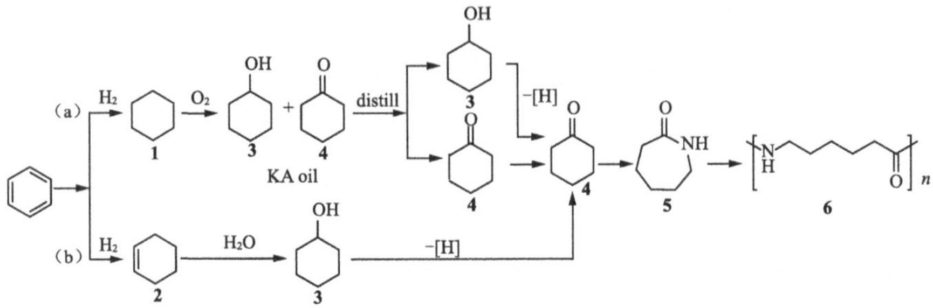

Fig. 1.3 Comparison of benzene total hydrogenation with selective hydrogenation into cyclohexene

with the catalysis of Ru catalyst, alkaline solution, and additives under 177 °C and 7.0 MPa. Then a large number of patents were applied by other chemical companies. The system of benzene selective hydrogenation was described in detail by many researchers especially Odenbrand. In 1970s, the Japanese companies including Mitsubishi, Toray, Sumitomo, and Mitsui prepared series of Ru-supported catalysts, the yield, and selectivity of cyclohexene were enhanced. The technical of benzene selective hydrogenation system was innovated by Asahi Chemical Industry Co., Ltd., and it was industrialized in the end of 1980s. In 2010, China, industrialized the whole technologies and broke out the abroad monopoly, becoming the second country in industrializing this technology and hold the proprietary intellectual property rights.

Figure 1.3 shows the comparison of benzene total hydrogenation with selective hydrogenation into cyclohexene.

In traditional benzene hydrogenation process (a), benzene is fully hydrogenated into cyclohexane (**1**) on Raney nickel catalyst, then cyclohexane is oxidized to cyclohexanol (**3**) and cyclohexanone (**4**) mixtures (KA oil). Cyclohexanol (**3**) could be obtained through the decomposition of KA oil under alkaline conditions, followed by waste base separation and distillation. Cyclohexanone (**4**) could also be obtained through the dehydrogenation of cyclohexanol. And KA oil could be oxidized into adipic acid with nitric acid. But the oxidation of cyclohexane belongs to the radical reaction, there exists safety hazards although this reaction could be controlled by inhibitor. The single-pass conversion of cyclohexane was controlled as 3–5%, because there are serious accidents happening occasionally in the world. This route has the disadvantages of long procedure, multistage, high energy consumption, and low yield. In addition, a large amount of wastewater and gas would be produced during manufacture, the utilization of carbon atom is around 85%, which could lead to resource waste and environmental pollution.

In benzene selective hydrogenation route (b), the selective hydrogenation into cyclohexene avoids the cyclohexane oxidation, which promotes safety production. Otherwise, the by-product of cyclohexane could be fully utilized. The production process of downstream stocks, including cyclohexanone (**4**), caprolactam (**5**), nylon 6 (**6**) and nylon 66 from benzene, would be shortened through the hydration of cyclohexene into cyclohexanol (**3**) and dehydration of cyclohexanol into cyclohexanone (**4**). The utilization of carbon atom of this new route is 100%, and the hydrogen consumption could be saved about 1/3. Thus, this is a giant technical innovation to traditional benzene fully-hydrogenation.

By comparison in Table 1.1, the utilization efficiency of benzene and water of traditional benzene fully-hydrogenation is 76.5% and 75.3%, while the selective hydrogenation is 100% and 96.2%, respectively. The consumption of hydrogen, water, and electricity, and the production of wastewater, waste gas, and waste solid could be decreased by about 32%, 41%, 50%, 85%, 86%, and 55% , respectively. The cost of preparing one ton products could be reduced by about 1,000–1,500 RMB.

Take the Chinese universally applied benzene (100,000 t/a) selective hydrogenation into cyclohexene as an example, the yearly consumption of benzene, water, and electricity could be saved about 8,300 t, 1,030,000 t, 2.82×10^7 kW·h. The wastewater, gas, and solid could be decreased by about 253,000 t, 1.01×10^9 Nm^3 and 24,000 t, respectively. And another 21,400 t cyclohexane could be produced.

Thus, the generalization and application of benzene selective hydrogenation technologies have not only economic benefit, but also social benefit in safety production, natural resource and ecological environment protection, pollution release and working condition improvement.

1.2 Foreign Developing History and Status in Catalytic Selective Hydrogenation Technologies of Benzene

In 1934, Horiuti and Polanyi developed the stepwise hydrogenation mechanism, benzene was hydrogenated into cyclohexane via the intermediates of cyclohexadiene and cyclohexene [5].

In 1957, Anderson firstly adopted a nickel membrane to catalyze the hydrogenation of benzene and detected the cyclohexene [6]. Cyclohexadiene is very active and would transform into cyclohexene as soon as it is produced, so researchers can't detect cyclohexadiene in hydrogenation products. But it could be detected in the products of benzene dehydrogenation products. In the following days, researchers reported the existence of cyclohexene under the low conversion of benzene catalyzed by Ru catalysts. And the Horiuti–Polanyi mechanism was generally accepted by researchers. Thus, it stimulated the interests of business circles and the academic community for the possibility of benzene selective hydrogenation into cyclohexene, though the selectivity and yield was not high.

In 1963, Hartog utilized ruthenium black catalysts to catalyze the hydrogenation of benzene in the water solution of aliphatic alcohol (50 wt.%) under room temperature and atmosphere, the yield of cyclohexene was only 2%. The selective hydrogenation of benzene into cyclohexene was beginning to made progress [7–9].

In 1972, the selective hydrogenation of benzene into cyclohexene was achieved breaking progress by Drinkard from Dupont company. The yield of cyclohexene could reach up to 30% with the catalysis of $RuCl_3$ under 450 K, 7 MPa H_2, and the

Table 1.1 Comparison of two processes for prepare cyclohexanone (60,000 t/a)

Entry	Class (unit)		Traditional fully hydrogenation			Selective hydrogenation		
			Consumption/t	Time/h	Yearly/8000 h	Consumption/t	Time/h	Yearly/8000 h
1	Stock	Benzene /t	1.02	7.65	61,200	0.962	7.215	57,720
2		H_2/Nm^3	1,000	7,500	6×10^7	680	5,100	4.08×10^7
3	Product	$C_6H_{10}O$/t	0.98	7.5	6×10^4	1.005	7.5	6×10^4
4		C_6H_{12}/t	—	—	—	0.215	1.6	1.28×10^4
5	Energy consumption	Water/t	25.2	189	1.51×10^6	14.9	112	8.96×10^5
6		Electric /(kW·h)	562	4,215	3.37×10^7	280	2,100	1.68×10^7
7		Steam/t	4.2	31.5	2.52×10^5	5.8	43.5	3.48×10^5
8	Three wastes	Water/t	2.96	22.2	1.78×10^5	0.43	3.24	2.59×10^5
9		Gas/Nm^3	11,783	88,370	7.07×10^8	1,680	12,600	1.01×10^8
10		Solid/kg	0.44	3.31	26,480	0.2	1.5	12,000
11	Water utilization efficiency/%		75.3			96.2		

reaction systems including benzene, basic aqueous solution, and additives of $ZnCl_2$, $TiCl_4$, and carbonyl cobalt [10, 11].

In 1974 and 1975, the selective hydrogenation of benzene into cyclohexene with the catalysis of Ru-supported catalysts in weak acidic water solution was conducted by Johnson and Nowack, from Philips petroleum company. It was considered that the ions of iron, nickel, and chromium that were corroded from the surface of the steel reactor could promote the selectivity [12–14].

The yield and selectivity of cyclohexene were both enhanced by Mitsubishi Chem. Ind., Toray Ind., Sumitomo Chem. Co., and Mitsui Petrochem. Ind., through the support studies ranging from general Al_2O_3, SiO_2 to other oxides, hydroxides, zeolites, rare earth oxides, and insoluble salts in combination with salt additives during 1975 and 1977 [15–23].

From 1980 to 1983, the selective hydrogenation of benzene into cyclohexene on Ru catalysts in steel autoclave was researched by Odenbrand. The catalysts, reaction systems, and mechanisms were described well. The Ru particles with the diameter range of 3–30 μm could be obtained through in-situ reduction of Ru hydroxides (the sodium hydroxide was added into $RuCl_3$ solution). The Ru catalyst and hydrogen were distributed in the emulsion of benzene and water. Ru particles, benzene, and hydrogen were, respectively, distributed in the water phase, oil drop with the diameter range of 0.05–0.12 mm and bubble with the diameter range of 0.1–0.8 mm. The limited mass transfer of hydrogen was one of the reasons for the high yield of cyclohexene. Two independent routes were assumed for benzene hydrogenation: Step-by-step hydrogenation to produce cyclohexane via cyclohexene, and one-step direct hydrogenation to produce cyclohexane. These metals which could partly poison Ru could block the active center which was helpful for one-step direct hydrogenation of benzene, thus cyclohexene selectivity is increased. They also thought the drop of ions of Fe, Cr, and Ni from the corroded inner wall of the equipment was one of the reasons for high cyclohexene selectivity [24–27].

From 1984 to 1987, the Ru catalyst without using additives was prepared by Niwa through "chemical mixture method", and the catalytic performance could be matched with the Ru-supported catalysts in cooperation with additives. The chemical mixture method is also called sol–gel method, the complexes of $RuCl_3$ with ethanol and tetraethoxysilane (or aluminum tri-sec-butoxide) were used to prepare the Ru/SiO_2 (or Ru/Al_2O_3) catalyst. The diameter of Ru microcrystalline was smaller than 20 Å, the specific area could reach up to 850 m^2/g, and these catalysts also had a micropore structure ($d_{pore} < 2$ nm). It was considered the Ru element in the catalyst was mixed with Si (or Al) in the status of the molecule. The high distribution of Ru and the strong interaction between Ru and support were the reason for the high yield of cyclohexene [28–30].

It has been considered that the selective hydrogenation of benzene into cyclohexene is very difficult for industrialization. From 1985 to 1990, the Asahi Kasei Company developed the catalytic technology for high selective hydrogenation of benzene. The company also had several technological innovations including preparation of special Ru particles (the diameter of microcrystalline was less than 20 nm) contained catalysts with *ex-situ* method, the introduction of zinc salts ($ZnSO_4 \cdot 7H_2O$)

to control the pH between 4 and 6.5, the addition of distribution agent to avoid the coalescence and adhesion of catalyst particles. The distribution agent could be the oxides, hydroxide or hydrous oxides of Zr, Hf, Ti, Nb, Cr, Fe, Co, Al, Ga, and Si that with the particles of 0.005–10 μm. The reaction system contained four phases of gas (H_2), oil, water, and solid (catalyst), which was a relatively comprehensive multiphase catalytic system. The reaction occurred in the water phase, the reactant and product transferred in different phases through dissolution, diffusion, and extraction, during which the diffusion process played an important role in enhancing the selectivity of cyclohexene. The zinc complexes could stabilize cyclohexene, avoiding further hydrogenation into cyclohexane. In 1988, the Asahi Kasei Company found a 60,000 t/a device on Mizushima for the production of cyclohexanone from cyclohexene through benzene selective hydrogenation. This device was put into operation in 1989, which indicates the first realization of the production of cyclohexene and its downstream products through benzene selective hydrogenation [31–34].

Before 1990s, the catalytic technical for benzene selective hydrogenation mostly existed in the type of patents, and the key technologies for catalyst preparation were in the status of confidentiality and blockage. After 1990s, many countries carried out such researches and the representatives were Netherland, Germany, Italy, India, Switzerland, Spain, Brazil, and China.

From 1992 to 1993, Struijk et al. conducted the selective hydrogenation of benzene into cyclohexene in the catalytic system of Ru catalysts and $ZnSO_4$ solution under 423 K and 5 MPa H_2. The content of the research including the effect of catalyst preparation and characterization, process parameters such as temperature, pressure, stirring speed, catalyst amount, Ru microcrystalline size, salt additives, pH, and water/benzene ratio on the selectivity and yield of cyclohexene. And the major conclusions were obtained as follows:

The hydrophilicity of catalyst is very important and the diffusion control leads to the high selectivity and yield of cyclohexene. As the hydrophilic catalyst is surrounded by a film of backwater membrane, the hydrogen controlled by such backwater layer has two important performances. One is to decrease the reaction rate of each elemental step of benzene hydrogenation, leading to the enhancement of cyclohexene selectivity. The other is to inhibit the inner diffusion of cyclohexene and the further hydrogenation into cyclohexane. The diffusion of benzene and cyclohexene that was controlled by the backwater layer also has two functions. On one hand, the dissolution difference between benzene and cyclohexene in water (benzene: 125 mol/m^3 and cyclohexene: 21 mol/m^3) and the distribution of benzene and cyclohexene leads to the concentration difference on the hydrophilic surface; on the other hand, as soon as the cyclohexene produces, it is more difficult for adsorption than desorption. In addition, when cyclohexene desorbs and dissolves in the oil phase, the cyclohexane only forms after prolonged collision and contact. Thus, with the formation of backwater and the increase in mass transfer resistance, the relatively high selectivity of cyclohexene could be observed. And this conclusion is also supported through the calculation of Wheeler-Weisz group ($\eta\varphi^2$) and Carberry number (Ca). The better hydrophilicity and thicker backwater film could be contributing to the higher selectivity of cyclohexane [35].

The inorganic or organic additives could make the hydrophobicity of catalyst particles surface transfer into hydrophilicity, leading to the formation of a layer of backwater film around the catalyst. Otherwise, the strongest active centers on the Ru catalyst surface could be selectively covered, which would be a benefit for reducing the adsorption enthalpy of benzene and cyclohexene and enhancing the desorption rate. The salts as additives should have sufficiently strong adsorption on the Ru catalyst surface, but the equilibrium coverage should be suitable, or the catalytic activity would reduce sharply. Moreover, the metal cation is difficult to be reduced under such hydrogenation conditions and would be the toxicant of Ru catalyst. In theory, the coverage percentage of salt should be adjusted to such a degree that the physical process of substance migration is the rate-determining step [36].

The factors of temperature, hydrogen pressure, stirring speed, catalyst amount, Ru microcrystalline size, salt additives, pH, and water/benzene volume ratio have an effect on the yield and selectivity of cyclohexene. For the adsorption of cyclohexene need to be activated, while the hydrogenation of cyclohexene needn't, the appropriate high temperature (150–160 °C) is favorable for promoting the selectivity of cyclohexene. The selective hydrogenation of benzene should proceed under suitable high hydrogen pressure (4–5 MPa). The relatively high stirring speed could eliminate the external diffusion restriction of benzene and hydrogen in the interface of oil–water and gas–water, and accelerate the mass transfer. The mass ratio of catalyst to distribution agent is 1:5, the total percentage of solids in the slurry is 3–5%, or the cyclohexene would be further hydrogenated into cyclohexane. The Ru microcrystalline, pH, and water/benzene volume ratio should be around 5 nm, 5.5–6.5, and 2–3, respectively [35, 36].

In 1996, the gaseous reaction kinetics of benzene hydrogenated into cyclohexene over Ru/La_2O_3 was conducted by Döbert in the alkaline solution containing $ZnCl_2$. It was considered that the cyclohexane only produces during the continuous hydrogenation of benzene, the conversion rate of benzene and cyclohexene were, respectively, independent on the partial pressure of benzene and cyclohexene. So it could be concluded that the adsorption and hydrogenation of benzene and cyclohexene occurred over different active sites [37].

During 1996 and 2001, the conclusions obtained by Milone, Hronec, and Ronchin from benzene selective hydrogenated into cyclohexene were as follows:

Milone et al. adopted the relatively low temperature (333 K) and high-pressure autoclave with teflon lining to release the diffusion and the metal ion pollution from the reactor surface. The effects of reaction condition, $RuCl_3$ precursor, chloridion, and catalyst pretreatment on yield and selectivity of cyclohexene were also investigated over Ru/Al_2O_3 in the presence of water. The diffusion controlling between hydrogen and liquid could be eliminated by increasing the stirring speed. But the calculation results of $\eta\varphi^2$ and Ca indicated the selectivity of cyclohexene was almost independent on hydrogen diffusion. In the range of all benzene conversion, the ignorance ($Ca(H_2) = 0.08$) and non-ignorance ($Ca(H_2) = 0.30$) of hydrogen diffusion to the selectivity of cyclohexene was the same to all catalysts, which was not similar to the Netherland researchers. The benzene hydrogenation occurs under the kinetic controlling and strong mass transfer restriction. Although the mass controlling is beneficial for the cyclohexene selectivity under high benzene conversion, the formation of cyclohexene couldn't be attributed to the substrate transfer. The Ru could be completely reduced under the hydrogen atmosphere, and the surface of the catalyst could be covered with a layer of hydrogen, leading to improving cyclohexene selectivity. However, no effect is observed when the catalyst is pretreated under the hydrogen atmosphere. The precursor of the Ru component with Cl^- is better than that without Cl^- for preparation of the catalyst. Firstly, the residue Cl^- could modify the electronic structure of Ru to form Ru(δ^+) species. Secondly, the existence of Cl^- could take up the strongest active sites of Ru, which is disadvantageous for the one-step hydrogenation of benzene into cyclohexane. Thirdly, the Cl^- anchored on the catalyst surface could bond with water through hydrogen bonding, promoting the interaction strength between water and catalyst surface and the hydrophilicity of catalyst. The function of water has the following two sides. On one hand, water has stronger adsorbability than benzene on the hydrophilic surface, so the water could easily replace the formed cyclohexene through competing adsorption. On the other hand, the water could bond with intermediate cyclohexene through hydrogen bonding, reducing the adsorbability of cyclohexene, thus the cyclohexene could be easily produced on the Ru active site absorbed with water molecule [38].

The nonpolar activated carbon and polar anionic cross-linked polymers was selected by Hronec as support for the preparation of Ru/C and Ru/ACLP catalysts. The effect of support polarity, water, and zinc salt additives on cyclohexene selectivity was conducted in a clave with glass lining under 100–110 °C and 1.5 MPa hydrogen. And the obtained conclusions are as follows: the polar support could promote the selectivity of cyclohexene and the Ru/ACLP is better than Ru/C under the conversion of 42–47%. Water can not only form a suitable chemical environment around Ru particles but also enhance the accessibility of metal particles distributed in hydrophilic polymers [39].

Ronchin et al. used 1 L PTFE-lined high-pressure reactor to study the effect of mass transfer, carrier and alkaline additive on cyclohexene selectivity and yield based on supported and unsupported Ru catalyst, under the reaction conditions of 150 °C, 5 MPa hydrogen pressure and 0.6 mol/L $ZnSO_4$ solution. Main conclusions were that, as to unsupported Ru catalyst, gas–liquid mass transfer resistance was not the control step when stirring rate was above 1000 r/min, and the reaction order of hydrogen was above one. Apparent activation energy was 19 kJ/mol for 130–150 °C, which was so small and meant that the reaction was limited by external diffusion or internal diffusion, but the relation between catalyst selectivity and the value of $\eta\varphi^2$ and Ca of benzene and hydrogen was not found. So, benzene selective hydrogenation was partly controlled by external and internal diffusion of benzene and hydrogen, while catalyst activity and selectivity depended strongly on the preparation procedure, especially on the properties of alkali and alkaline-earth metal hydroxides used for catalyst precursor preparation. Turnover frequency (TOF) of benzene increased with the increasing of molecular weight of hydroxides in the same group. Based on double isothermal line method, they studied the reversible chemical absorption formed between Ru atom exposured on the catalyst surface, and hydrogen molecule poorly bonded on the catalyst surface. The amount of hydrogen on the catalyst surface by poor chemical absorption depended on the property of alkali during precipitation. It was the above amount of hydrogen that affected catalyst activity and selectivity [40, 41]. As to supported Ru catalyst, the reaction order of hydrogen was zero, and the reaction rate was not affected by hydrogen pressure. Catalyst selectivity was mainly affected by the nature of the carrier and the interaction between metal and carrier. The carrier with the best hydrophilicity gave the best results. With the best catalyst of Ru/ZrO_2 and the best precipitant of KOH, cyclohexene selectivity and yield was 80% and 41%, respectively. Strongly chemical absorbed hydrogen was linked to the exposed metal atom, while poorly chemical absorbed hydrogen was linked to highly unsaturated active center, which had relation with initial reaction rate. Catalysts containing TiO_2 and Fe_2O_3 gave very low hydrogen and high selectivity, which may be explained by that strong metal-support interaction reduced the absorption ability of catalyst surface to hydrogen. For the same group of alkali used for catalyst preparation, the initial reaction rate increased with the increasing of amount of reversible absorbed hydrogen. The $\eta\varphi^2$ and Ca of catalyst had the same behavior using IA metal hydroxide as precipitant; catalyst activity using IIA metal hydroxide as precipitant was lower than that using IA metal hydroxide, but both have similar selectivity. The aging time during catalyst preparation became much longer, initial activity of catalyst became much lower because the reaction rate was reduced, but initial selectivity and cyclohexene yield aroused. Turnover frequency of benzene increased with the increasing of molecular weight of hydroxide in the same group, so alkali can be used as additive or modifier. When catalyst was treated by pure water, benzene hydrogenation reaction rate was fast with a low cyclohexene selectivity, while treated by NaOH, reaction order of cyclohexene hydrogenation reduced from 1.5–2 to 1, and hydrohexene selectivity increased obviously. This could be explained from the saturation effect of NaOH-absorbed active center, increasing of average diameter of catalyst and diameter of catalyst microcrystal. Catalyst actvity reduced obviously when it is treated with N_2H_2 or HCOOH [42].

During 1997 and 2000, the influence of organic additives and support on the yield and selectivity of cyclohexene was studied by Suryawanshi and Mallat. It was discovered that the additive of monoethanolamine (MEA) was better than zinc sulfate, because the latter would lead to the acidity of the reaction system and the pollution of reboiler. Among the supports of Al_2O_3, ZrO_2, TiO_2, and Nb_2O_3, Al_2O_3 was the best [43]. The selectivity of cyclohexene could be promoted by mixing appropriate organic or inorganic compounds with reactants [44].

From 2003 to 2005, the Ru/Al_2O_3 catalysts were prepared by Mazzieri from $RuCl_3$ and hydrochloride solution with different concentrations using the impregnation method. And the catalysts were characterized with FT-IR, TPR, and so on, and the conclusions were obtained as follows:

The aftertreatment of catalysts including calcination and reduction after calcination all could not remove the chloride ion, but it could be decreased sharply by washing with NH_4OH. The electronic state of Ru species including Ru^0, $RuCl_3$, Ru oxides, and Ru oxychlorides could be affected by the treatment methods. The Ru species would also influence the catalytic activity and selectivity, and the catalyst with the highest chloride amount had the best selectivity of cyclohexene. Because the chloride ion could keep Ru in the status of electron deficiency through modifying the electronic property, leading to the decrease of cyclohexene adsorption. The basic sites of Al_2O_3 could induce the hydrolysis products of $RuCl_3$, such as ruthenium oxychloride, ruthenium hydroxide, co-precipitate on Al_2O_3. It would affect the distribution of Ru, but it had no effect on catalytic performance [45–47].

From 2003 to 2010, Estevam, da-Silva, da-Costa, and Rodrigues researched the supported catalysts, and the conclusions were obtained as follows:

The Ru–Fe/TiO_2 and Ru–Fe/SiO_2 catalysts were prepared by Estevam and da-Silva with co-impregnation method. And the impact of the support structure and Fe additives on catalytic activity was studied. It was proved by XRD, oxygen chemisorption, and TPD that the reducibility and distribution of Ru depended on the support structure. And the influence of Fe on Ru species and its relationship with catalytic activity was also revealed. The Ru supported on SiO_2 was easier to be reduced than that on TiO_2, the addition of Fe could promote this process. The addition of Fe could also enhance the distribution and oxidation of Ru. Although this effect was more obvious on SiO_2, it would not influence the stability of Ru/TiO_2. In summary, Ru–Fe/TiO_2 had higher cyclohexene selectivity than Ru–Fe/SiO_2 [48, 49]. While the impact of preparation condition (calcination temperature and reduction) on the catalytic performance of Ru/CeO_2 was investigated by da-Costa, and it was proved that Ru/CeO_2 was a good catalytic system under $TiCl_3$ solution. The catalyst reduced directly would have better activity, cyclohexene selectivity, and yield than that reduced after calcination [50]. The support natural property of Al_2O_3 and Nb_2O_5 on the activity of Ru catalysts that with different particle sizes and specific areas were researched by Rodrigues. It was found that the support natural property had no effect on cyclohexene selectivity. However, the decrease in particle size and specific area could promote the yield and selectivity of cyclohexene [51].

In 2011, a very simple catalysis system only containing Ru/Al_2O_3 and ionic liquid solution (ppm) were reported by Schwab. The characterization results of XPS revealed that the chemical adsorption of ionic liquid didn't lead to the electronic state (binding energy) change of Ru. So the reason for lowering activation should be attributed to the group effect of a second metal or ligand influence on the kinetic interference of benzene hydrogenation. The anion of ionic liquid would interact strongly with Ru center, blocking active sites and changing the geometrical arrangement of active sites. Thus, the active sites for benzene hydrogenation into cyclohexane were diluted or decreased. It was a new ligand effect and could be treated as the decrease of adsorption enthalpy of hydrogen. In general, the adsorption enthalpy of water on Ru would decrease to half in the presence of hydrogen. The coverage of hydrogen would also be decreased in the existence of an ionic liquid, which could promote the hydrophily of catalyst and inhibit excessive hydrogenation of benzene [52].

After 1990s, many countries were engaged in the catalyst research for benzene selective hydrogenation into cyclohexene. On one hand, the catalysts with high activity and selectivity were wished to be developed, on the other hand, the catalytic mechanism and the scientific nature of high yield and selectivity of cyclohexene were to be investigated. For the catalyst development, the researches were mainly focused on the supported catalysts and simple catalytic systems, but the yield and selectivity was generally low, and there was no industrial value. As to the catalytic mechanism and the scientific nature of high yield and selectivity of cyclohexene, although muchimportant progress had been made, there still existed different academic views that needed to be proved with the development of science and technology.

1.3 Domestic Research and Progress on Selective Hydrogenation Technologies of Benzene

In the year of 1991, the article entitled "partial hydrogenation of benzene into cyclohexene" was published in *Petrochemical Technology* by Wang. This article mentioned that the Asahi Kasei Company found a 60,000 t/a device on Mizushima for the production of cyclohexanone from cyclohexene through benzene selective hydrogenation in November 1988. And this device was put into operation in 1989. The cyclohexene has an active double bond and is widely used as organic chemical stocks. The cyclohexene, especially, could be directly oxidized into cyclohexanone and adipic acid, hydrolysis into cyclohexanol, which could not only shorten the production process of ε-caprolactam and adipic acid, but also has a relatively high economic profit. During the recent 20 years, researches about the new preparation process of cyclohexene have been conducted abroad, and the hydrogenation catalyst development is the key technology [53].

In 1992, the article entitled "Preparation of cyclohexanol from cyclohexene" was published in *China Synthetic Fiber Industry* by Shi. This article summarized the catalysts and process of benzene selective hydrogenation into cyclohexene via Asahi method, then the cyclohexene was further hydrolyzed into cyclohexanol. The industrial production of cyclohexanol from cyclohexene had induced the attention from China and abroad. Compared with the cyclohexane oxidation method, the cyclohexene method had advantages of less by-product (by-product cyclohexane could be further utilized), high carbon yield (close to 100%), energy saving, low material consumption, and pollution. So the intermediate cyclohexene is an important fine chemical stock and has a wider application perspective [54].

In 1992, the article entitled "The new preparation process of cyclohexanol and cyclohexanone" was published in *Modern Chemical Industrial* by Su, Ye, and Wu from Changchun Institute of Applied Chemistry, Chinese Academy of Sciences. This article introduced the process of benzene selective hydrogenation into cyclohexene, cyclohexene hydrolysis into cyclohexanol, cyclohexanol dehydrogenation into cyclohexanone, and benzene conversion into cyclohexene or phenylcyclohexane. Due to the advantages of the process, the conversion of benzene into cyclohexanol and cyclohexanone via cyclohexene has been paid more attention to, and it has been the strategic issue in the field of petrochemical industry. Cyclohexanol and cyclohexanone are all important chemical stocks, they could be used to produce nylon 6, caprolactam, adipic acid, civetone, rubber anti-aging agent (diphenol cyclohexane), and fruit antiseptic (phenylphenol). It is a new route for benzene selective hydrogenation into cyclohexanol and cyclohexanone through the conversion of benzene into cyclohexene and phenylcyclohexane [2].

In the same year, the article entitled "The research on benzene selective hydrogenation into cyclohexene over surface modified catalyst" was published in *Chemical Reaction Engineering and Technology* by Ye, Pang, Huang, and Yi. This article stated that benzene would be completely hydrogenated into cyclohexane via the traditional gaseous catalytic method, and there's no intermediate cyclohexene. In the view of thermodynamics, the hydrogenation of benzene into cyclohexane was easier to proceed than into cyclohexene. In addition, the cyclohexene was also easily hydrogenated into cyclohexane. But the selective hydrogenation product cyclohexene could be obtained on the modified catalysts. Cyclohexene was very active as it contained the double bond that could be applied in medicine, food, farming chemicals, fodder, polyester, and fine chemicals. Especially in the field of petrochemical industry, the chemicals produced from benzene could be simplified and efficient using cyclohexene as stock [55].

In 1993, the article entitled "New production technology of polyamide—partial hydrogenation of benzene into cyclohexene" was published in *Chinese Polymer Bulletin* by Ye, Pang, and Huang from South China University of Technology. This article introduced and commented on the new rout for polyamide production through benzene selective hydrogenation into cyclohexene. It illustrated the application perspective in industrial manufacture by comparing it with the existing process. It also reviewed the developing history, current status, and future tendency in the reaction process, catalyst preparation, reaction mechanism, separation, and application of cyclohexene. Compared with the cyclohexane oxidation route, the benzene selective hydrogenation method could save 1/3 hydrogen, the carbon yield could reach up to 100%. It was a natural resource and energy saving, procedure shorting and yield increasing, there was no environmental pollution step, and the reaction condition was relatively mild [1]. Successful development of selective catalytic hydrogenation of benzene to cyclohexene will give cyclohexene a wide application in the field of chemical industry and fine chemicals, etc.

In 1996, the article entitled "The Ru catalyst developing progress for benzene hydrogenation into cyclohexene" was published in *Speciality Petrochemicals* by Tang. This article pointed out that the production of adipic acid from the oxidation of cyclohexene, which was prepared through hydrogenation of benzene, was the universally applied traditional process in the world. The cyclohexane oxidation route existed the problems of serious side reactions, low conversion, complicated process, and poor security. The selective hydrogenation of benzene into cyclohexene, followed by hydrolyzing into cyclohexanol and oxidizing into adipic acid, was an alternative route for adipic acid production. But the key problem was the development of catalysts for selective hydrogenation of benzene. While this study in our country only on the start-up stage, to quicken the research work and industrial step had significant on our adipic acid and polyamide industry [56].

In 1997, the article entitled "The research progress on benzene selective hydrogenation into cyclohexene" was published in *Speciality Petrochemicals* by Jia, Zhang,

Xu, and Lin that from Dalian Institute of Chemical Physics, Chinese Academy of Sciences. This article reviewed the research progress on selective hydrogenation of benzene into cyclohexene in liquid phase over Ru catalysts and introduced the characterization of Ru catalysts prepared by the chem-mixed method and its effect on benzene selective hydrogenation into cyclohexene in liquid phase. As early as 1901, it was well-known that cyclohexene could be prepared through benzene hydrogenation, but cyclohexene was easily hydrogenated into cyclohexane. So it was a dream for preparing cyclohexene through benzene hydrogenation at that time. The selective hydrogenation of benzene into cyclohexene had stimulated researchers' interests for it was an economic industrial route. The selective hydrogenation of benzene into cyclohexene had been industrialized in Japan, the corrosion problems of additives had been resolved, and the key point was to enhance the yield of cyclohexene. This reaction had a good application perspective in our country [4].

In sum, it is easy to figure out, at the beginning of 1990s, the scientists and scholars from petrochemical industry, petrochemical fiber companies, Chinese Academy of Sciences, and universities focused their attention on benzene selective hydrogenation technologies. And they introduced the developing history and current status for our readers. By comparing the benzene selective hydrogenation with traditional benzene completely hydrogenation technologies, it was of great significance for the development of selective hydrogenation technologies.

It was reported that the studies on catalyst preparation and related fundamental theory for benzene selective hydrogenation into cyclohexene have been conducted by "Taiwan National Central University", Fudan University, Dalian Institute of Chemical Physics (Chinese Academy of Sciences), East China University of Science and Technology, Sichuan University, Hebei University of Technology, Institute of Chemistry (Chinese Academy of Sciences), and so on.

During 1997 and 2001, the La and Zn binary oxides supported Ru catalyst was researched in aqueous sodium hydroxide by Taiwan scholars. The results showed that the binary oxides La_2O_3–ZnO were better than La_2O_3 or ZnO for the enhancement of yield and selectivity of cyclohexene. Because the ZnO component could promote the hydrophily of Ru catalyst, leading to the slow adsorption of benzene on the catalyst and the fast desorption of cyclohexene from the catalyst. The further hydrogenation rate was decreased and the selectivity of cyclohexene was increased. This slow adsorption mechanism explained all the binary support could reduce the hydrogenation rate of benzene on Ru catalyst and promote the selectivity of cyclohexene. The scholars also proposed that the benzene, hydrogen, and cyclohexene adsorb on the same active sites competitively. With the increase in hydrogen pressure, the hydrogenation rate of benzene was accelerated. When the coverage of hydrogen was equal to benzene, the hydrogenation rate of benzene reaches the maximum [57–61].

In 1999, the amorphous alloy catalyst Ru–B reported by Fudan University was used for benzene selective hydrogenation into cyclohexene. The amorphous alloy could provide more active sites due to the short-range order, long-range disorder, isotropy, and highly unsaturated coordinate on the surface. In the amorphous alloy structure of Ru–B, Ru was metallic state and B was elemental and salt states, the electron transferred from B to Ru. With the increasing percentage of B, the yield and selectivity of cyclohexene was all promoted [62, 63]. In 2004, it was proved by XPS that the Zn^{2+} could be reduced into Zn due to hydrogen spillover. The percentage of Zn was 1.1 wt.% in the reduced catalyst, the Zn formed alloy with Ru, covering the highest active sites reversely. The electron transferred partially from Ru, changing the electronic structure of Ru and affecting the activity and selectivity of the catalyst [64–66]. From 2008 to 2009, the Ru-Ba/SBA-15 and Ru-Ce/SBA-15 catalysts were developed by Fudan University using a double solvent method. The $CdSO_4$ and $ZnSO_4$ were used to modify the catalytic performance of Ru-La/SBA-15 [67–69]. From 2012 to 2014, Fudan University designed a Ru/B–ZrO_2 catalyst and proved the additive active sites was the Lewis acid sites on the support. The proposal of additive active sites could provide the new method for the development of selective hydrogenation catalyst [70, 71].

From 2006 to 2009, Dalian Institute of Chemical Physics (Chinese Academy of Sciences) reported a SiO_2 stabilized colloid Ru catalyst by micro-emulsion method. This catalyst has high activation and selectivity due to the higher distribution of Ru (4 nm), and the more exposure active sites than SiO_2 supported Ru catalyst. This research also reported another Ru-cordierite catalyst used for fixed bed process, after the modification with ZrO_2–Al_2O_3 or Al_2O_3 coating, the relative high yield of cyclohexene could be obtained under low space velocity. The core–shell distribution of active components, large aperture, and the formation of Taylor flow were the determining factors for the high activation and selectivity [72–76].

During 2007 and 2010, East China University of Science and Technology researched the benzene selective hydrogenation into cyclohexene over Ru–Zn/ZrO_2 through theory calculation and experiment. The precursor of Ru catalyst was reduced in $ZnSO_4$ solution, the reduced Zn formed alloy with Ru. Whether in the stage of reduction or hydrogenation, atomic Zn universally existed in the bulk phase or on the surface of Ru-based catalyst, leading to the decreasing the chemical adsorption energy of benzene and cyclohexene. Especially for the chemical adsorption of cyclohexene, the passivated surface of catalyst inhibited the further hydrogenation of cyclohexene. The experimental results and theory proved the existence of the optimal extent of Zn. The Zn atoms affected the catalytic activation and cyclohexene selectivity through adsorption and surface reactions [77–81].

In 2008, Sichuan University reported an amorphous alloy catalyst RuCoB/γ–Al_2O_3 through reduction-impregnation method, Co oxides and B element could promote the distribution of Ru. The additives of amine and alcohol could stabilize the monimolimnion around catalyst and decrease the adsorption of cyclohexene on active sites. The yield of cyclohexene could be obviously increased in the presence of $ZnSO_4$ and ethylenediamine [82, 83].

In 2009, Hebei University of technology reported a Ru–Zn/SiO_2 catalyst prepared by the water/oil microemulsion method. The size of Ru particles was controlled by adding hydrazine hydrate, and the Ru species were highly distributed to form microcrystalline, which would be a benefit for enhancing the selectivity of cyclohexene [84].

From 2011 to 2015, the Institute of Chemistry (Chinese Academy of Sciences) reported a Ru–Cu/ZnO catalyst prepared by deposition–precipitation and impregnation- co-precipitation method, the selectivity of cyclohexene could be obviously promoted in aqueous NaOH solution. In the structure of Ru/ZnO–$ZrO_x(OH)_y$ catalyst, the hydroxyl group, Zn, and Zr atoms existed synergistic effect, so the yield of cyclohexene could be high without any additives [85, 86]. It also reported another Ru-Cd/BEN catalyst [87, 88]. The Ru/HAP and Ru–Zn/HAP catalysts were also prepared by the ion exchanging method with the nontoxic, rich in nature support of HAP and the NaOH was selected as a modification agent, the yield of cyclohexene could be relatively high [89]. The selectivity of cyclohexene over Ru/TiO_2 catalyst in aqueous NaOH solution could be obviously enhanced through the addition of little ZnO. Because the ZnO would form $Na_2Zn(OH)_4$ in aqueous NaOH solution, which could efficiently inhibit the further hydrogenation of cyclohexene [90].

The selective hydrogenation of benzene into cyclohexene and its downstream products have an impressive market perspective. In 1995, Henan province spent 3 billion RMB for introducing the process of benzene selective hydrogenation into nylon 66 salt (60,000 t/a) and put it into production in 1998. This was the first apparatus in China, and the second in the world. But the catalyst preparation technology was monopolized, restricting the autonomously and sustainably development of Chinese companies. To solve the technology problem, Zhengzhou University has carried out the research on the catalyst preparation method since 1998.

During 1998 and 2007, Zhengzhou University developed the first generation amorphous alloy catalyst Ru-M-B/ZrO_2 (M = Fe and La) with the main indexes of 40% benzene conversion, 80% selectivity of cyclohexene, and above 32% yield of cyclohexene.

With the support of Educational Natural Science Foundation of Henan Province (The development of catalyst with high activity and selectivity for benzene selective hydrogenation, No. 2000100014) and Key Scientific and Technological Project of Henan Province (The development of catalyst for benzene selective hydrogenation into cyclohexene, No. 001090107), the later project passed provincial identification in 2001. And the catalyst and its preparation method for benzene selective hydrogenation into cyclohexene was applied for national invention patent in the same year [91]. In 2004, the catalyst and its preparation, modulation, and regeneration methods for benzene selective hydrogenation into cyclohexene were applied for the national invention patent [92].

Under the support of technical innovation found for small and medium enterprises in science and technology (The catalyst and process for benzene selective hydrogenation into cyclohexene, No. 02C26214100384), three pilot results "The catalyst and process for benzene selective hydrogenation into cyclohexene", "The catalyst preparation and application technologies for benzene partial hydrogenation into cyclohexene", and "The apparatus and process for benzene partial hydrogenation into cyclohexene" passed the identification of Henan department of Science and Technology. The catalyst and its preparation method, and the hydrogenation process were applied for the national invention patent [93].

The comment on "catalyst and catalytic process for benzene selective hydrogenation into cyclohexene" was that "The research on catalyst and process for benzene partial hydrogenation into cyclohexene was completed by Zhengzhou University in cooperation with Puyang Chemical Co. Ltd. The technical indexes of the pilot catalyst were that the selectivity of cyclohexene should be stabilized above 80% under the conversion of 40%. This catalyst had advantages in high utilization of Ru and low cost, and attain or surpass the foreign level. The pilot test of benzene selective hydrogenation into cyclohexene was finished in the interval apparatus. The effect of the component of the reaction system on catalytic activation and selectivity was investigated systematically. And the key problems involved in the production of cyclohexene were all solved, which was innovative, applicable, and advanced."

After 2005, the catalysts had been used in the industrial apparatus. In 2007, the import catalysts were all replaced by the self-produced and the localization of catalyst for benzene selective hydrogenation was realized."

In 2005, our country introduced the 100,000 t/a cyclohexanone production technology from cyclohexene that hydrogenated from benzene. The technical indexes of catalyst were 51% conversion of benzene, 77.7% selectivity of cyclohexene, and 40% yield of cyclohexene. Compared with the first generation in 1995, the conversion of benzene and the yield of cyclohexene was enhanced by 11 percentage points and 8 percentage points, respectively.

From 2008 to 2010, the second catalyst Ru–Zn–Na_2SiO_3-PEG-10000 for benzene selective hydrogenation was developed by Zhengzhou University in cooperation with the enterprise. The main indexes were 50% conversion of benzene, above 80% selectivity of cyclohexene, and 40% yield of cyclohexene. Based on the second generation catalyst, the catalytic process, key equipment, and apparatus were developed for the production of cyclohexene and its downstream products through benzene hydrogenation. And it was industrialized in 2010.

During 2011 and 2013, the third generation nano-bimetal catalysts Ru-M (M = Zn, Mn, Fe, La, and Ce) were developed by Zhengzhou University in cooperation with enterprises. Through the interaction between transition metal (Zn, Mn, Fe) or rare earth metal (La, Ce) oxides and $ZnSO_4$, the in-situ formed basic salt would adsorb on the surface of the catalyst. And the high yield and selectivity of cyclohexene would be obtained near the threshold value. The main technical indexes were 60% conversion of benzene, above 80% selectivity, and 48% yield of cyclohexene. The conversion of benzene and yield of cyclohexene was increased by 10 percentage points and 8 percentage points, respectively. The monolayer-dispersed catalyst Ru-M (M = Zn, Mn, Fe, La, and Ce) for benzene selective hydrogenation into cyclohexene passed the identification of Henan department of science and technology. And these research results were applied for the national invention patent [94–96].

From 2014 to 2015, the technology for in-situ formed basic salt was broken through and the fourth generation core–shell catalyst Ru–Zn@BZSS was developed by Zhengzhou University in cooperation with the enterprise. The main technical indexes were 70% conversion of benzene, above 80%, and 56% of selectivity and yield, respectively. Compared with the third generation catalyst, the yield of cyclohexene was increased by 8 percentage points. The project "The development of key technical and industrial application for benzene selective hydrogenation" passed the identification of China Petroleum and Chemical Industry Federation. And the research results were applied for the national invention patent [97–100].

In 2015, the review on benzene selective hydrogenation was published on *Chem. Soc. Rev.* (Royal Society of Chemistry), the 59 catalysts were listed and their catalytic performances were also compared with each other. The catalyst Ru–Zn developed by Zhengzhou University in 2012, took the highest level for the 83.8% conversion of benzene, 75.5% selectivity, and 63.3% yield of cyclohexene [101].

Among the production chain of benzene selective hydrogenation into cyclohexene and its downstream products, the preparation of catalyst takes the important status. Based on the four generations of self-produced catalysts, the catalytic process, key equipment, and apparatus are developed by Zhengzhou University and enterprise. And 11 invention patents are applied [102–112]. The results have been generalized and applied in Shandong, Shanxi, Jiangsu, Zhejiang, Fujian, Tianjin, Chongqing, Hebei, and so on. Four sets of apparatus for the industrial preparation of catalyst (1000–3000 kg/a) and 11 sets of equipment for preparing cyclohexene and its downstream products from benzene selective hydrogenation (100,000–200,000 t/a) were set up. The capacity could reach above 1400,000 t/a. Up to now, six research results have passed the provincial identification and 17 invention patents have been authorized.

From 1998 to 2015, more than 20 masters and doctors took part in the research on benzene selective hydrogenation [113–131], and more than 20 papers were published [132–158]. In recent years, with the support of the National Natural Science Foundation (No. 21273205) and China Postdoctoral Science Foundation (No. 2012M511589), the multiphase catalytic theory research was conducted and the catalytic mechanism was revealed.

In the field of benzene selective hydrogenation, the foreign monopoly was broken by Zhengzhou University and enterprises. So our country became the second one to industrialize and hold the proprietary intellectual property rights. In 2013, the project "The catalyst and catalytic process for benzene selective hydrogenation into cyclohexene" earned the first prize of the science and technology progress award of Henan province. In 2015, the project "Development and industrial application of catalytic technical for benzene selective hydrogenation into cyclohexene" gained the first prize of technical invention award of China Petroleum and Chemical Industry Federation.

1.4 Main Technology Index on Selective Catalysts of Hydrogenation of Benzene Over World

The main index of benzene selective hydrogenation into cyclohexene contains the conversion of benzene, the yield, and selectivity of cyclohexene. In 2015, the review on benzene selective hydrogenation was published on *Chem. Soc. Rev.* (Royal Society of Chemistry), the 59 catalysts were listed and the index were compared [101]. Table 1.2 presents the main indexes of the catalysts.

Table 1.2 Main indexes of catalysts for benzene selective hydrogenation

No.	Catalyst	Conversion/%	Selectivity/%	Yield/%	Reference
1	Ru-NPs	0.3	65.0	0.2	L. M. Rossi, G. Machado, *J. Mol. Catal. a-Chem.*, 2009, **298**, 69–73
2	Ru-NPs	2.0	34.0	0.7	E. T. Silveira, A. P. Umpierre, L. M. Rossi, et al. *Chem. Eur. J.*, 2004, **10**, 3734–3740
3	Ru/C	43.0	8.6	3.7	C. Zanutelo, R. Landers, W. A. Carvalho, et al. *Appl. Catal. A*, 2011, **409**, 174–180
4	Ru/Polymer	47.2	8.1	3.8	M. Hronec, Z. Cvengrošová, M. Králik, et al. J. Mol. Catal. a-Chem., 1996, **105**, 25–30
5	Ru/Al_2O_3	35.0	12.3	4.3	M. F. F. Rodrigues, A. J. G. Cobo, Catal. Today, 2010, **149**, 321–325
6	Ru/Al_2O_3	62.2	7.3	4.5	C. Milone, G. Neri, A. Donato, et al. J. Catal., 1996, **159**, 253–258
7	Ru–Fe/TiO_2	20.0	25.0	5.0	J. W. Da-Silva, A. J. G. Cobo, Appl. Catal. a-Gen., 2003, **252**, 9–16
8	Ru–Cd/Bentonite	57.4	43.1	24.8	W. T. Wang, H. Z. Liu, G. D. Ding, et al. Chem. Cat. Chem., 2012, **4**, 1836–1843
9	Ru–Co-B/Al_2O_3	62.7	45.7	28.7	G.-Y. Fan, W.-D. Jiang, J.-B. Wang, et al. Catal. Commun., 2008, **10**, 98–102
10	Ru–Cu/SiO_2	83.3	37.7	31.4	S. I. Niwa, F. Mizukami, M. Kuno, et al. J. Mol. Catal., 1986, **34**, 247–249

(continued)

Table 1.2 (continued)

No.	Catalyst	Conversion/%	Selectivity/%	Yield/%	Reference
11	Ru/La_2O_3–ZnO	77.8	42.6	33.1	S. C. Hu , Y. W. Chen, Ind. Eng. Chem. Res., 1997, **36**, 5153–5159
12	RuB–Zn/$ZrO_2 \cdot xH_2O$	73.1	62.3	45.6	Z. Liu, S. H. Xie, B. Liu, et al. New J. Chem., 1999, **23**, 1057–1058
13	Ru/ZnO–ZrOx(OH)y	77.5	72.3	56.0	H. Z. Liu, T. Jiang, B. X. Han, et al. Green Chem., 2011, **13**, 1106–1109
14	Ru-[bmim]BF4	12.2	40.5	4.9	Y. Quin, W. Xue, F. Li, et al. Chin. J. Catal., 2011, **32**, 1727–1732
15	Ru/Al_2O_3	18.3	60.0	11.0	F. Schwab, M. Lucas, P. Claus, Angew. Chem. Int. Ed., 2011, **50**, 10453–10456
16	Ru/Al_2O_3	38.6	33.3	12.9	P. T. Suryawanshi, V. V. Mahajani, J. Chem. Technol. Biotechnol., 1997, **69**, 154–160
17	Ru/SiO_2	60.0	23.3	14.0	E. V. Spinacé, J. M. Vaz, Catal. Commun., 2003, **4**, 91–96
18	$Ru/A12O_3$	60.0	26.7	16.0	R. S. Suppino, R. Landers, A. J. G. Cobo, Appl. Catal. a-Gen., 2013, **452**, 9–16
19	Ru/CeO_2	65.0	26.2	17.0	P. da Costa Zonetti, R. Landers, A. J. G. Cobo, Appl. Surf. Sci., 2008, **254**, 6849–6853
20	Ru-[bmim]BF4	49.5	34.1	17.0	W. Xue, Y. F. Qin, F. Li, et al. Chin. J. Catal., 2012, **33**, 1913–1918
21	Ru/SiO_2	64.9	33.1	21.5	J. Bu, Y. Pei, P. Guo, et al. Stud. Surf. Sci. Catal., 2007, **165**, 769–772

(continued)

Table 1.2 (continued)

No.	Catalyst	Conversion/%	Selectivity/%	Yield/%	Reference
22	Ru–La/ZrO_2	35.0	70.0	24.5	S. Liu, Y. Wu, Z. Liu, et al. J. Nat. Gas. Chem., 2005, **14**, 226–232
23	Ru-B	60.4	45.4	27.4	Z. Liu, W. L. Dai, B. Liu, et al. J. Catal., 1999, **187**, 253–256
24	Ru/Bentonite	61.5	45.2	27.8	W. T. Wang, H. Z. Liu, T. B. Wu, et al. J. Mol. Catal. a-Chem., 2012, **355**, 174–179
25	Ru	69.0	40.6	28.0	L. Ronchin, L. Toniolo, Appl. Catal. a-Gen., 2001, **208**, 77–89
26	Ru/Ga_2O_3–ZnO	75.0	38.0	28.5	S. C. Hu,Y. W. Chen, J. Chem. Technol. Biotechnol., 2001, **76**, 954–958
27	Ru–Zn/SiO_2	62.7	47.7	29.9	W. Xue, Y. Song, Y. Wang, et al. Catal. Commun., 2009, **11**, 29–33
28	Ru–Zn/SiO_2	55.0	56.4	31.0	S. C. Hu,Y. W. Chen, Ind. Eng. Chem. Res., 2001, **40**, 6099–6104
29	Ru–Zn/HAP	69.8	47.3	33.0	P. Zhang, T. B. Wu, T. Jiang, et al. Green Chem., 2013, **15**, 152–159
30	Ru-B/SiO_2	68.4	49.7	34.0	S. H. Xie, M. H. Qiao, H. X. Li, et al. Appl. Catal. a-Gen., 1999, **176**, 129–134
31	Ru–Co-B/Al_2O_3	72.6	48.0	34.8	G. Y. Fan, R. X. Li, X. J. Li et al. Catal. Commun., 2008, **9**, 1394–1397
32	Ru-B/$Al_2O_3 \cdot xH_2O$	77.4	51.2	39.6	J. Wang, P. Guo, S. Yan, M. et al. J. Mol. Catal. a-Chem., 2004, **222**, 229–234

(continued)

Table 1.2 (continued)

No.	Catalyst	Conversion/%	Selectivity/%	Yield/%	Reference
33	Ru/SiO_2	68.0	63.0	42.0	J. B. Ning, J. Xu, J. Liu, et al. Catal. Lett., 2006, **109**, 175–180
34	$Ru–Zn/m\text{-}ZrO_2$	69.2	62.7	43.4	J. Wang, Y. Wang, S. Xie, et al. Appl. Catal. a-Gen., 2004, **272**, 29–36
35	$Ru\text{-}B/ZrO_2$	83.0	56.0	47.0	G. B. Zhou, J. L. Liu, X. H. Tan, et al. Ind. Eng. Chem. Res., 2012, **51**, 12205–12213
36	$Ru/B\text{-}ZrO_2$	80.0	60.0	48.0	G. B. Zhou, Y. Pei, Z. Jiang, K. N. et al. J. Catal., 2014, **311**, 393–403
37	Ru–Cu/ZnO	72.3	68.3	49.4	H. Liu, S. Liang, W. Wang, et al. J. Mol. Catal. a-Chem., 2011, **341**, 35–41
38	Ru/ZrO_2	83.0	62.0	51.0	G. Zhou, X. Tan, Y. Pei, et al. ChemCatChem, 2013, **5**, 2425–2435
39	$Ru–La\text{-}B/ZrO_2$	80.8	66.5	53.7	S. Liu, Z. Liu, Z. Wang, et al. Appl. Catal. a-Gen., 2006, **313**, 49–57
40	$Ru–Ce/SiO_2$	85.0	63.3	53.8	J. L. Liu, L. J. Zhu, Y. Pei, et al. Appl. Catal. a-Gen., 2009, **353**, 282–287
41	$Ru–La/SiO_2$	82	69	57	J. Liu, Y. Zhu, J. Liu, et al. J. Catal., 2009, **268**, 100–105
42	$Ru–Fe\text{-}B/ZrO_2$	80.6	71.1	57.3	Z. Y. Liu, H. J. Sun, D. B. Wang, et al. Chin. J Chem, 2010, **28**, 1927–1934
43	Ru–Zn	82.0	71.8	58.9	H. J. Sun, H. X. Wang, H. B. Jiang, et al. Appl Catal a-Gen, 2013, **450**, 160–168

(continued)

Table 1.2 (continued)

No.	Catalyst	Conversion/%	Selectivity/%	Yield/%	Reference
44	Ru–La	88.0	67.6	59.5	H. J. Sun, Y. Y. Dong, S. H. Li, et al. J. Mol. Catal. a-Chem., 2013, **368**, 119–124
45	Ru–Zn–Mn	89.0	67.3	59.9	X. L. Zhou, H. J. Sun, W. Quo, et al. J. Nat. Gas. Chem., 2011, **20**, 53–59
46	Ru–Zn	80.0	75.0	60.0	H. Nagahara, M. Ono, M. Konishi, et al. Appl. Surf. Sci., 1997, **121**, 448–451
47	Ru–Zn	80.0	75.0	60.0	H. Nagahara, M. Ono,Y. Fukuoka, Stud. Surf. Sci. Catal., 1995, **92**, 375–378
48	Ru–Zn	68.7	78.9	61.4	H. J. Sun, Z. H. Chen, W. Guo, et al. Chin. J. Chem., 2011, **29**, 369–373
49	Ru–Zn	83.8	75.5	63.3	H. J. Sun, Y. J. Pan, H. X. Wang, et al. Chin. J. Catal., 2012, **33**, 610–620
50	Ru/Al_2O_3	–	33.0	–	V. Mazzieri, F. Coloma-Pascual, A. Arcoya, P. L' Argentiere, et al. Appl. Surf. Sci., 2003, **210**, 222–230
51	Ru/Al_2O_3	–	33.0	–	V. Mazzieri, N. Figoli, F. C. Pascual, et al. Catal. Lett., 2005, **102**, 79–82
52	Ru/Al_2O_3	–	32.6	–	V.A.Mazzieri,P.C. L'Argentiere, F. Coloma- Pascual, et al. Ind. Eng. Chem. Res., 2003, **42**, 2269–2272
53	$Ru–Ba/SiO_2$	–	–	50.8	J. Bu, J. L. Liu, X. Y. Chen, et al. Catal. Commun., 2008, **9**, 2612–2615

(continued)

Table 1.2 (continued)

No.	Catalyst	Conversion/%	Selectivity/%	Yield/%	Reference
54	Ru–Zn/ZrO_2	–	–	44.0	P.-Q. Yuan, B.-Q. Wang, Y.-M. Ma, et al. J. Mol. Catal. a-Chem., 2009, **301**, 140–145
55	Ru/ZrO_2	–	–	41.0	L. Ronchin, L. Toniolo, Catal. Today, 2001, **66**, 363–369
56	Ru	–	–	34.0	L. Ronchin, L. Toniolo, Catal. Today, 1999, **48**, 255–264
57	Ru/Yb_2O_3	–	–	30.1	L. Ronchin, L. Toniolo, React. Kinet. Catal. Lett., 2003, **78**, 281–289
58	Ru/La_2O_3	–	–	14.0	F. Schwab, M. Lucas, P. Claus, Green Chem., 2013, **15**, 646–649
59	Ru Black	–	–	0.2	F. Haretog, J. Catal., 1963, **2**, 79–81

In Table 1.2, the catalysts with underline were developed by Zhengzhou University. The number 49 catalyst was developed in 2012, the conversion of benzene could reach up to 83.8% with the 63.3% yield and 75.5% selectivity of cyclohexene. The main index of other self-produced catalysts is all ranked firstly.

References

1. Ye, D.Q., Pang, X.C., Huang, Z.T.: New technology for production of polyamide—the selective hydrogenation of benzene to cyclohexene. Chin. Polym. Bull. **3**, 170–177 (1993)
2. Su, W.P., Ye, X.K., Wu, Y.: New process for production of cyclohexanol and cyclohexanone. Mod. Chem. Ind. **1**, 37–44 (1992)
3. Li, G., Huang, K.L., Wu, G. et al.: Studies on Process Intensification for Hydrogenation of Benzene to cyclohexene in Liquid Phase. In: Hunan Sci-tech Forum-Symposium for Low Carbon and Green Chemical Technology, pp. 49–52 (2010).
4. Jia, J.F., Zhang, T., Xu, Z.S., et al.: The progress of partial hydrogenation of benzene to cyclohexene. Spec. Chem. **3**, 46–50 (1997)
5. Horiuti, I., Polanyi, M.: The mechanism of benzene hydrogenation. Trans. Faraday Soc. **30**, 1164–1172 (1934)
6. Anderson, J.R.: The catalytic hydrogenation of benzene and toluene over evaporated films of nickel and tungsten. Aust. J. Chem. **10**, 409–416 (1957)
7. Hartog, F., Zwietering, P.: Olefins as intermediates in the hydrogenation of aromatic hydrocarbons. J. Catal. **2**, 790–881 (1963)
8. Hartog, F., Tebben, J.H., Weterings, C.A.M.: Proc. 3rd Int. Congr. Catal., Amsterdam, 1210 (1964).

9. Hartog, F.: Preparation of cyclic alkenes. US Patent 3391206 (1968).
10. Drinkard, W.C.: Catalysts and catalysis system for selective partial hydrogenation. DE Patent 2221137A (1972).
11. Drinkard, W.C.: A heterogeneous system for selective partial hydrogenation. NL Patent 7205832 (1972).
12. Johnsonm, M.M., Nowack, G.P.: Selective hydrogenation of aromatics. US Patent 3793383 (1974).
13. Nowack, G.P., Johnson, M.M.: Selective partial hydrogenation of aromatics. US Patent 3912787 (1975).
14. Nowack, G.P., Johnson, M.M.: Hydrogenation catalysts. US Patent 4401640 (1975).
15. Kayeyama, Y.: Method for separating cyclohexene. GB Patent 1454717 (1975).
16. Aizawa, H., Kuroda, A.: Process for preparing cyclic olefins by selective partial hydrogenation of aromatic hydrocarbons. US Patent 4197415 (1976).
17. Yasuhara, Y., Nishino, M., Matauhira, S.: Method for producing cyclohexane derivatives directly from aromatic hydrocarbons. US Patent 4212990 (1977).
18. Yasuhara, Y., Nishino, M., Matauhira, S.: Method for partially hydrogenating a monocyclic aromatic hydrocarbon. JP Patent 78144543 (1977).
19. Ichihashi, H., Yoshioka, H.: A method for producing cycloolefins. Eur. Patent 170915 (1985).
20. Ichihashi, H., Yoshioka, H.: A method for producing cycloolefins. Eur. Patent 214530 (1987).
21. Ichihashi, H., Yoshioka, H.: A method for producing cycloolefins. US Patent 4575572 (1986).
22. Matsunaga, F., Fukuhara, H., Mitsuki, Y.: Process for producing a cycloolefin. Eur. Patent 316142 (1987).
23. Fukuhara, H., Matsunaga, F., Nakashima, Y.: Method for partial nuclear hydrogenation of aromatic hydrocarbon compounds and a hydrogenation catalyst. Eur. Patent 323192 (1989).
24. Odenbrand, C.U.I., Lundin, S.T.J.: Hydrogenation of benzene to cyclohexene on a ruthenium catalyst: influence of some reaction parameters. J. Chem. Technol. Biotechnol. **30**, 677–687 (1980)
25. Odenbrand, C.U.I., Lundin, S.T.J.: Hydrogenation of benzene to cyclohexene on an unsupported ruthenium catalyst: effect of poisons. J. Chem. Technol. Biotechnol. **31**, 660–669 (1981)
26. Odenbrand, C.U.I., Andersson, S.L.T.: Hydrogenation of benzene to cyclohexene on a ruthenium catalyst: physical and chemical characterization of the catalyst and its precursor. J. Chem. Technol. Biotechnol. **32**, 365–375 (1982)
27. Odenbrand, C.U.I., Lundin, S.T.J.: Hydrogenation of benzene to cyclohexene on an unsupported ruthenium catalyst: an ESCA study of titanium poisoned catalysts. J. Chem. Technol. Biotechnol. **33**, 151–156 (1983)
28. Niwa, S.I., Mizukami, F., Isoyama, S., et al.: Method for partial nuclear hydrogenation of aromatic hydrocarbon compounds and a hydrogenation catalyst therefor. US Patent 4495373 (1984).
29. Niwa, S.I., Mizukami, F., Isoyama, S., et al.: Partial hydrogenation of benzene with ruthenium catalysts prepared by a chemical mixing procedure: preparation and properties of the catalysts. J. Chem. Technol. Biotechnol. **36**, 236–246 (1986a)
30. Niwa, S.I., Mizukami, F., Isoyama, S., et al.: Selective hydrogenation of benzene to cyclohexene with ruthenium catalysts prepared by a chemical mixing procedure: preparation and properties of the catalyst. J. Mol. Catal. **34**, 247–249 (1986b)
31. Mitsui, O., Fukuoka, Y.: Process for producing a cycloolefin. US Patent 4678861 (1985).
32. Nagahara, H., Kohishi, M.: Method for partially hydrogenating a monocyclic aromatic hydrocarbon. Eur. Patent 220525 (1986).
33. Nagahara, H., Kohishi, M.: Process for producing cyclic alcohol. JP Patent 6281332 (1986).
34. Nagahara, H., Ono, M., Konishi, M., et al.: Partial hydrogenation of benzene to cyclohexene. Appl. Surf. Sci. **121**, 448–451 (1997)
35. Struijk, J., d'Angremond, M., Lucas-de Regt, W.J.M., et al.: Partial liquid phase hydrogenation of benzene to cyclohexene over ruthenium catalysts in the presence of an aqueous salt solution I. preparation, characterization of the catalyst and study of a number of process variables. Appl. Catal. A: Gen., **83** 263–295 (1992).

36. Struijk, J., Moene, R., van der Kamp, T., et al.: Partial liquid-phase hydrogenation of benzene to cyclohexene over ruthenium catalysts in the presence of an aqueous salt solution II. Influence of various salts on the performance of the catalyst. Appl. Catal. A: Gen., **89**, 11–102 (1992).
37. Döbert, F., Gaube, J.: Kinetics and reaction engineering of selective hydrogenation of benzene towards cyclohexene. Chem. Eng. Sci. **51**, 167–171 (1996)
38. Milone, C., Neri, G., Donato, A., et al.: Selective hydrogenation of benzene to cyclohexene on Ru/γ-Al_2O_3. J. Catal. **159**, 253–258 (1996)
39. Hronec, M., Cvengrošová, Z., Králik, M., et al.: Hydrogenation of benzene to cyclohexene over polymer-supported ruthenium catalysts. J. Mol. Catal. A: Chem. **105**, 25–30 (1996)
40. Ronchin, L., Toniolo, L.: Selective hydrogenation of benzene to cyclohexene using a suspended. Ru catalyst in a mechanically agitated tetraphase reactor. Catal. Today, **48**, 255–264 (1999).
41. Ronchin, L., Toniolo, L.: Selective hydrogenation of benzene to cyclohexene using a Ru catalyst suspended in an aqueous solution in a mechanically agitated tetraphase reactor: a study of the influence of the catalyst preparation on the hydrogenation kinetics of benzene and of cyclohexene. Appl. Catal. A: Gen. **208**, 77–89 (2001)
42. Ronchin, L., Toniolo, L.: Selective hydrogenation of benzene to cyclohexene catalyzed by ru supported catalysts influence of the alkali promoters on kinetics, selectivity and yield. Catal. Today **66**, 363–369 (2001)
43. Suryawanshi, P.T., Mahajani, V.V.: Liquid-phase hydrogenation of benzene to cyclohexene using ruthenium-based heterogeneous catalyst. J. Chem. Tech. Biotech. **69**, 154–160 (1997)
44. Mallat, T., Baiker, A.: Selectivity enhancement in heterogeneous catalysis induced by reaction modifiers. Appl. Catal. A: Gen. **200**, 3–22 (2000)
45. Mazzieri, V., Fígolia, N., Pascualb, F.C., et al.: Effect of the support on the selective hydrogenation of benzene over ruthenium catalysts. 1. Al_2O_3 and SiO_2. Catal. Lett. **102**, 1–2 (2005).
46. Mazzieri, V.A., L'Argentière, P.C., Coloma-Pascual, F., et al.: Effect of chlorine on the properties of Ru/Al_2O_3. Ind. Eng. Chem. Res. **42**, 2269–2272 (2003)
47. Mazzieri, V., Coloma-Pascual, F., Arcoya, A., et al.: XPS, FTIR and TPR characterization of Ru/Al_2O_3. Catal. Appl. Surf. Sci. **210**, 222–230 (2003)
48. Estevam, V.S., Jorge, M.V.: Liquid-phase hydrogenation of benzene to cyclohexene catalyzed by ru/sio_2 in the presence of water-organic mixtures. Catal. Commun. **4**, 91–96 (2003)
49. da Silva, J.W., Cobo, A.J.G.: The role of the titania and silica supports in Ru–Fe catalysts to partial hydrogenation of benzene. Appl. Catal. A: Gen. **252**, 9–16 (2003).
50. da Costa, P.Z., Landers, R., Cobo, A.J.G.: Thermal treatment effects on the Ru/CeO_2 catalysts performance for partial hydrogenation of benzene. Appl. Surf. Sci. **254**, 6849–6853 (2008).
51. Rodrigues, M.F.F., Cobo, A.J.G.: Influence of the support nature and morphology on the performance of ruthenium catalysts for partial hydrogenation of benzene in liquid phase. Catal. Today **149**, 321–325 (2010)
52. Schwab, F., Lucas, M., Claus, P.: Ruthenium-catalyzed selective hydrogenation of benzene to cyclohexene in the presence of an ionic liquid. Angew. Chem. Int. Ed. **50**, 1–5 (2011)
53. Wang, D.S.: Partial hydrogenation of benzene to cyclohexene. Petrochem. Tech. **20**(11), 786–791 (1991)
54. Shi, Z.P.: Cyclohexanol process via cyclohexene. Syn. Fiber Ind. **15**(6), 49–53 (1992)
55. Ye, D.Q., Pang, X.C., Huang, Z.T., et al.: A study on the selective hydrogenation of benzene to cyclohexene by catalytic surface modification. Chem. React. Eng. Tech. **8**(2), 210–213 (1992)
56. Tang, Z.Z.: Development progress of ruthenium catalysts for cyclohexene from hydrogenation of benzene. Spec. Chem. **1**, 1–4 (1996)
57. Hu, S.C., Chen, Y.W.: Partial hydrogenation of benzene to cyclohexene on ruthenium catalysts supported on La_2O_3-ZnO binary oxides. Ind. Eng. Chem. Res. **36**, 5153–5159 (1997)
58. Hu, S.C., Chen, Y.W.: Partial hydrogenation of benzene: a review. J. Chin. Inst. Chem. Engrs. **29**, 387–396 (1998)

59. Hu, S.C., Chen, Y.W.: Effect of preparation on Ru-Zn ultrafine catalysts in partial hydrogenation of benzene. Ind. Eng. Chem. Res. **40**, 3127–3132 (2001)
60. Hu, S.C., Chen, Y.W.: Partial hydrogenation of benzene on Ru-Zn/SiO_2 catalysts. Ind. Eng. Chem. Res. **40**, 6099–6104 (2001)
61. Hu, S.C., Chen, Y.W.: Liquid phase hydrogenation of benzene to cyclohexene on ruthenium catalysts supported on zinc oxide-based binary oxides. J. Chem. Technol. Biotechnol. **76**, 954–958 (2001)
62. Liu, Z., Dai, W.L., Liu, B., et al.: The effect of boron on selective benzene hydrogenation to cyclohexene over ruthenium boride powders. J. Catal. **187**, 253–256 (1999)
63. Xie, S.H., Qiao, M.H., Li, H.X., et al.: A novel Ru-B/SiO_2 amorphous catalyst used in benzene-selective hydrogenation. Appl. Catal. A: Gen. **176**, 129–134 (1999)
64. Li, H.X., Dai, W.L., Qiao, M.H.: Preparation of the Ni-B amorphous alloys with variable boron extent and its correlation to the hydrogenation activity. Appl. Catal. A: Gen. **238**, 119–130 (2003)
65. Wang, J.Q., Wang, Y.Z., Xie, S.H., et al.: Partial hydrogenation of benzene to cyclohexene on a Ru-Zn/m-ZrO_2 nanocomposite catalyst. Appl. Catal. A: Gen. **272**, 29–36 (2004)
66. Wang, J.Q., Guo, P.J., Yan, S.R., et al.: Colloidal RuB/$Al_2O_3{\cdot}xH_2O$ catalyst for liquid phase hydrogenation of benzene to cyclohexene. J. Molec. Ctal, A: Chem. **222**, 229–234 (2004).
67. Bu, J., Liu, J.L., Chen, X.Y., et al.: Ru/SBA-15 catalysts for partial hydrogenation of benzene to cyclohexene: tuning the ru crystallite size by Ba. Cataly. Commun. **9**, 2612–2615 (2008)
68. Liu, J.L., Zhu, L.J., Pei, Y., et al.: Ce-promoted Ru/SBA-15 catalysts prepared by a "two solvents" impregnation method for selective hydrogenation of benzene to cyclohexene. Appl. Catal. A: Gen. **353**, 282–287 (2009)
69. Liu, J.L., Zhu, Y., Liu, J., Pei, Y., et al.: Discrimination of the roles of $CdSO_4$ and $ZnSO_4$ in liquid phase hydrogenation of benzene to cyclohexene. J. Catal. **268**, 100–105 (2009)
70. Zhou, G.B., Liu, J.L., Tan, X.H., et al.: Effect of support acidity on liquid-phase hydrogenation of benzene to cyclohexene over Ru-B/ZrO_2 catalysts. Ind. Eng. Chem. Res. **51**, 12205–12213 (2012)
71. Zhou, G.B., Pei, Y., Jiang, Z., Fan, K.N., et al.: Doping effects of B in ZrO_2 on structural and catalytic properties of Ru/B-ZrO_2 catalysts for benzene partial hydrogenation. J. Catal. **311**, 393–403 (2014)
72. Ning, J.B., Xu, J., Liu, J., Lu, F.: Selective hydrogenation of benzene to cyclohexene over colloidal ruthenium catalyst stabilized by silica. Catal. Lett. **109**, 175–179 (2006)
73. Zhao, Y.J., Zhou, J., Zhang, J.G., et al.: Monolithic Ru-based catalyst for selective hydrogenation of benzene to cyclohexene. Catal. Commun. **9**, 459–464 (2008a)
74. Zhao, Y.J., Zhou, J., Zhang, J.G., et al.: Selective hydrogenation of benzene to cyclohexene on a Ru/Al_2O_3/cordierite monolithic catalyst: effect of mass transfer on the catalytic performance. Ind. Eng. Chem. Res. **47**, 4641–4647 (2008b)
75. ZhaoY, J., Zhou, J., Zhang, J.G., et al.: Preparation and characterization of Ru/Al_2O_3/ cordierite monolithic catalysts for selective hydrogenation of benzene to cyclohexene. Catal. Lett. **131**, 597–605 (2009)
76. Zhao, Y.J., Zhou, J., Zhang, J.G., et al.: Liquid-phase selective hydrogenation of benzene to cyclohexene on Ru/Al_2O_3-ZrO_2/cordierite monolithic catalysts. J. Mol. Ctal. A: Chem. **309**, 35–39 (2009)
77. Yuan, P.Q., Ma, Y.M., Cheng, Z.M., et al.: Effect of Zn^{2+}/Zn layer on H_2 dissociation on ruthenium (0001) surface: a first principles density functional stud. J. Mol. Struct: Theochem. **807**, 185–189 (2007)
78. He, H.M., Yuan, P.Q., Ma, Y.M., et al.: Theoretical and experimental study on the partial hydrogenation of benzene over Ru-Zn/ZrO_2 catalyst. Chin. J. Catal. **30**, 312–318 (2009).
79. Yuan, P.Q., Wang, B.Q., Ma, Y.M., et al.: Hydrogenation of cyclohexene over Ru-Zn/Ru (0001) surface alloy: a first principles density functional study. J. Mol. Catal. A: Chem. **301**, 140–145 (2009a)
80. Yuan, P.Q., Wang, B.Q., Ma, Y.M., et al.: Partial hydrogenation of benzene over the metallic Zn modified Ru-based catalyst. J. Mol. Catal. A: Chem. **309**, 124–130 (2009b)

81. Yuan, P.Q., Ma, Y.M., Li, X.K., et al.: Hydrophilic property of Ru (0001) with and without H adatoms. J. Mol. Struct.: Theochem. **942**, 77–82 (2010)
82. Fan, G.Y., Li, R.X., Li, X.J., et al.: Effect of organic additives on partial hydrogenation of benzene. Catal. Commun. **9**, 1394–1397 (2008)
83. Fan, G.Y., Jiang, W.D., Wang, J.B., et al.: Selective hydrogenation of benzene to cyclohexene over RuCoB/γ-Al_2O_3 without additives. Catal. Commun (2008). https://doi.org/10.1016/j.catcom.2008.08.002
84. Xue, W., Song, Y., Wang, Y.J., et al.: Effect of hydrazine hydrate on the Ru-Zn/SiO_2 catalysts performance for partial hydrogenation of benzene. Catal. Commun. **11**, 29–33 (2009)
85. Liu, H.Z., Liang, S.G., Wang, W.T., et al.: The partial hydrogenation of benzene to cyclohexene over Ru-Cu catalyst supported on ZnO. J. Mol. Ctal. A: Chem. **341**, 35–41 (2011)
86. Liu, H.Z., Jiang, T., Han, B.X., Liang, S.G., et al.: Highly selective benzene hydrogenation to cyclohexene over supported ru catalyst without additives. Green Chem. **13**, 1106–1109 (2011)
87. Wang, W.T., Liu, H.Z., Wu, T.B., et al.: Ru catalyst supported on bentonite for partial hydrogenation of benzene to cyclohexene. J. Mol. Ctal. A: Chem. **355**, 174–179 (2012)
88. Wang, W.T., Liu, H.Z., Ding, G.D., et al.: Ru-Cd/bentonite for the partial hydrogenation of benzene: a catalyst without additives. Chem. Cat. Chem. **4**, 1836–1843 (2012)
89. Zhang, P., Wu, T.B., Jiang, T., et al.: Ru-Zn Supported On Hydroxyapatite As An Effective Catalyst For Partial Hydrogenation Of Benzene. Green Chem., 2012, View article online, 10.1039/c2gc36596k.
90. Wu, T.B., Zhang, P., Jiang, T., et al.: Enhancing the selective hydrogenation of benzene to cyclohexene over Ru/TiO_2 catalyst in the presence of a very small amount of ZnO. Sci. China. Chem. **58**, 93–100 (2015)
91. Liu, S.C., Li, L.M., Wang, X.Y.: The catalyst prepared for benzene selective hydrogenation to cyclohexene. CN patent, 0112208.5 (2001).
92. Liu, S.C., Li, L. M., Wang, X.Y.: The preparation, modulation and regeneration method of catalyst for benzene selective hydrogenation to cyclohexene. CN patent, 200410060451.0 (2004).
93. Liu, S.C., Li, L.M., Wang, X.Y., et.al.: Production technology, catalyst and its preparation method for hydrogenation of benzene to cyclohexene. CN patent, 2005100175793 (2005).
94. Liu, Z.Y., Liu, S.C., Li, Z.J., et.al.: The catalytic system and method for benzene selective hydrogenation to cyclohexene. CN patent, 201210255301X (2012).
95. Liu, Z.Y., Liu, S.C., Sun, H.J., et.al.: Mono-dispersed catalyst and its preparation method for benzene selective hydrogenation to cyclohexene. CN patent, 201210060765.5 (2012).
96. Liu, Z.Y., Liu, S.C., Li, Z.J., et.al.: A catalytic system containing nano Ru catalyst and basic $ZnSO_4$ salt for benzene selective hydrogenation to cyclohexene., CN patent, 2012102553024 (2012).
97. Liu, Z.Y., Li, Z.J., Liu, S.C., et.al.: The catalyst and its preparation method for benzene selective hydrogenation to cyclohexene. CN patent, 2015104816954 (2015).
98. Liu, Z.Y., Li, Z.J., Liu, S.C., et.al.: A catalytic activation and selectivity modulation method of Ru-Zn catalyst for benzene selective hydrogenation to cyclohexene. CN patent, 2015104799094 (2015).
99. Liu, Z.Y., Li, Z.J., Liu, S.C., et.al.: The catalyst production system and method for benzene selective hydrogenation to cyclohexene. CN patent, 2015104799840 (2015).
100. Liu, Z.Y., Li, Z.J., Liu, S.C., et.al. An in-situ regeneration method of Ru–Zn catalyst for benzene selective hydrogenation to cyclohexene. CN patent, 201510479707X (2015).
101. Foppa, L., Dupont, J.: Benzene partial hydrogenation: advances and perspectives. Chem. Soc. Rev. **44**, 1886–1897 (2015)
102. Liu, Z.Y., Liu, S.C., Li, Z.J., et.al.: The apparatus for selective hydrogenation of benzene to cyclohexene in liquid phase. CN patent, 201110253889.0 (2011).
103. Yuan, X.M., Wang, Z.W., Song, H., et al.: The apparatus for selective hydrogenation of benzene to cyclohexene. CN patent, 201010613726.4 (2010).

104. Yang, K.J., Yuan, X.M., Sun, S.Q., et al.: The apparatus for selective hydrogenation of benzene. CN patent, 201010210982.9 (2010).
105. Meng, Q.G., Liu, X.A., Chen, F.H., et al.: The apparatus and process for selective hydrogenation of benzene to cyclohexene. CN patent, 201110127661.7 (2011).
106. Meng, Q.G., Liu, X.A., Chen, F.H., et al.: The apparatus for selective hydrogenation of benzene to cyclohexene. CN patent, 201120157770.9 (2011).
107. Li, Y., Yuan, X.M., Yang, K.J., et al.: A recycle catalyst for benzene partial hydrogenation to cyclohexene. CN patent, 201210121986.9 (2012).
108. Yang, K.J., Yuan, X.M., Liu, X.W., et al.: An apparatus for gas-liquid-liquid-solid reaction. CN patent, 201210074362.6 (2012).
109. Liu, X.W., Yang, K.J., Yuan, X.M., et al.: A method for cyclohexene production from highly pure benzene. CN patent, 201210176338.3 (2012).
110. Liu, X.W., Yuan, X.M., Wang, G., et al.: A method for continuous production of cyclohexene. CN patent, 2012103774518 (2012).
111. Yang, K.J., Yuan, X.M., Liu, X.W., et al.: A method for production of caprolactam from highly pure benzene. CN patent, 2012101763203 (2012).
112. Liu, X.W., Li, Y., Xu, H., et al.: A method for high production of cyclohexene. CN patent, 2012103774787 (2012).
113. Wang, H.R.: Study on the novel Ru catalysts for selective hydrogenation of benzene to cyclohexene. Master thesis, Zhengzhou University, Zhengzhou (2001).
114. Luo, G.: Study on preparation and characterization of Ru-Zn catalyst for selective hydrogenation of benzene to cyclohexene. Master thesis, Zhengzhou University, Zhengzhou (2002).
115. Ji, Y.L.: Study on Ru-type amorphous alloy catalysts for selective hydrogenation of benzene to cyclohexene. Master thesis, Zhengzhou University, Zhengzhou (2003).
116. Han, M.L.: Study on novel Ru catalyst for selective hydrogenation of benzene to cyclohexene. Master thesis, Zhengzhou University, Zhengzhou (2004).
117. Shi, R.J.: Study on catalyst pilot for interval selective hydrogenation of benzene to cyclohexene. Master thesis, Zhengzhou University, Zhengzhou (2005).
118. Wang, H.: Study on catalyst pilot for continuous selective hydrogenation of benzene to cyclohexene. Master thesis, Zhengzhou University, Zhengzhou (2005).
119. Wu, Y.M.: Study on additives in Ru catalyst for selective hydrogenation of benzene to cyclohexene. Master thesis, Zhengzhou University, Zhengzhou (2006).
120. Huang, Z.X.: Studies on novel Ru-Zn catalyst for selective hydrogenation of benzene to cyclohexene. Master thesis, Zhengzhou University, Zhengzhou (2006).
121. Wang, D.W.: Effect of Zn extent on catalytic performance and age of Ru-Zn catalyst. Master thesis, Zhengzhou University, Zhengzhou (2010).
122. Zhou, X.L.: Study on Ru-Zn catalyst for selective hydrogenation of benzene to cyclohexene. Master thesis, Zhengzhou University, Zhengzhou (2011).
123. Guo, W.: Study on reaction additives and surface modification of catalyst for selective hydrogenation of benzene. Master thesis, Zhengzhou University, Zhengzhou (2011).
124. Chen, Z.H.: The Ru catalyst modified with basic $ZnSO_4$ salt for selective hydrogenation of benzene to cyclohexene. Master thesis, Zhengzhou University, Zhengzhou (2012).
125. Pan, Y.J.: Ru catalysts modified with the fourth transition metal for selective hydrogenation of benzene to cyclohexene. Master thesis, Zhengzhou University, Zhengzhou (2013).
126. Dong, Y.Y.: Ru catalysts modified with rare earth element La and Ce for selective hydrogenation of benzene to cyclohexene. Master thesis, Zhengzhou University, Zhengzhou (2013).
127. Wang, H.X.: Study on Ru-type catalysts for selective hydrogenation of benzene. Master thesis, Zhengzhou University, Zhengzhou (2013).
128. Sun, H.J.: Selective hydrogenation of benzene to cyclohexene over Ru catalyst modified by $(Zn(OH)_2)_3(ZnSO_4)(H_2O)_x$ [D]. Doctor thesis, Zhengzhou University, Zhengzhou (2013).
129. Jiang, H.B.: Deactivation mechanism and kinetic studies on Ru-Zn catalyst for selective hydrogenation of benzene to cyclohexene. Master thesis, Zhengzhou University, Zhengzhou (2014).

130. Li, S.H.: Effect of Zn assistant precursor on Ru-Zn catalyst for selective hydrogenation of benzene. Master thesis, Zhengzhou University, Zhengzhou (2015).
131. Zhang, Y.Q.: The activation and selectivity modulation method and mechanism for selective hydrogenation of benzene. Master thesis, Zhengzhou University, Zhengzhou (2015).
132. Liu, S.C., Luo, G., Han, M.L., et al.: Characterization of the catalyst prepared by impregnation for selective hydrogenation of benzene to cyclohexene. Chin. J. Catal. **22**, 559–562 (2001)
133. Liu, S.C., Luo, G., Wang, H.R., et al.: Study on operation conditions for liquid phase selective hydrogenation of benzene to cyclohexene over Ru-M-B/ZrO_2 catalyst. Chin. J. Catal. **23**, 317–320 (2002)
134. Liu, S.C., Luo, G., Xie, Y.L., et al.: Study on Ru-Zn catalyst prepared by precipitation method for selective hydrogenation of benzene to cyclohexene. J. Mol. Catal. (China) **16**, 349–354 (2002a)
135. Liu, S.C., Guo, Y.Q., Yang, X.L., et al.: Kinetic equations for liquid-phase selective hydrogenation of benzene to cyclohexene. Chin. J. Catal. **24**, 42–46 (2003)
136. Liu, S.C., Luo, G., Xie, Y.L.: Study on Ru-Zn catalyst prepared by precipitation method for selective hydrogenation of benzene to cyclohexene. Chin. J. Mol. Catal. **16**(5), 349–354 (2002)
137. Liu, S.C., Zhu, B.Z., Luo, G., et al.: Characterization of amorphous Ru-M-B/ZrO_2 catalysts for partial hydrogenation of benzene to cyclohexene. Chin. J. Mol. Catal. **16**(3), 217–222 (2002)
138. Liu, S.C., Guo, Y.Q., Yang, X.L., et al.: Kinetic equations for liquid-phase selective hydrogenation of benzene to cyclohexene. Chin. J. Chem. **24**, 41–46 (2003b)
139. Liu, S.C., Wu, Y.M., Wang, Z., et al.: Study of a Ru-La/ZrO_2 catalyst prepared by precipitation method for selective hydrogenation of benzene to cyclohexene. J. Natur. Gas Chem. **14**, 226–232 (2005)
140. Si, R.J., Liu, S.C., Liu, Z.Y., et al.: Study on Ru-M_1-M_2 catalyst prepared by precipitation method for selective hydrogenation of benzene to cyclohexene. Chin. J. Mol. Catal. **19**, 141–145 (2005)
141. Liu, S.C., Liu, Z.Y., Zhao, S.H., et al.: Study on the nanosized amorphous Ru-Fe-B/ZrO_2 alloy catalyst for benzene selective hydrogenation to cyclohexene. J. Nat. Gas Chem. **15**, 319–326 (2006a)
142. Huang, Z.X., Liu, Z.Y., Wu, Y.M., et al.: Selective hydrogenation of benzene to cyclohexene by a novel Ru-Zn catalyst. Chin. J. Mol. Catal. **20**(3), 226–229 (2006)
143. Liu, S.C., Liu, Z.Y., Liu, Y.L., et al.: Effect of lanthanum on performance of RuB amorphous alloy catalyst for benzene selective hydrogenation. J. Rare Earths **24**, 456–460 (2006)
144. Liu, S.C., Liu, Z.Y., Wang, Z., et al.: A novel amorphous alloy Ru-La-B/ZrO_2 catalyst with high activity and selectivity for benzene selective hydrogenation. Appl. Catal. A: Gen. **313**, 49–57 (2006)
145. Liu, S.C., Liu, Z.Y., Zhao, S.H., et al.: Study on the amorphous alloy Ru-Fe-B/ZrO_2 nanosized catalyst for benzene selective hydrogenation to cyclohexene. J. Nat. Gas Chem. **150**, 1–9 (2006b)
146. Liu, S.C., Liu, Z.Y., Wang, Z., Wu, Y.M., et al.: Characterization and study on performance of the Ru-La-B/ZrO_2 amorphous alloy catalysts for benzene selective hydrogenation to cyclohexene under pilot conditions. Chem. Eng. J. **139**, 157–164 (2008)
147. Liu, Z.Y., Sun, H.J., Wang, D.B., et al.: The modifiable character of a novel Ru-Fe-B/ZrO_2 catalyst for benzene selective hydrogenation to cyclohexene. Chin. J. Chem. **28**, 1927–1934 (2010)
148. Liu, Z.Y., Sun, H.J., Wang, D.B., et al.: Selective hydrogenation of benzene to cyclohexene over Ru-Zn catalyst with nanosized zirconia as dispersant. Chin. J. Catal. **21**, 150–152 (2010)
149. Sun, H.J., Guo, W., Zhou, X.L., et al.: Process in Ru-based amorphous alloy catalysts for benzene selective hydrogenation to cyclohexene. Chin. J. Catal. **32**, 1–16 (2011)
150. Sun, H.J., Zhang, X.D., Chen, Z.H., et al.: Monolayer dispersed Ru-Zn catalyst and its performance in the selective hydrogenation of benzene to cyclohexene. Chin. J. Catal. **32**, 224–230 (2011)

151. Sun, H.J., Chen, Z.H., Guo, W., et al.: Effect of organic additives on the performance of nano-sized Ru-Zn catalyst. Chin. J. Chem. **29**, 369–373 (2011)
152. Zhou, X.L., Sun, H.J., Guo, W., et al.: Selective hydrogenation of benzene to m cyclohexene on Ru-based catalysts promoted with Mn and Zn. J. Nat. Gas Chem. **20**, 53–59 (2011)
153. Sun, H.J., Jiang, H.B., Li, S.H., et al.: Selective hydrogenation of benzene to cyclohexene over nanocomposite Ru-Mn/ZrO_2 catalysts. Chin. J. Catal. **34**, 684–694 (2013)
154. Sun, H.J., Pan, Y.J., Li, S.H., et al.: Selective hydrogenation of benzene to cyclohexene over Ce-promoted Ru catalysts. J. Eng. Chem. **22**, 710–716 (2013)
155. Sun, H.J., Pan, Y.J., Jiang, H.B., et al.: Effect of transition metals (Cr, Mn, Fe Co, Ni, Cu and Zn) on the hydrogenation properties of benzene over Ru-based catalyst. Appl. Catal. A: Gen. **464–465**, 1–9 (2013)
156. Sun, H.J., Dong, Y.Y., Li, S.H., et al.: The role of La in improving the selectivity to cyclohexene of Ru catalyst for hydrogenation of benzene. J. Mol. Catal. A: Chem. **368–369**, 119–124 (2013)
157. Sun, H.J., Jiang, H.B., Li, S.H., et al.: Effect of alcohols as additives on the performance of a nano-sized Ru-Zn (2.8%) catalyst for selective hydrogenation of benzene to cyclohexene. Chem. Eng. J., **218**, 415–424 (2013)
158. Sun, H.J., Wang, H.X., Jiang, H.B., et al.: Effect of $(Zn(OH)_2)_3(ZnSO_4)(H_2O)_5$ on the performance of Ru-Zn catalyst for benzene selective hydrogenation to cyclohexene. Appl. Catal. A: Gen. **450**, 160–168 (2013)

Chapter 2
Benzene Selective Hydrogenation Thermodynamics, Heterogeneous Catalytic Kinetics Catalysis Mechanism and Scientific Essence

It is disadvantageous in thermodynamics for benzene selective hydrogenation to cyclohexene. The reaction route and rate-controlling step could be changed through multiphase catalysis and the high selectivity of cyclohexene under the high conversion of benzene. The thermodynamic results reveal that the selective hydrogenation of benzene should proceed under suitable temperature and pressure.

The kinetic study shows that the appropriate temperature is 135–145 °C and hydrogen pressure is 4.5–5.0 MPa. The relatively high stirring speed could release the diffusion restriction of gas/liquid interface and liquid/liquid interface, which is necessary for the high yield and selectivity of cyclohexene. The suitable pH of the slurry is 5.6–5.8 and the concentration of H^+ would affect the existing status, surface species, and property and active sites of Zn^{2+}. Of which, the insoluble basic salts could obviously promote the selectivity of cyclohexene. The optimal ratio of water to benzene is 2:1, catalyst disperses in the water phase, and the salt additives adsorb on the catalyst surface. The strongest active sites are selectively covered by salt additives and the hydrophobic Ru particles would be transformed into hydrophilic. The kinetic results revealed that the reaction order of benzene conversion is first-order to concentration, second-order to hydrogen (low pressure), fraction order to hydrogen (medium pressure), and zero-order to hydrogen (high pressure). The further hydrogenation of cyclohexene to cyclohexane is zero-order to concentration, second-order to hydrogen (low pressure), fraction order to hydrogen (medium pressure), and zero-order to hydrogen (high pressure). The apparent activation energy is 58.5 kJ/mol.

In the multiphase catalytic system of benzene selective hydrogenation, the benzene and hydrogen are all have strong adsorption and dissociation ability on the pure Ru component. Around the activated benzene molecules, the concentration of hydrogen atom is relatively high and benzene is easily hydrogenated to cyclohexane. The electron enrichment of Ru caused by the electronic and geometry effect of the second metal decreased the space density of Ru active sites. The hydrogen concentration could be decreased through the hydrophilic modification and the diffusion

Z. Liu et al., *Catalytic Technology for Selective Hydrogenation of Benzene to Cyclohexene*,
https://doi.org/10.1007/978-981-15-6411-6_2

restriction of monimolimnion, inhibiting the hydrogenation of benzene to cyclohexane directly. As soon as the cyclohexene produces, it would desorb quickly and the yield and selectivity of cyclohexene are increased.

This chapter introduces the multiphase catalytic kinetics of benzene selective hydrogenation to cyclohexene and its mechanism.

2.1 Thermodynamics of Benzene Selective Hydrogenation

2.1.1 Thermodynamic Data

Figure 2.1 shows the thermodynamic data of each benzene hydrogenation step. In Fig. 2.1, it is proposed that the hydrogenation of benzene contains two lines, one is the hydrogenation of benzene to cyclohexene, and further into cyclohexane. The other is the direct hydrogenation of benzene to cyclohexane. Based on the data of $\Delta_r G_m^0$ and $\Delta_r H_m^0$, the selective hydrogenation of benzene to cyclohexene is disadvantageous in thermodynamic theory, and the cyclohexane is easier to produce than cyclohexene.

Due to the restriction of thermodynamics, it has been a dream for selective hydrogenation of benzene to cyclohexene though it is well known [1]. The benzene would be completely hydrogenated to cyclohexane without cyclohexene through the traditional gaseous catalytic hydrogenation method. But the cyclohexene could be obtained via multiphase catalysis, during which the reaction route is changed [2].

2.1.2 Effect of Temperature

The values of $\Delta_r H_m^0$ of each benzene hydrogenation step are all in minus, which means the reaction is exothermal. According to the reaction isobar, the increasing temperature is adverse for each reaction step. By comparing the absolute value of $\Delta_r H_m^0$ between benzene hydrogenation to cyclohexane and selective hydrogenation to cyclohexene, the high temperature could inhibit the production of cyclohexane

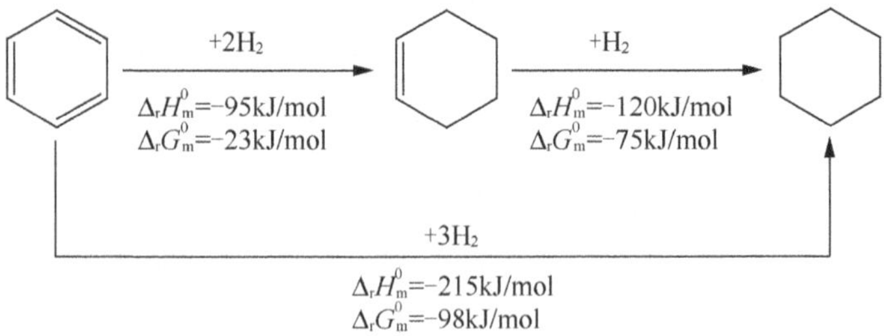

Fig. 2.1 Thermodynamic data of benzene selective hydrogenation steps

and promote the selectivity of cyclohexene. Thus, the selective hydrogenation of benzene to cyclohexene should proceed under the suitable temperature.

$$\frac{\mathrm{d}\ln K^0}{\mathrm{d}T}=\frac{\Delta_r H_m^0}{RT^2}$$

2.1.3 *Effect of Pressure*

According to the relationship between equilibrium constant and temperature, the relatively high hydrogen pressure is better for cyclohexene production. Because there is hydrogen gas involving in the selective hydrogenation of benzene and the volume change ΔV_m of each reaction is less than zero. In addition, the hydrogen consumption for selective hydrogenation of benzene to cyclohexene among the one-step and sub-step hydrogenation to cyclohexane.

$$\left(\frac{\partial \ln K_x}{\partial p}\right)_T=-\frac{\Delta V_m}{RT}$$

2.1.4 *Effect of Inert Gas*

During the selective hydrogenation of benzene to cyclohexene, the air in the clave is often replaced with nitrogen, so the partial nitrogen would remain in hydrogen. When the total pressure is constant, the existence of inert gas would dilute hydrogen, which is similar to the release of total pressure.

For the detailed chemical reaction, not only thermodynamics, but also kinetics should be considered. As to the exothermal reaction, the effect of thermodynamic and kinetic factor on chemical equilibrium are reverse. And for multiphase catalysis, it is more complicated.

2.2 Heterogeneous Catalytic Kinetics of Benzene Selective Hydrogenation

After the determination of thermodynamics, the most important problems are focus on kinetics and the problems occurs in industry almost belong to the range of kinetics.

2.2.1 *Macroscopic Kinetics*

(1) Effect of hydrogen pressure

The selective hydrogenation of benzene to cyclohexene should proceed under the relatively high hydrogen pressure. But in the different reaction systems (different reaction temperatures and catalysts), the effect of hydrogen pressure on the hydrogenation rate of benzene and selectivity of cyclohexene is also different. For example, in the catalytic system of 40 °C, Ru black catalyst and water, the hydrogenation rate of benzene increases with the enhancement of hydrogen pressure. It attains the maximum at 4 MPa and then decreases between 4–9 MPa [3]. In the reaction system of 130 °C, Ru/La_2O_3-ZnO catalyst and NaOH aqueous solution, the hydrogenation rate of benzene climbs the peak at 2.34 MPa, and it would be inhibited at 2.34–5.10 MPa [4]. Under the condition of 150 °C, Ru-Zn catalyst and $ZnSO_4$ additive, the selectivity of cyclohexene is constant at the hydrogen pressure of 2.0–5.0 MPa [5].

The effect of hydrogen pressure could be explained from the view of competing for adsorption on active sites. The adsorption rate of hydrogen, benzene, and cyclohexene on active sites could be ranked as hydrogen>cyclohexene>benzene [4]. The adsorption and desorption are dynamic equilibrium and the surface reaction is the rate-controlling step. The hydrogenation rate of benzene is restricted due to the low coverage of hydrogen on catalyst surface under low hydrogen pressure. When the coverage of hydrogen is equal to benzene, the hydrogenation rate attains the maximum. And then it would decrease with the further increasing hydrogen pressure due to the low coverage of benzene.

The selectivity of cyclohexene is mainly dependent on the surface property of catalyst, desorption rate of intermediate cyclohexene, and the production rate of cyclohexane. The catalytic activity and selectivity could be promoted after the catalyst surface modified and the cyclohexene could be obtained on the modified hydrophilic catalyst [2]. The effect of hydrogen pressure on the selectivity of cyclohexene is not obvious and the suitable pressure is 4.5–5.0 MPa.

(2) Effect of mass transfer

The multiphase catalytic system of benzene selective hydrogenation contains hydrogen, benzene, water, solid catalyst and dispersion agent, existing gas/liquid, liquid/liquid, and liquid/solid interface. The hydrogenation reaction occurs on the catalyst surface, so the activity of catalyst and selectivity of cyclohexene depends on the mass transfer. The suitable stirring speed could eliminate the diffusion restriction of gas/liquid and liquid/liquid interface. The thinner stranded membrane of liquid/solid interface would be a benefit for the activation of reactant molecules, which is necessary for the high yield and selectivity of cyclohexene.

During the selective hydrogenation of benzene to cyclohexene, the increasing stirring speed (below 1000 r/min) would obviously accelerate the reaction rate. When the stirring speed above 1000 r/min, the reaction rate increases slowly, the yield and selectivity of cyclohexene attain the maximum when the stirring speed is 1500 r/min. The catalyst would be smashing and stick to the reactor surface when the stirring speed is 2000 r/min. Thus, the suitable stirring speed should be controlled in the range of 1000–1500 r/min, and the diffusion resistance of hydrogen and benzene could be lowered as far as possible.

The treatment for the mass transfer phenomenon in the liquid/solid interface during the liquidus hydrogenation is provided by the experiment and theory.

In the experiment, the effect of mass transfer in liquid/solid interface on the reaction rate could be determined through the measure of reaction rate change with the amount of catalyst. According to the correlation results of $1/r_{0,\,H_2}$ to the $1/w_{cat.}$, it could be obtained that the slope of the line is proportional to the summary resistance of liquid/solid and cogne diffusion. As for the system containing Ru catalyst and $ZnSO_4$ solution, the relatively large amount of catalyst would decrease the coverage of $ZnSO_4$. When the ratio of $ZnSO_4$ to catalyst is kept constant, the coverage of $ZnSO_4$ on the catalyst surface would not change. The experimental results show that with the increase of catalyst amount, the selectivity of cyclohexene would decrease from 70 to 35% when the concentration of $ZnSO_4$ is constant. While the selectivity of cyclohexene would not change (50–55%) under the constant ratio of $ZnSO_4$ to the catalyst. The increasing coverage of $ZnSO_4$ on the catalyst surface would control the diffusion of hydrogen, decreasing the benzene hydrogenation rate of each step. And the decreasing hydrogenation rate of cyclohexene would directly enhance the selectivity of cyclohexene. The more hydrophilic of catalyst surface is, the more monimolimnion around catalyst, and the higher selectivity of cyclohexene is [5–7].

In theory, the restriction extent of external and internal diffusion could be judged through the calculation of the Carberry number (Ca) [8] and Wheeler-Weisz group ($\eta\varphi^2$) [9, 10].

The Ca of H_2 could be calculated from Eq. (2.2.1).

$$Ca = \frac{r_{\text{obs},\,H_2}}{k_{\text{ls}} \frac{6 \cdot w}{d_{\text{p}} \cdot \rho_{\text{p}}} C^{*}_{\text{H}_2 \cdot \text{l}}} \tag{2.2.1}$$

where $r_{\text{obs},\text{H}_2}$ is the apparent reaction rate of hydrogen (mol/s), k_{ls} is the mass transfer coefficient (m/s), w is the amount of catalyst (g), d_{p} and ρ_{p} is the particle diameter (m) and density (g/m^3), respectively, $C^{*}_{\text{H}_2 \cdot \text{l}}$ is the equilibrium concentration of hydrogen in the liquid phase (mol/cm^3). The Carberry number (Ca) represents the ratio of apparent reaction rate to the maximum mass transfer or the concentration difference in liquid/solid interface, characterizing the restriction extent of external diffusion. When the value of Ca below 0.05, the external diffusion could be ignored.

The $\eta\varphi^2$ could be calculated through Eq. (2.2.2).

$$\eta\varphi^2 = \frac{d_p^2}{4D_{eff}}\left(\frac{1}{V_p}\cdot\frac{dn}{dt}\right)\frac{1}{C_s} \tag{2.2.2}$$

where D_{eff} is the efficient diffusion coefficient (m^2/s), and it could be replaced with the diffusion coefficient of hydrogen in water, 4 is the bending factor τ, C_s is the surface concentration of catalyst. $\eta\varphi^2$ characterizes the restriction extent of internal diffusion. When it bellows 0.1, the internal diffusion could be ignored.

While the different conclusions would be obtained under different temperatures, catalysts, and catalytic systems for selective hydrogenation of benzene.

Under the reaction condition of 150 °C, 5 MPa hydrogen, 1500 r/min, 75 mL water, water/benzene = 2.67, 0.2 g catalyst, and 3.5 g $ZnSO_4 \cdot 7H_2O$, the calculation results from Eq. 2.2.1 show that with the proceeding of reaction, the value of *Ca* of hydrogen decreases from 1.0 to 0.2, while the benzene increases from 0.3 to 0.5. That means the external diffusion of hydrogen decreases gradually while that of benzene increases. Based on the calculation results of Eq. (2.2.2), the value of $\eta\varphi^2$ characterizes the internal diffusion restriction above 100 in the beginning time, which means that the control of internal diffusion of hydrogen is heavy and the reaction only occurs on the external surface of the catalyst. In the beginning time of hydrogenation, the conversion of benzene is low and the selectivity of cyclohexene is high. With the proceeding of reaction, the conversion increases, which the selectivity decreases. So it could be determined that the selective hydrogenation of benzene to cyclohexene is seriously restricted by the diffusion of liquid/solid interface [5]. And such restriction could decrease the production of cyclohexane and promote the selectivity of cyclohexene.

The activity and selectivity of Ru/Al_2O_3 catalysts with different Ru contents from precursors were tested in the presence of water at 60 °C. The calculation results of *Ca* and $\eta\varphi^2$ reveal that in the existence of water, the production of cyclohexene could not be simply attributed to the diffusion restriction of liquid/solid interface, but such restriction plays an important role in enhancing the yield and selectivity of cyclohexene [11, 12].

When the Ru/La_2O_3-ZnO catalyzed selective hydrogenation of benzene to cyclohexene in NaOH aqueous solution under 125–175 °C and 2.34–5.10 MPa hydrogen pressure, the values of *Ca* and $\eta\varphi^2$ are all small, but the activation energy is relatively high (120–150 °C, 79 kJ/mol). So it could be considered that the reaction proceeds in the range of kinetics [4].

Under the conditions of 150 °C, 5.0 MPa hydrogen pressure, 1.96 g Ru-Zn catalyst (2.8 wt%, microcrystalline diameter: 3–5 nm, particle size distribution: 0.5–5 μm), 45.7 g $ZnSO_4 \cdot 7H_2O$, 140 mL benzene and 280 mL H_2O, the high yield and selectivity of cyclohexene is obtained when the catalyst is pretreated for 22 h. It is concluded that the surface property and the diffusion restriction of hydrogen in liquid/solid interface is crucial for the high selectivity of cyclohexene through the calculation of *Ca* and $\eta\varphi^2$ in combination with the catalyst component, slurry condition, and pretreatment process [13].

(3) Effect of pH of the slurry

In industry, the pH of the reaction system is an important controlling index.

$$Zn^{2+} + H_2O \rightleftharpoons [Zn(OH)]^+ + H^+ \quad (2.2.3)$$

$$Zn^{2+} + 2H_2O \rightleftharpoons Zn(OH)_2 + 2H^+ \quad (2.2.4)$$

$$3Zn(OH)_2 + ZnSO_4 + 3H_2O \rightleftharpoons (Zn(OH)_2)_3 \cdot ZnSO_4 \cdot 3H_2O \downarrow \quad (2.2.5)$$

The catalytic system of benzene selective hydrogenation contains a large amount of $ZnSO_4$ and the pH of slurry would affect the following equilibrium and determine the existence of Zn^{2+}. Although the species of $(Zn(OH)_2)_3 \cdot ZnSO_4 \cdot 3H_2O$, $Zn(OH)_2$, $[Zn(OH)]^+$, and Zn^{2+} could all adsorb on the surface of Ru catalyst, occupy the strongest active sites preferentially, transfer the hydrophobic surface of Ru catalyst into hydrophilic, decrease the activity and promote the selectivity of cyclohexene. But the different existences of Zn^{2+} species have different impact extents and the insoluble basic salt compound $(Zn(OH)_2)_3 \cdot ZnSO_4 \cdot 3H_2O$ could obviously promote the selectivity of cyclohexene. This salt compound, which is a benefit for the formation of backwater film, could not only control the diffusion of hydrogen and decrease the coverage of hydrogen on the catalyst surface, but also promote the desorption of cyclohexene due to the dissolution difference between benzene and cyclohexene in water. In addition, the effect of zinc species on the selectivity of cyclohexene could be ranked as follows:

$$(Zn(OH)_2)_3 \cdot ZnSO_4 \cdot 3H_2O > Zn(OH)_2 > [Zn(OH)]^+ > Zn^{2+}$$

The experimental results show that the suitable pH of the slurry is 5.6–5.8.

(4) Effect of water and salt additives

The existence of water and the addition of salt is necessary for the increase of yield and selectivity of cyclohexene, and the experiments show the suitable ratio of water to benzene is 2:1. The catalyst disperses in the water phase and the reaction occurs on the catalyst surface. The additional salt additives could selectively cover the strongest active sites and transfer the hydrophobicity of catalyst particles into hydrophilicity, leading to the formation of backwater film around Ru particles.

The salt additives, such as $ZnSO_4$, $CoSO_4$, $FeSO_4$, $MnSO_4$, and so on, should have adsorption ability on the Ru catalyst surface and be difficult to be reduced to metal, and the $ZnSO_4$ is the best. The additives could decrease the catalytic activity and enhance the selectivity of cyclohexene. As $MgSO_4$ and Na_2SO_4, these salts could not be adsorbed by Ru, so it only exists in the slurry and has no effect. For $CuSO_4$,

$Pb(NO_3)_2$ and $Fe_2(SO_4)_3$, the metal ions not only have strong adsorption ability, but also are easy to be reduced. The reduced metal atom would block the active sites of Ru. As to $CdSO_4$ and $In_2(SO_4)_3$, the low concentration of salt could almost adsorb on Ru completely, leading to the deactivation of the catalyst. In theory, salt has an optimal concentration in water to keep the equilibrium coverage on the catalyst surface, and it should be modulated to the extent that the diffusion of hydrogen is the rate-controlling step to obtain the highest yield and selectivity of cyclohexene [6].

The additive $ZnSO_4 \cdot 7H_2O$ has the cation reduce potential of $\varphi^0_{Zn^{2+}/Zn} = -0.7628V$, the Zn^{2+} is difficult to be reduced to metal Zn under the hydrogenation condition, so the $ZnSO_4$ is considered to be one of the best additives. But the research results show that little amount of Zn^{2+} is reduced to metal Zn under hydrogenation condition due to the overflow phenomenon of hydrogen.

$$Zn^{2+}_{ads}(n - x)H_2O + 2Hads = Zn^0_{ads} + 2H^+ + (n - x)H_2O \qquad (2.2.6)$$

where n is the water molecule number of first Zn^{2+} hydration shell, x is the outgoing water molecule number during chemical adsorption. The quantum chemistry calculation results show that the chemical adsorption ability of benzene and cyclohexene would be decreased due to the existence of Zn. The Zn-center catalyst has a passivated surface and it would inhibit the further hydrogenation of cyclohexene to cyclohexane [14, 15].

The combination of macrokinetics research and industrial operation condition would provide the reference and guidance.

2.2.2 *Heterogeneous Catalytic Kinetic Equations*

In 2003, the highly selective amorphous Ru-M-B/ZrO_2 catalysts were prepared by chemical reduction method. The experimental results show that the diffusion effect of reactants could be eliminated when the temperature is below 150 °C, the stirring speed is lower than 900 r/min, and the catalyst particle size is smaller than 0.15 μm. And the kinetic parameter and rate equations of each hydrogenation step were obtained in the ranging of kinetic reaction [16, 17].

The conclusions could be obtained as follows if the hydrogenation of benzene is a simple and continuous reaction, namely sub-step hydrogenation of benzene to cyclohexane via cyclohexene, ignoring the direct hydrogenation to cyclohexane.

The reaction order of benzene conversion is first-order to the benzene concentration, is second-(fraction and zero) order to hydrogen under low (medium and high) pressure. The reaction order of further hydrogenation of cyclohexene into cyclohexane is zero-order to cyclohexene concentration, is second (fraction and zero)-order to hydrogen under low (medium and high) pressure.

The hydrogenation kinetic equations of benzene hydrogenation are as follows:

Under low pressure

$$-\mathrm{d}c(\mathrm{BZ})/\mathrm{d}t = k_1 c(\mathrm{BZ}) \cdot p^2(H_2) \tag{2.2.7}$$

Under medium pressure

$$-\mathrm{d}c(\mathrm{BZ})/\mathrm{d}t = \frac{k_1 c(\mathrm{BZ}) \cdot \sigma p^2(H_2)}{1 + \sigma p^2(H_2)} \tag{2.2.8}$$

Under high pressure

$$-\mathrm{d}c(\mathrm{BZ})/\mathrm{d}t = k_1 c(\mathrm{BZ}) \tag{2.2.9}$$

The σ in Eq. (2.2.8), is equal to the Henry coefficient of hydrogen dissolution in the reaction system, the low, medium, and high pressure of hydrogen are corresponding to Eq. (2.2.7), Eq. (2.2.8), and Eq. (2.2.9), respectively. When the $p(H_2)$ is medium, the reaction order is a fraction, so Eq. (2.2.8), has universality.

And the formation rates of cyclohexane are as follows:

Under low pressure

$$\mathrm{d}c(\mathrm{HA})/\mathrm{d}t = k_2 p^2(H_2) \tag{2.2.10}$$

Under medium pressure

$$\mathrm{d}c(\mathrm{HA})/\mathrm{d}t = \frac{k_2 \sigma p^2(H_2)}{1 + \sigma p^2(H_2)} \tag{2.2.11}$$

Under high pressure

$$\mathrm{d}c(\mathrm{HA})/\mathrm{d}t = k_2 \tag{2.2.12}$$

And the formation rates of cyclohexene are as follows:

Under low pressure

$$\frac{\mathrm{d}c(\mathrm{HE})}{\mathrm{d}t} = -\frac{\mathrm{d}c(\mathrm{BZ})}{\mathrm{d}t} + \frac{\mathrm{d}c(\mathrm{HE})}{\mathrm{d}t} = k_1 c(\mathrm{BZ}) p^2(H_2) - k_2 p^2(H_2) \tag{2.2.13}$$

Under medium pressure

$$\frac{\mathrm{d}c(\mathrm{HE})}{\mathrm{d}t} = -\frac{\mathrm{d}c(\mathrm{BZ})}{\mathrm{d}t} + \frac{\mathrm{d}c(\mathrm{HE})}{\mathrm{d}t} = \frac{k_1 c(\mathrm{BZ}) \sigma p^2(H_2)}{1 + \sigma p^2(H_2)} - \frac{k_2 \sigma p^2(H_2)}{1 + \sigma p^2(H_2)} \tag{2.2.14}$$

Under high pressure

$$\frac{\mathrm{d}c(\mathrm{HE})}{\mathrm{d}t} = -\frac{\mathrm{d}c(\mathrm{BZ})}{\mathrm{d}t} + \frac{\mathrm{d}c(\mathrm{HE})}{\mathrm{d}t} = k_1 c(\mathrm{BZ}) - k_2 \tag{2.2.15}$$

If the original concentration of benzene, cyclohexene, and cyclohexane are x_0, 0, and 0, respectively, at the beginning time, and the concentration would change to x, y, and z after the reaction time of t. So the Eqs. (2.2.9), (2.2.15), and (2.2.12) could be transformed into Eqs. (2.2.16), (2.2.17), and (2.2.18).

$$x = x_0 \cdot e^{-k_1 t} \tag{2.2.16}$$

$$y = x_0(1 - e^{-k_1 t}) - k_2 t \tag{2.2.17}$$

$$z = k_2 t \tag{2.2.18}$$

According to formulas of Eqs. (2.2.16) to (2.2.18), the calculated data are in accordance with the experimental results, the relative average deviation is ± 10%.

The explanation for kinetic equations: due to the existence of water in the reaction system, the concentration of hydrogen in water is proportional to hydrogen pressure. And the concentration of benzene, cyclohexene, and cyclohexane in water is proportional to their molar fraction in the organic phase. The surface of the hydrophilic catalyst is covered with a film of backwater, so the diffusion and adsorption of benzene and hydrogen through the liquid/solid interface are restricted. Owing to the occurrence of hydrogenation on the catalyst surface, the reaction rate and adsorption is related to the coverage of hydrogen and benzene. The different reaction orders reflect the adsorption extent of benzene, hydrogen, and cyclohexene on the catalyst surface. The zero-order means the saturated adsorption on the catalyst surface and the reaction rate is independence on the concentration in liquid phase. The second-order means low adsorption and coverage on the catalyst surface, the reaction rate is exponential to the concentration. And the first-order means relatively strong adsorption and coverage on the catalyst surface, the reaction rate is in line with the concentration in liquid phase.

The multiphase catalytic kinetics of benzene selective hydrogenation on Ru catalysts still have the kinetic characterization of simple continuous reactions. With the extension of reaction time, the concentration of benzene decreases, the concentration of product cyclohexane increases, but the concentration of intermediate cyclohexene decreases after climbing the peak. And the reaction time when cyclohexene attains the highest concentration could be calculated by Eq. (2.2.19).

$$t_m = \frac{\ln k_2 - \ln k_1}{k_2 - k_1} \tag{2.2.19}$$

where k_1 and k_2 represents the rate constant of benzene hydrogenation to cyclohexene and further hydrogenation of cyclohexene to cyclohexane, respectively. The intermediate cyclohexene is the main product and cyclohexane is the by-product. The t_m in Eq. (2.2.19), is the residue time of benzene in industry, and it is a very important controlling parameter. The benzene and slurry flow could be determined by t_m, efficient volume of reaction still and the ratio of water to benzene.

2.2.3 Apparent Activation Energy

The selective hydrogenation of benzene proceeds under the conditions of 5.0 MPa hydrogen pressure, 1.96 g Ru-Zn catalyst with the microcrystalline of 3–5 nm and particle size of 0.5–5 μm, 9.8 g ZrO_2, 45.7 g $ZnSO_4 \cdot 7H_2O$, 140 mL benzene, 280 mL H_2O, and 1400 r/min.

Table 2.1 lists the benzene conversion and cyclohexene selectivity at different temperatures.

The curve of C_{BZ}-t is figured with the data of Table 2.1, then the rate constant k could be calculated from C_{BZ}-t curves at different temperatures. The macro-activation energy and frequency coefficient could be obtained as $E_a = 58.5$ kJ/mol and $A = 3.17 \times 10^6$ from the slope and intercept of ln$k$-1/T figure. The relation between rate constant and reaction temperature of benzene hydrogenation could be expressed as

$$k = 3.17 \times 10^6 \exp(-7304/T)$$

The macro-activation energy could be used to judge whether control of diffusion or reaction. In general, the activation energy is 15–20 kJ/mol under diffusion control, while it is above 40 kJ/mol under reaction control. At such conditions, the hydrogenation of benzene belongs to reaction control, but the activation energy is low, meaning the crucial status of diffusion control.

Table 2.1 Benzene conversion and cyclohexene selectivity at different temperature

t/min	C_{BZ}/%, S_{HE}/%	T/K				
		423	413	403	393	373
5	C_{BZ}	30.94	24.55	15.76	13.27	3.89
	S_{HE}	76.44	78.70	75.29	67.01	63.76
15	C_{BZ}	79.42	65.61	44.78	35.01	11.16
	S_{HE}	63.02	70.64	72.93	73.41	69.24
30	C_{BZ}	98.86	89.03	74.53	62.98	24.31
	S_{HE}	45.25	58.63	63.93	66.37	68.30
45	C_{BZ}	99.46	96.78	88.59	79.49	34.01
	S_{HE}	31.38	45.96	54.78	58.66	66.04
60	C_{BZ}	99.85	99.14	95.58	88.50	43.00
	S_{HE}	21.22	34.48	44.99	51.46	63.73

2.2.4 Selectivity and Yield of Cyclohexene

Due to the different effects of temperature on each hydrogenation rate, the temperature has a crucial impact on the yield and selectivity of cyclohexene. The experimental results show that the hydrogenation rate could be increased by 4 times, and the selectivity of cyclohexene would be enhanced from 3.4 to 71.4% when the reaction temperature rise from 50 to 150 °C [5].

According to Horiuti–Polanyi mechanism, the selectivity of cyclohexene could be expressed as Eq. (2.2.20) [6].

$$S_{\mathrm{HE}} = \frac{r_{\mathrm{des}} - r_{\mathrm{ads}}}{r_{\mathrm{H}}} \tag{2.2.20}$$

where the S_{HE} means the selectivity of cyclohexene, r_{des} and r_{ads} represents the adsorption and desorption rate of cyclohexene, respectively, r_{H} means the hydrogenation rate of cyclohexene. It could be known from Eq. (2.2.20), that to promote the selectivity of cyclohexene, the desorption rate (r_{des}) should be increased, while the adsorption (r_{ads}) and the further hydrogenation rate (r_{H}) of cyclohexene should be decreased.

The adsorption rate of cyclohexene (r_{ads}) would be very slow if the catalyst surface is covered with a film of backwater. The cyclohexene would have very low potential energy, dissolution, and diffusion rate in water when the cyclohexene and cyclohexane drops dissolve in benzene, thus $r_{\mathrm{ads}} \approx 0$.

The desorption rate of cyclohexene r_{des} could be expressed as

$$r_{\mathrm{des}} = k_{\mathrm{des}} \theta_{\mathrm{C_6H_{10}}} \mathrm{e}^{\left(\frac{-\Delta H^{\neq}}{RT}\right)} \tag{2.2.21}$$

where k_{des} is the desorption rate of cyclohexene, $\theta_{\mathrm{C_6H_{10}}}$ is the coverage of cyclohexene, $\Delta H^{\neq}$ is the desorption enthalpy, the value is approximately equal to the negative value of adsorption enthalpy, $-\Delta H_{\mathrm{ads}} = 25$–$40$ kJ/mol.

The hydrogenation rate of cyclohexene could be expressed as

$$r_{\mathrm{H}} = k_{\mathrm{H}} \theta_{\mathrm{H}}^2 \theta_{\mathrm{C_6H_{10}}} \mathrm{e}^{\left(\frac{-E_{\mathrm{H}}^{\neq}}{RT}\right)} \tag{2.2.22}$$

where k_{H} is the hydrogenation rate constant, θ_{H} and $\theta_{\mathrm{C_6H_{10}}}$ are the coverage of hydrogen and cyclohexene, respectively, the $E_{\mathrm{H}}^{\neq}$ is the hydrogenation activation energy of cyclohexene.

According to the selectivity definition of cyclohexene, the Eq. (2.2.20), could be simplified as

$$S_{\mathrm{HE}} = \frac{k_{\mathrm{des}}}{k_{\mathrm{H}} \theta_{\mathrm{H}}^2} \mathrm{e}^{\left(\frac{-(\Delta H^{\neq} - E_{\mathrm{H}}^{\neq})}{RT}\right)} \tag{2.2.23}$$

The hydrogenation rate of cyclohexene is high although the temperature is very low, so it could be considered that $E_{\mathrm{H}}^{\neq} \approx 0$, then the Eq. (2.2.23), could be changed to (2.2.24).

$$S_{\mathrm{HE}} \approx \frac{k_{\mathrm{des}}}{k_{\mathrm{H}}\theta_{\mathrm{H}}^{2}} \mathrm{e}^{\left(\frac{-\Delta H^{\neq}}{RT}\right)} \tag{2.2.24}$$

$\Delta H^{\neq} = 25 - 40\ \mathrm{kJ/mol}$. It could be seen from Eq. (2.2.24) that with the increase of temperature, the selectivity of cyclohexene enhances heavily, which is agreed with the experimental results.

The influence of temperature on the yield and selectivity of cyclohexene could also be discussed from the continuous reaction kinetic characterization. As to the simple continuous reaction

$$C_6H_6(\mathrm{BZ}) \xrightarrow{2\mathrm{H}_2, k_1} \mathrm{C_6H_{10}(HE)} \xrightarrow{\mathrm{H}_2, k_2} \mathrm{C_6H_{12}(HA)} \tag{2.2.25}$$

$$y_{\mathrm{m}} = a\left(\frac{k_1}{k_2}\right)^{\frac{k_2}{k_2 - k_1}} \tag{2.2.26}$$

where k_1 and k_2 are, respectively, the rate constant of benzene hydrogenation to cyclohexene and cyclohexene further hydrogenation to cyclohexane, y_{m} is the maximum concentration of intermediate cyclohexene, a is the initial concentration of reactant benzene. Obviously, y_{m} is related to a, k_1 and k_2. If $k_1 \gg k_2$, the value of y_m is relatively high, or is relatively low.

According to the Arrhenius formula for continuous reaction, raising the temperature is a benefit for the reaction with high activation energy. The activation energy of benzene hydrogenation to cyclohexene is above 40 kJ/mol, while the activation energy of cyclohexene further hydrogenation to cyclohexane is almost 0. Thus, the k_1 increases more than k_2 with the rising temperature. And it could be seen from Eq. (2.2.26), that the increasing y_{m} leads to the promoting yield and selectivity of cyclohexene.

The impact of temperature on yield and selectivity of cyclohexene could also be explained from the influence of temperature on adsorption and desorption rate. With the increasing temperature, the hydrogenation rate of benzene reaches the maximum and then decreases. Because the initial increasing temperature could promote the coverage of benzene and hydrogen on the catalyst surface, but desorption of benzene and hydrogen on the catalyst surface occurs and the coverage reduces with the further enhancement of temperature.

The experiments show that the suitable temperature of benzene selective hydrogenation is around 150 °C. In such temperature range, the high conversion of benzene, the relatively high yield and selectivity of cyclohexene could be all obtained. In industry, in order to prevent the agglutination of Ru nanoparticles and extend the catalyst age, the operating temperature is often controlled as 135–145 °C. In the later period, the temperature could be enhanced to 150 °C, to fully utilize the potential of the catalyst.

The kinetic equations can be seen as the medium of quantitative chemistry, on one hand, it could provide a reference for the reactor design and condition determination. For zero-order reaction, the rate is independent of the reactor and stir. As to first-order, the plug flow reactor is advantageous for the high selectivity of cyclohexene. On the other hand, it could also be a benefit for the proposal of reaction route and mechanism.

2.2.5 Microscopic Kinetics

According to the Langmuir-Hinshelwood route, the reaction occurs between the activated benzene and hydrogen that adsorbed on the surface of solid catalysts. Without considering the diffusion restriction, the proposed elemental reactions are as follows: the benzene and hydrogen molecules adsorb on the surface, the benzene and hydrogen get converted into intermediate cyclohexene, benzene hydrogenates directly to cyclohexane, cyclohexene further hydrogenates to cyclohexane, the cyclohexene and cyclohexane get desorbed. The benzene, cyclohexene, and cyclohexane are all adsorbed on the same active sites one by one, and the hydrogenation route is as follows: [18, 19].

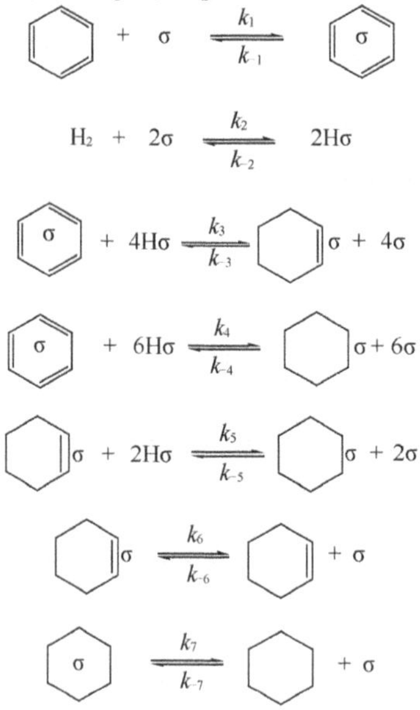

①
②
③
④
⑤
⑥
⑦

The reaction rate of the first step could be listed as Eq. (2.2.27).

$$k_1 C_{BZ} \theta_0 = k_{-1} \theta_{BZ}$$

$$K_1 = \frac{k_1}{k_{-1}} = \frac{\theta_{BZ}}{C_{BZ}\theta_0}$$

$$\theta_{BZ} = K_1 C_{BZ}\theta_0 \tag{2.2.27}$$

where C_{BZ} is the concentration of benzene in water, θ_0 and θ_{BZ} are the blank active sites and the coverage of benzene, respectively, k_1 and k_{-1} are the absorption and desorption rate constant of benzene, respectively, and K_1 is the adsorption equilibrium constant of benzene.

The reaction rate of the second step could be listed as Eq. (2.2.28).

$$k_2 p_{H_2}\gamma\theta_0^2 = k_{-2}\theta_H^2$$

$$K_2 = \frac{k_2}{k_{-2}} = \frac{\theta_H^2}{p_{H_2}\gamma\theta_0^2}$$

$$\theta_H = \theta_0\sqrt{K_2\gamma p_{H_2}} \tag{2.2.28}$$

where p_{H_2} is the hydrogen pressure, γ is the Henry coefficient of hydrogen in the liquid phase, θ_H is the coverage of hydrogen, k_2 and k_{-2} are the absorption and desorption rate constant of hydrogen, respectively, and K_2 is the adsorption equilibrium constant of hydrogen.

According to the measurement result of activation energy, if the third step is the rate-controlled step, the rate equation could be expressed as

$$r = -\mathrm{d}C_{BZ}/\mathrm{d}t = k_3\theta_{BZ}\theta_H^4 - k_{-3}\theta_{HE}\theta_0^4 \tag{2.2.29}$$

where θ_{HE} is the coverage of cyclohexene.

As for the fourth step

$$k_4\theta_{BZ}\theta_H^6 = k_{-4}\theta_{HA}\theta_0^6$$

$$K_4 = \frac{k_4}{k_{-4}} = \frac{\theta_{HA}\theta_0^6}{\theta_{BZ}\theta_H^6}$$

For the fifth step

$$k_5\theta_{HE}\theta_H^2 = k_{-5}\theta_{HA}\theta_0^2$$

$$K_5 = \frac{k_5}{k_{-5}} = \frac{\theta_{HA}\theta_0^2}{\theta_{HE}\theta_H^2} \tag{2.2.30}$$

For the sixth step

$$k_6\theta_{\rm HE} = k_{-6}C_{\rm HE}\theta_0$$

$$K_6 = \frac{k_6}{k_{-6}} = \frac{C_{\rm HE}\theta_0}{\theta_{\rm HE}}$$

$$\theta_{\rm HE} = \frac{C_{\rm HE}\theta_0}{K_6} \quad (2.2.31)$$

For the seventh step

$$k_7\theta_{\rm HA} = k_{-7}C_{\rm HA}\theta_0$$

$$K_7 = \frac{k_7}{k_{-7}} = \frac{C_{\rm HA}\theta_0}{\theta_{\rm HA}}$$

$$\theta_{\rm HA} = \frac{C_{\rm HA}\theta_0}{K_7} \quad (2.2.32)$$

The coverage summary of each substrate and blank active sites θ_0 on the catalyst surface is equal to 1

$$\theta_0 + \theta_{\rm H} + \theta_{\rm BZ} + \theta_{\rm HE} + \theta_{\rm HA} = 1$$

The $\theta_{\rm BZ}$ [Eq. (2.2.27)], $\theta_{\rm H}$ [Eq. (2.2.28)], $\theta_{\rm HE}$ [Eq. (2.2.31)], and $\theta_{\rm HA}$ [Eq. (2.2.32)] are substituted in the above formula

$$\left(1+K_1C_{\rm BZ}+\sqrt{K_2\gamma p_{\rm H_2}}+\frac{C_{\rm HE}}{K_6}+\frac{C_{\rm HA}}{K_7}\right)\theta_0 = 1 \quad (2.2.33)$$

The dissolution of cyclohexene and cyclohexane is low in water, and the formed cyclohexane and cyclohexene on catalyst surface are easy to condense into oil drop, diffusing into the oil phase fast, so K_6, $K_7 \gg K_2$, K_1. Due to the high hydrogen pressure and the small dissolution of benzene in water, $\sqrt{K_2\gamma p_{\rm H_2}} \gg K_1C_{\rm BZ}$, and the Eq. (2.2.33), could be simplified as

$$(1+\sqrt{K_2\gamma p_{\rm H_2}})\theta_0 = 1$$

$$\theta_0 = \frac{1}{1+\sqrt{K_2\gamma p_{\rm H_2}}} \quad (2.2.34)$$

The $\theta_{\rm BZ}$, $\theta_{\rm H}$, $\theta_{\rm HE}$, and θ_0 are substituted in Eq. (2.2.29)

$$r = -{\rm d}C_{\rm BZ}/{\rm d}t = k_3\theta_{\rm BZ}\theta_{\rm H}^4 - k_{-3}\theta_{\rm HE}\theta_0^4$$
$$= \left(k_3K_1K_2^2\gamma^2C_{\rm BZ}p_{\rm H_2}^2 - k_{-3}\frac{C_{\rm HE}}{K_6}\right)\theta_0^5$$

$$= \left(k_3 K_1 K_2^2 \gamma^2 C_{BZ} p_{H_2}^2 - k_{-3} \frac{C_{HE}}{K_6}\right) \left(\frac{1}{1+\sqrt{K_2 \gamma p_{H_2}}}\right)^5$$

If the reverse reaction rate could be ignored, the rate equation could be converted into Eq. (2.2.35).

$$r = -\mathrm{d}C_{BZ}/\mathrm{d}t = k_3 K_1 K_2^2 \gamma^2 C_{BZ} p_{H_2}^2 \frac{1}{\left(1+\sqrt{K_2 \gamma p_{H_2}}\right)^5} \qquad (2.2.35)$$

It is defined that $k = k_3 K_1 K_2^2 \gamma^2$, the denominator is unfolded through binomial series. Because the values of γ and K_2 are low, so the higher term could be ignored.

$$r = -\mathrm{d}C_{BZ}/\mathrm{d}t \approx k \frac{C_{BZ} p_{H_2}^2}{1 + 4\sqrt{K_2 \gamma p_{H_2}}}$$

The δ is defined to $4\sqrt{K_2 \gamma} = \delta$, so

$$r = -\mathrm{d}C_{BZ}/\mathrm{d}t \approx k \frac{C_{BZ} p_{H_2}^2}{1 + \delta p_{H_2}^{1/2}} \qquad (2.2.36)$$

Under low pressure, $1 + \delta p_{H_2}^{1/2} \approx 1$

$$r = -\mathrm{d}C_{BZ}/\mathrm{d}t \approx k C_{BZ} p_{H_2}^2 \qquad (2.2.37)$$

Equation (2.2.37) is totally in accordance with the conversion rate of benzene in Eq. (2.2.27), under low pressure in Sect. 2.2.2. Under medium and high pressure, the simplified Eq. (2.2.36),is identical to Eqs. (2.2.8) and (2.2.9).

If the hydrogen adsorption is the rate-determined step

$$r = k_2 p_{H_2} \theta_0^2 - k_{-2} \theta_H^2$$

$$r = (k_2 p_{H_2} - k_{-2} K_2 p_{H_2}) \left(\frac{1}{(1+\sqrt{K_2 \gamma p_{H_2}})}\right)^2$$

The rate equation could be transformed into Eq. (2.2.38), if the reverse reaction rate is ignored. And it is different from Eqs. (2.2.7) to (2.2.9) in the Sect. of 2.2.2.

$$r = \frac{k_2 p_{H_2}}{1+2\sqrt{K_2 \gamma p_{H_2}} + K_2 \gamma p_{H_2}} \qquad (2.2.38)$$

If the benzene adsorption is the rate-determined step

$$r = (k_1 C_{BZ} - k_{-1} K_1 C_{BZ}) \frac{1}{1+\sqrt{K_2 \gamma p_{H_2}}}$$

If the reverse reaction rate could be ignored and $\delta = \sqrt{K_2 \gamma}$

$$r \approx \frac{k_1 C_{BZ}}{1+\sqrt{K_2 \gamma p_{H_2}}} = \frac{k_1 C_{BZ}}{1+\delta p_{H_2}^{1/2}} \quad (2.2.39)$$

Similarly, Eq. (2.2.39), is also different from the kinetic rate of Eqs. (2.2.7)–(2.2.9) in Sect. 2.2.2.

In conclusion, the obtained rate equations is in accordance with the macro-rate equations based on the measurement of activation energy. But it is different if the adsorption of benzene and hydrogen is the rate-determined step, which means the proposal of reaction route is reasonable. The micro-kinetic research provides the reference for the catalytic mechanism understanding of benzene selective hydrogenation to cyclohexene.

2.3 Heterogeneous Catalytic Mechanism and Scientific Essence for Selective Hydrogenation of Benzene into Cyclohexene

2.3.1 Heterogeneous Catalytic Mechanism

In 1934, the reaction mechanism of the sub-step hydrogenation of benzene was firstly proposed by Horiuti and Polanyi. The activated benzene and hydrogen converted into cyclohexane on the VIII group metals via cyclohexadiene and cyclohexene.

Figure 2.2 presents the hydrogenation Horiuti–Polanyi mechanism of benzene on VIII group metals [20].

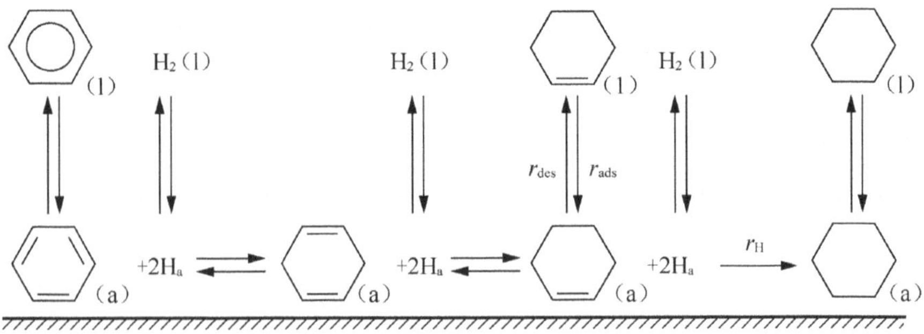

Fig. 2.2 Hydrogenation Horiuti–Polanyi mechanism of benzene on VIII group metals, the subscript of (l) and (**a**), respectively, represent liquid and adsorption status

In 1957, the cyclohexene was detected by Anderson when the catalytic hydrogenation of benzene on Ni membrane [21], proving cyclohexene was the intermediate of benzene hydrogenation. So the Horiuti–Polanyi mechanism was accepted by researchers. Up to now, the selective hydrogenation of benzene have been industrialized, the Horiuti–Polanyi mechanism has played an important role in the development of benzene selective hydrogenation. Although cyclohexadiene is not detected in the hydrogenation product of benzene, its existence has been determined with the surface vibration spectrum during the dehydrogenation of cyclohexene on Pt catalyst. Thus, the Horiuti–Polanyi mechanism has been universally accepted.

In 1983, the parallel reaction mechanism was proposed by Prasad during the research on selective hydrogenation of benzene to cyclohexene via gas-solid phase catalysis [22].

Figure 2.3 shows the parallel reaction mechanism proposed by Prasad.

In Fig. 2.3a, the benzene molecules adsorb vertically on the catalyst surface and bond with the σ active sites, the hydrogen molecules adsorb on the different sites with benzene. The sub-step hydrogenation of benzene to cyclohexadiene, cyclohexene, and cyclohexane occurs on the catalyst surface. In Fig. 2.3b, the benzene molecules adsorb horizontally on the catalyst surface and bond with the π active sites, forming the van der Waals complexes with six hydrogen atoms to produce cyclohexane. Under low benzene/hydrogen ratio, it proceeds as the first route, while under high hydrogen pressure as the second one.

The mechanism of benzene hydrogenation to cyclohexane in one-step was queried in the view of homogeneous catalysis, because the probability for seven particles' touch at one time was very low. But based on the interaction between aromatic

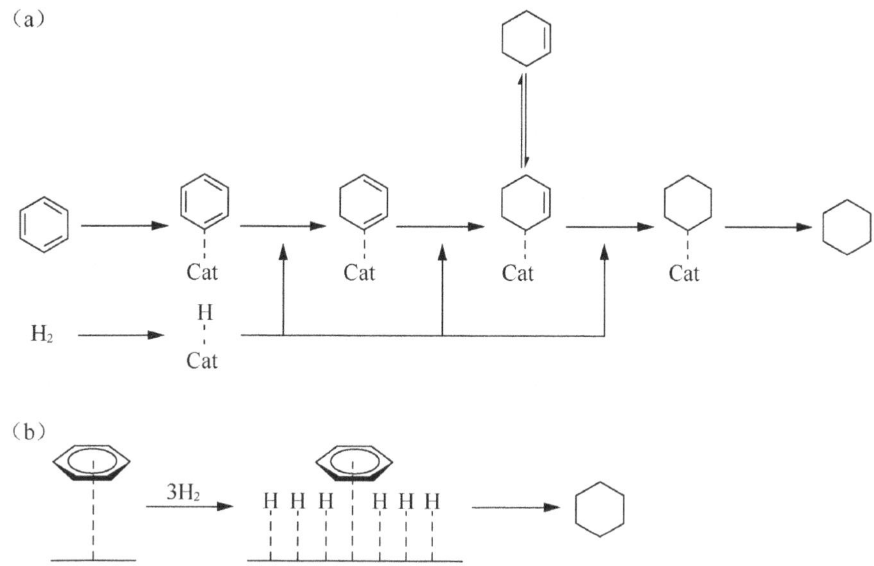

Fig. 2.3 The parallel reaction mechanism proposed by Prasad

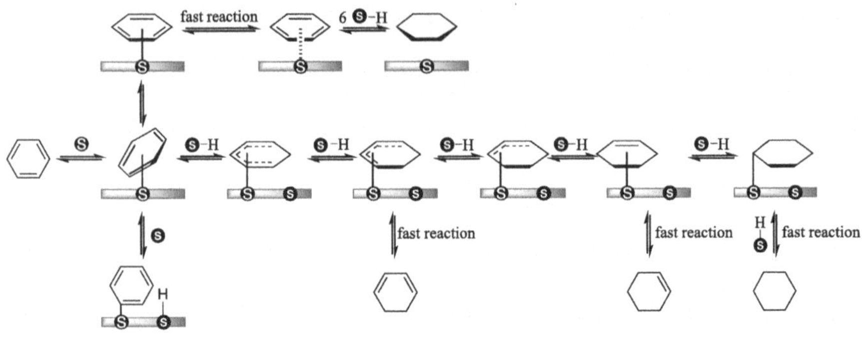

Fig. 2.4 Mechanism of one-step and multi-step hydrogenation of benzene. S, s and imaginary line mean the benzene coordination sites, hydrogen coordination sites and van der Waals force, respectively

molecules and metal surface and the chemical adsorption researches, if the coverage of disassociated hydrogen atoms is relatively high, there exists the probability to form cyclohexane from one benzene molecule and six hydrogen atoms. This fact was also proved by the gaseous hydrogenation of benzene on Ni catalysts under high pressure, namely, benzene was hydrogenated to cyclohexane on Ni catalysts fast. Although on the selective Ru catalyst surface, if the surface is hydrophobic, benzene would also be hydrogenated to cyclohexane quickly in the liquid phase.

In regard to the adsorption and reactions of benzene, hydrogen, and cyclohexene on Ru catalyst surface, it is considered that benzene forms the active species through π and σ/π bonds, so it has two parallel reaction routes.

Figure 2.4 shows the mechanism of one-step and multi-step hydrogenation of benzene [23].

Figure 2.4 also explains the hydrogenation mechanism of benzene in detail. On one hand, benzene adsorbs on the catalyst surface through the complanate π bonds and has high activity. Hydrogen stays in the status of adsorption and disassociation. If there are enough disassociated hydrogen atoms around the activated benzene molecules, the benzene is easily hydrogenated to cyclohexane. On the other hand, benzene molecules adsorb on the Ru catalyst surface in a suitable angle through σ/π bonds. The activated benzene molecules occur reaction with the closed hydrogen atoms to form cyclohexadiene, cyclohexene, and finally convert to cyclohexane via sub-step hydrogenation. Or benzene molecules adsorb vertically on the active sits of Ru, and it is considered inactive. The bond types of benzene with Ru catalyst surface are determined by the number, geometry structure, and electronic property of Ru active sites. And it provides the reference for utilizing additives and surface modification to increase the yield and selectivity of cyclohexene.

2.3.2 *Scientific Essence of High Selectivity and Yield of Cyclohexene*

(1) The electronic and geometry effect of the second metal

Through the characterization of XRD, XPS, AES-Ar^+, TEM, and H_2-TPD, it is proved that the M species enriched on Ru-M (M is Zn, Mn, Fe, La, and Ce) nano-bimetal catalyst surface would form an insoluble basic salt $(Zn(OH)_2)_3 \cdot ZnSO_4 \cdot 3H_2O$ with $ZnSO_4$ in slurry under the pretreatment and hydrogenation. Such basic salt would chemically adsorb on the surface of Ru. When the M stays in the optimal status, the basic salt disperses in monolayer, the electron transfer from M species to Ru, and Ru becomes the electron enrichment site $Ru^{(\delta-)}$, which is a benefit for the desorption of cyclohexene and could avoid further hydrogenation to cyclohexane. The M species change the spatial arrangement of Ru active sites through geometry effect, dilute the density of Ru active sites, decrease the coverage of hydrogen atom around activated benzene molecules, and release the opportunities for direct hydrogenation of benzene to cyclohexane.

Figure 2.5 presents the electron cloud density of carbon atoms in benzene and cyclohexene molecules.

It could be seen from Fig. 2.5, that the electron density of six carbons in benzene is -0.239×10^6, the electron density of six carbons is -0.230×10^2、-0.494×10^2, and -0.464×10^2, which means that the partial electron density of carbons in cyclohexene is higher than that in benzene. The repulsive interaction of $Ru^{(\delta-)}$ active sites to cyclohexene is stronger than to benzene, and the adsorption enthalpy

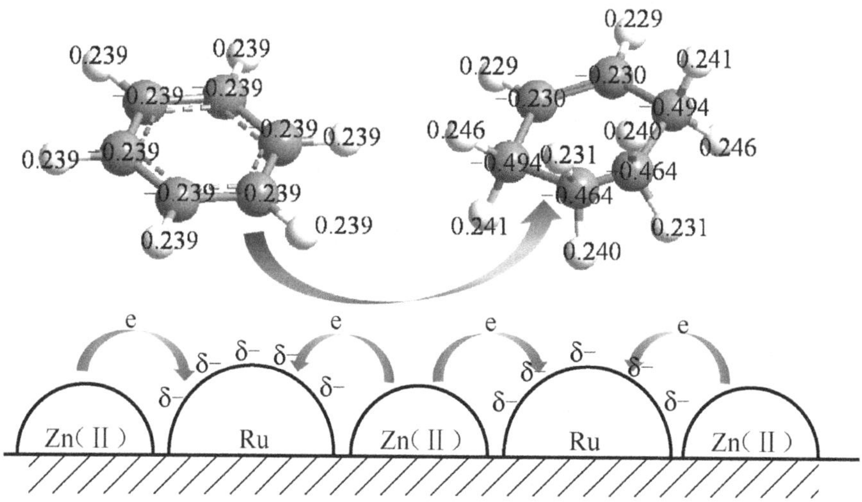

Fig. 2.5 The electron interaction between Ru and Zn (II) species and the electron density of benzene and cyclohexene

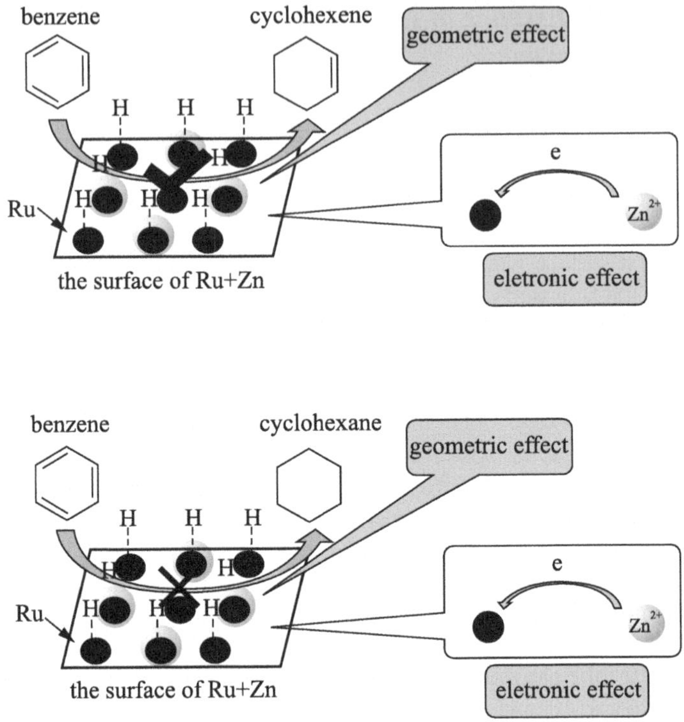

Fig. 2.6 Electron and geometry effect of additives M

of cyclohexene is lower than benzene. The rising Zn content is a benefit for the increase of $Ru^{(\delta-)}$ active sites. Benzene and cyclohexene adsorb on the same active sites competitively and the selectivity of cyclohexene could be enhanced due to the easy desorption from active sites.

Figure 2.6 displays the electron and geometry effect of additives M.

In Fig. 2.6, the black balls represent the hexagonal close packed lattice (H. C. P.) Ru atoms, gray balls mean the second metal atoms. Without the second metal, only Ru atoms as the active sites and hydrogen molecules adsorb preferentially on the Ru active sites and disassociate into hydrogen atoms. The catalyst surface has high coverage and density of hydrogen atoms, the benzene is easily hydrogenated to cyclohexane in one-step. While the addition of the second metal could not only change the electron enrichment active sites $Ru^{(\delta-)}$, but also occupy partial Ru lattices to change the geometry arrangement and dilute the density of Ru active sites. So the direct hydrogenation of benzene to cyclohexane could be inhibited and the sub-step hydrogenation of benzene to cyclohexene is dominant. If the intermediate cyclohexene could desorb fast, the direct hydrogenation to cyclohexane would be avoided, the yield and selectivity of cyclohexene could be enhanced.

(2) The compatibility change on the catalyst surface

The surface of Ru catalysts is hydrophobic and has strong compatibility with organic molecules. Benzene could be completely hydrogenated to cyclohexane within 5 min under the conditions of 150 °C, 5 MPa, and without modification in the slurry. The Ru-M catalysts could interact with $ZnSO_4$ that in slurry under pretreatment and hydrogenation, the hydrophobic surface would be changed to hydrophilic.

Figure 2.7 shows the compatibility change on Ru-Zn catalyst surface.

Figure 2.8 shows the compatibility change on Ru-La catalyst surface.

Figure 2.9 shows the compatibility change on Ru-Ce catalyst surface.

In Figs. 2.7, 2.8 and 2.9, the surface of Ru-M catalysts forms backwater film, the hydrophobic surface would change to hydrophilic. The better solubility of benzene in water than cyclohexene lead to the dominant adsorption of benzene on Ru active sites, the intermediate cyclohexene is easy to be expelled from catalyst surface, and is transferred to oil phase in the form of droplets. The energy is decreased, and it is hard to diffuse on the catalyst surface through backwater layer, avoiding the further hydrogenation to cyclohexane.

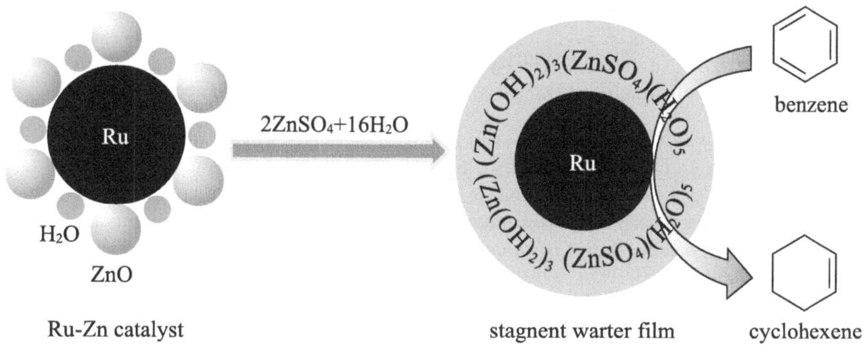

Fig. 2.7 The compatibility change on Ru-Zn catalyst surface

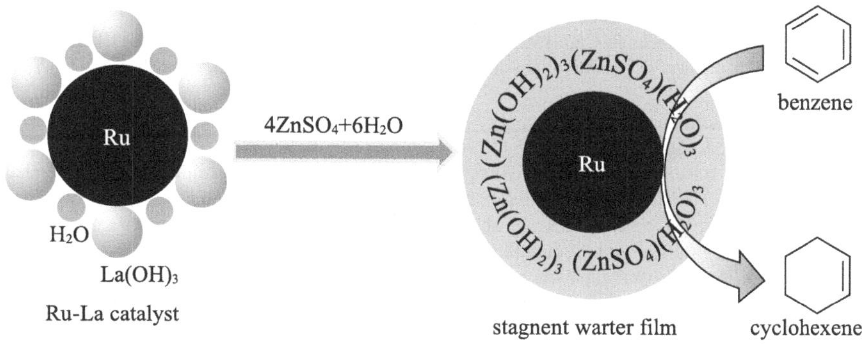

Fig. 2.8 The compatibility change on Ru-La catalyst surface

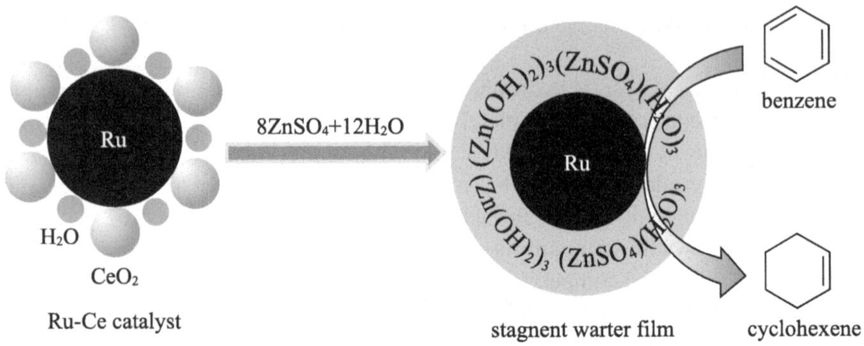

Fig. 2.9 The compatibility change on Ru-Ce catalyst surface

References

1. Jian, J.F., Zhang, T., Xu, Z.S., et al.: The progress of partial hydrogenation of benzene to cyclohexene. Spec. Petrochem. **3**, 46–50 (1997)
2. Ye, D.Q., Pang, X.X., Huang, Z.T., et al.: A study on the selective hydrogenation of benzene to cyclohexene by catalytic surface modification. Chem. React. Eng. Technol. **8**(2), 210–213 (1992)
3. Odenbrand, C.U.I., Lundin, S.T.J.: Hydrogenation of benzene to cyclohexene on an unsupported ruthenium catalyst: effect of poisons. J. Chem. Technol. Biotechnol. **31**, 660–669 (1981)
4. Hu, S.C., Chen, Y.W.: Partial hydrogenation of benzene to cyclohexene on ruthenium catalysts supported on La_2O_3-ZnO binary oxides. Ind. Eng. Chem. Res. **36**, 5153–5159 (1997)
5. Struijk, J., d'Angremond, M., Lucas-de Regt, W.J.M., et al.: Partial liquid phase hydrogenation of benzene to cyclohexene over ruthenium catalysts in the presence of an aqueous salt solution I. Preparation, characterization of the catalyst and study of a number of process variables. Appl. Catal. A: Gen. 83: 263–295 (1992)
6. Struijk, J., Moene, R., van der Kamp, T., et al.: Partial liquid-phase hydrogenation of benzene to cyclohexene over ruthenium catalysts in the presence of an aqueous salt solution II. Influence of various salts on the performance of the catalyst. Appl. Catal. A: Gen. 89, 11–102 (1992)
7. Roberts, G.W., Rylander, P.N., Greenfield, H.: Catalysis in Organic Synthesis, p. 1. Academic Press, New York (1976)
8. Carberry, J.J.: Physico-Chemical Aspects of Mass and Heat Transfer in Heterogeneous Catalysis (Catalysis, Science and Technology, Vol. 8), p. 131. Springer, Berlin (1987)
9. Wheeler, A.: Advances in Catalysis. Vol. 3, p. 249. Academic Press, New York (1951)
10. Weisz, P.B., Prater, D.C.: Advances in Catalysis. Vol. 6, p. 143. Academic Press, New York (1954)
11. Milone, C., Neri, G., Donato, A., et al.: Selective hydrogenation of benzene to cyclohexene on Ru/γ-Al_2O_3. J. Catal. **159**, 253–258 (1996)
12. Ronchin, L., Toniolo, L.: Selective hydrogenation of benzene to cyclohexene using a suspended ru catalyst in a mechanically agitated tetraphase reactor. Catal. Today **48**, 255–264 (1999)
13. Sun, H.J., Pan, Y.J., Wang, H.X., et al.: Selective hydrogenation of benzene to cyclohexene over a Ru-Zn catalyst with diethanolamine as an additive. Chin. J. Catal. **33**, 610–629 (2012)
14. Yuan, P.Q., Wang, B.Q., Ma, Y.M., et al.: Hydrogenation of cyclohexene over Ru–Zn/Ru (0001) surface alloy: a first principles density functional study. J. Molec. Catal. A: Chem. **301**, 140–145 (2009)
15. Yuan, P.Q., Wang, B.Q., Ma, Y.M., et al.: Partial hydrogenation of benzene over the metallic zn modified Ru-based catalyst. J. Molec. Catal. A: Chem. **309**, 124–130 (2009)

16. Liu, S.C., Guo, Y.Q., Yang, X.L., et al.: Kinetic equations for liquid-phase selective hydrogenation of benzene to cyclohexene. Chin. J. Catal. **24**, 42–46 (2003)
17. Yang, X.L.: Study on Selective Hydrogenation Kinetic of Benzene to Cyclohexene on Ru-M-B/ZrO_2. Zhengzhou University, Zhengzhou (2003)
18. Song, Y.: Study on the Selective Hydrogenation of Benzene and Its Kinetic over Ru Catalysts. Hebei University of Technology, Tianjin (2009)
19. Jiang, H.B.: Studies on Deactivation Mechanism and Kinetic of Selective Hydrogenation of Benzene to Cyclohexene over Ru Catalysts. Zhengzhou University, Zhengzhou (2014)
20. Horiuti, I., Polanyi, M.: The mechanism of benzene hydrogenation. Trans. Faraday Soc. **30**, 1164–1172 (1934)
21. Anderson, J.R.: The catalytic hydrogenation of benzene and toluene over evaporated films of nickel and tungsten. Aust. J. Chem. **10**, 409–416 (1957)
22. Prasad, K.H.V., Prasad, K.B.S., Mallikarjunan, M.M., Vaidyeswaran, R.: Self-poisoning and rate multiplicity in hydrogenation of benzene. J. Catal. **84**, 65–73 (1983)
23. Foppa, L., Dupont, J.: Benzene partial hydrogenation: advances and perspectives. Chem. Soc. Rev. **44**, 1886–1897 (2015)

Chapter 3
The First-Generation Catalyst for Selective Hydrogenation of Benzene to Cyclohexene-Ru–M–B/ZrO_2(M=Fe, La) Amorphous Alloys

In 1989, Japan, firstly industrialized the selective hydrogenation of benzene to cyclohexene and the downstream products catalytic technology. Their technologies were transferred to our country in 1995 and 2005, respectively. The main technical indexes of the first-generation catalyst were as follows: the conversion of benzene was 40%, the selectivity of cyclohexene was 80%, and the yield of cyclohexene was 32%. For the second-generation catalyst: benzene conversion was 51%, cyclohexene selectivity was 77.7%, and cyclohexene yield was 40%. However, due to the lack of technology for catalyst preparation, the autonomic and sustainable development of domestic enterprise has been restricted seriously.

In the face of the practical needs of enterprises, Zhengzhou University has conducted the research on preparation of catalyst for selective hydrogenation of benzene to cyclohexene. Firstly, a breakthrough was made in the research of catalyst, where the first-generation of Ru–M–B/ZrO_2 (M=Fe, La) amorphous alloy system was developed to selectively generate benzene, and the researches on its pilot and industrial applications were also carried out. The main technical indexes were as follows: benzene conversion of 40%, cyclohexene selectivity of over 80%, cyclohexene yield of more than 32%, which reached a similar level of first-generation catalyst abroad.

In Ru–M–B/ZrO_2, Ru is the active component and provides an active center for benzene hydrogenation. Agent M restrains the coalescence of Ru active components, improves the structural stability of Ru–M–B amorphous alloy, affects Ru active center electronic properties through electronic effect, and makes influences on the crystallite size and geometric configuration of active components through the geometric effect. Element B exists in the form of both free and oxidation state, partial electrons of free state B transfer to Ru, which make Ru to be a rich electronic active center, while the oxidation state of B can form an intermolecular hydrogen bond with water molecules, improves the catalyst surface hydrophilicity. ZrO_2 increases the dispersion and utilization of active components. By adding inorganic or organic additives to the reaction system, preprocessing the catalyst can significantly improve the selectivity and yield of cyclohexene.

Z. Liu et al., *Catalytic Technology for Selective Hydrogenation of Benzene to Cyclohexene*,
https://doi.org/10.1007/978-981-15-6411-6_3

The pilot study of Ru–M–B/ZrO_2 catalyst in 50 L batch reactor and 100 L continuous device was carried out, and we investigated the catalyst important performance indexes such as activity, selectivity, adjustable character, regeneration and lifetime, as well as the technological process and operating conditions, etc.

The contents in this chapter are as follows: preparation and characterization of Ru–M–B/ZrO_2 amorphous alloy catalyst, the function of each component in Ru–M–B/ZrO_2, operating conditions and pilot study of Ru–M–B/ZrO_2 amorphous alloy catalyst.

3.1 Preparation and Characterization of Ru–M–B/ZrO_2 Amorphous Alloy Catalyst

Amorphous alloys have structural features of "long range disorder", "short range order", and rich surface defects, so they have good catalytic activity and special selectivity [1]. Since the first report in 1981 [2], it has always been the hot spot of the catalytic materials research. Ru–M–B/ZrO_2 (M=Fe, La) catalyst is prepared by the chemical reduction method and applied to selective hydrogenation of benzene to cyclohexene [3–7]. It expands the application areas of amorphous alloys.

Ru–M–B/ZrO_2 (M=Cr, Mn, Fe, Co, Ni, Cu, zinc) catalyst is prepared by the chemical reduction method as follows: Take $RuCl_3 \cdot 3\ H_2O$ as Ru precursor, add transition elements and rare-earth elements, respectively, as additives, with $NaBH_4$ as a reducing agent, take ZrO_2 as the carrier.

Table 3.1 presents the catalyst composition and the conversion of benzene, selectivity, and yield of cyclohexene at 25 min, respectively.

Table 3.1 shows that Ru–M–B/ZrO_2 itself without additive M has high activity and cyclohexene selectivity. With the addition of M and the increase of content, a further reduction of conversion of benzene and increase of selectivity of cyclohexene can be observed. The second metal M can reduce the activity of the catalyst, and enhances the selectivity of cyclohexene significantly.

Temperature, pH, the concentration of $NaBH_4$, charging method, NaOH residue, etc., can affect the B content in the amorphous alloy, the microcrystal size, active center, surface properties, etc., and ultimately affect the activity and selectivity of the catalyst. The distribution of M on the catalyst is influenced by straight addition method or inverse addition method. The straight addition method is to add the $NaBH_4$ alkaline solution to the ionic solution with Ru^{3+} and M ions. Due to the gradual increase of pH, Ru^{3+} is reduced before M ions, and eventually, some active sites of Ru are covered by M, which leads to a significant decrease in the catalyst activity. The inverse addition method is to add Ru^{3+} and M ions to $NaBH_4$ alkaline solution. Because the quantity of $NaBH_4$ solution is much more than the required amount for reducing Ru^{3+} and M ions, Ru and M can be reduced at the same time and the constituent is almost uniform. Therefore, the benzene conversion and cyclohexene selectivity are better.

Table 3.1 Catalyst composition and conversion of benzene, selectivity and yield of cyclohexene at 25min[a]

Catalyst[b]	The content of B/%[c]	The content of M/%[d]	C_{BZ}/%[e]	S_{HE}/%	Y_{HE}/%
Ru–B/ZrO_2	0.065	0	83.9	72.8	61.1
Ru–Cr(6%)–B/ZrO_2	0.277	0.282	44.0	82.4	36.3
Ru–Cr(21%)–B/ZrO_2	0.331	0.655	24.1	84.3	20.3
Ru–Mn(6%)–B/ZrO_2	0.144	0.441	69.6	78.5	54.6
Ru–Mn(21%)–B/ZrO_2	0.205	0.985	64.3	80.0	51.4
Ru–Fe(6%)–B/ZrO_2	0.118	0.108	53.5	80.9	43.3
Ru–Fe(21%)–B/ZrO_2	0.221	1.196	28.6	89.9	25.7
Ru–Co(6%)–B/ZrO_2	0.152	0.124	75.8	82.8	62.8
Ru–Co(21%)–B/ZrO_2	0.181	0.761	68.8	81.4	56.0
Ru–Ni(6%)–B/ZrO_2	0.129	0.186	50.4	81.7	41.1
Ru–Ni(21%)–B/ZrO_2	0.207	0.734	31.5	85.9	27.1
Ru–Cu(6%)–B/ZrO_2	0.078	0.266	50.9	85.3	43.4
Ru–Cu(21%)–B/ZrO_2	0.100	1.337	22.9	90.1	20.6
Ru–Zn(6%)–B/ZrO_2	0.078	0.437	76.4	76.8	58.6
Ru–Zn(21%)–B/ZrO_2	0.321	1.813	60.4	82.4	49.7

[a]Reaction conditions: 423 K, 5 MPa H_2, 1000 r/min, 2 g catalyst, 45.7 g $ZnSO_4 \cdot 7H_2O$, 140 mL benzene, 280 mL H_2O

[b]The number inside parenthesis is the theoretical content of M

[c, d]The actual content of B and M determined by ICP-AES (mass fraction)

[e] The conversion of benzene C_{BZ} (mole fraction), the selectivity S_{HE} (mole fraction) and yield Y_{HE} (molar fraction of cyclohexene) at 25 min; following are the same

The filtrate pH of the catalyst washing solution is 7.5–8, which indicates that there exists a small amount of NaOH residue. The alkali metal has the effect of the electronic promoter, and OH^- reacts with Zn^{2+} in reaction slurry to produce minim $Zn(OH)_2$, which stays on the surface of the catalyst by chemical adsorption. It is beneficial for improving the selectivity of cyclohexene [8–15].

Table 3.1 also shows that, with the addition of M, the content of B in catalyst increases, indicating that M can regulate B content in the catalyst. With the increase of B content, the selectivity of cyclohexene grows. The second metal or metal oxide results in the geometrical rearrangement of the active component metal Ru, reducing or diluting the activity sites for further hydrogenation from cyclohexene to cyclohexane (i.e., the ensemble effect). The second metal or metal oxide selectively covers some of the strongest active centers, reducing the endothermic enthalpy of benzene and cyclohexene on catalyst surface (i.e., the coordination effect). Due to the ensemble effect and coordination effect of the second metal or metal oxide, the desorption ratio of cyclohexene increases, and the maximum cyclohexene yield with Ru–M–B/ZrO_2 catalyst is 62.8%.

Fig. 3.1 TEM (**a**) and SAED (**b**) images of Ru–M–B/ZrO_2 amorphous alloy catalyst

Figure 3.1 shows the images of transmission electron microscopy (TEM) and constituency electron diffraction (SAED) of the Ru–M–B/ZrO_2 amorphous alloy catalyst.

In Fig. 3.1a, the larger light-colored spherical or ellipsoidal spherical particles are micron-sized ZrO_2, particles darker in color are loosely-bounded nano-grade Ru–Fe–B active components. ZrO_2 has a dispersion effect on Ru–Fe–B. The micrinite diameter of Ru–Fe–B is 3.6–5.3 nm. In Fig. 3.1b, the electron diffraction ring gives the amorphous structure characteristics of Ru–Fe–B. On the catalyst, the benzene selective hydrogenation has good activity and selectivity. At 25 min, conversion of benzene is 50%, selectivity of cyclohexene is 80%, and yield is 40%. The high activity and selectivity come from the small size effect, surface effect, and quantum size effect of nano-grade Ru–Fe–B particles, the electronic effect and geometric effect of second metal Fe, as well as the dispersion of ZrO_2 and the possible effect from the additional active center.

From Fig. 3.2, all samples show hydrogen consumption peak at 320–400 K, corresponding to the reduction of RuO_x, shoulder peak is caused by the stepwise reduction of Ru oxide, produced in the process of washing and drying catalyst. No reduction peak of the promoter M is found within 550 K. The hydro-temperature of catalyst is 423 K, which indicates that Ru is at reduction state in the reaction condition, while promoter M is in the oxidation state.

Figure 3.3 presents N_2 low-temperature adsorption-desorption isotherm and pore size distribution curve of the Ru–Co–B/ZrO_2 sample. It can be seen from Fig. 3.3a, Ru–Co–B/ZrO_2 adsorption-desorption isotherm is close to the classification of fourth type isotherm and H1 hysteresis loop by International Federation of Pure and Applied Chemistry (IUPAC). The catalyst belongs to the middle hole structure, and the pore shape is the uniform hole formed by the spherical or capillary particles opening at both ends. The relative pressure of the adsorption-desorption isotherm is greater than 0.85, and when approaching 1.0, the isotherm rises sharply, indicating that there is

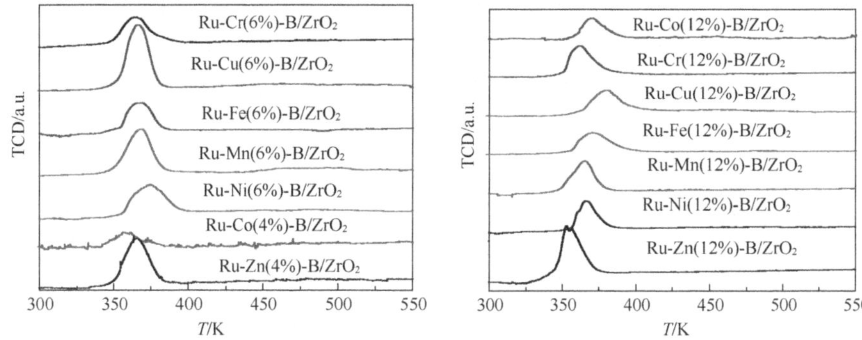

Fig. 3.2 H_2-TPR spectrum of the Ru–M–B/ZrO_2 catalyst

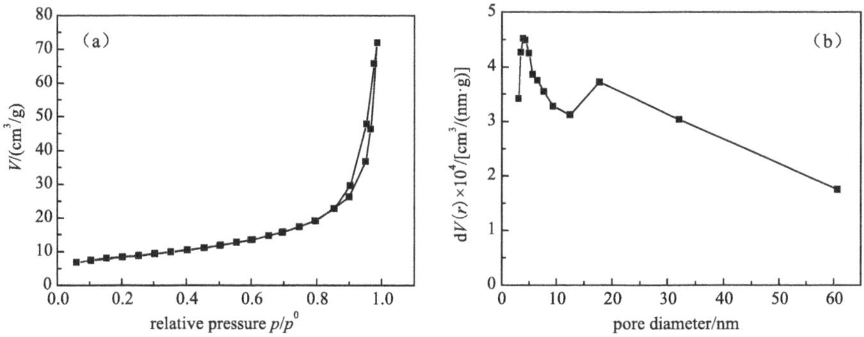

Fig. 3.3 Adsorption-desorption isotherm (**a**) and pore diameter distribution curve (**b**) of Ru–M–B/ZrO_2

not only containing an abundant middle hole, but also a large hole. As shown in Fig. 3.3b, the pore diameter of the catalyst is around 20 nm.

Table 3.2 shows the BET specific surface area S_{BET}, total pore volume V_p, and average pore diameter d_p of Ru–Co–B/ZrO_2 (atomic ratio Co/Ru = 6%).

According to the data in Table 3.2, the specific surface area of the catalyst is lower, indicating that the micropores are less and the pore diameter is larger, which is consistent with the pore diameter distribution curve. The larger pore diameter is advantageous to the transfer of the mass and desorption and evolution of the intermediates cyclohexene to avoid further hydrogenation.

Table 3.2 BET specific surface area $\boldsymbol{S}_{BET}$, total pore volume $\boldsymbol{V}_p$, and average pore diameter $\boldsymbol{d}_p$ of Ru–Co–B/ZrO_2

	$S_{BET}/(m^2/g)$	$V_P/(cm^3/g)$	d_P/nm
Ru–Co–B/ZrO_2	27.9	0.09	15.94

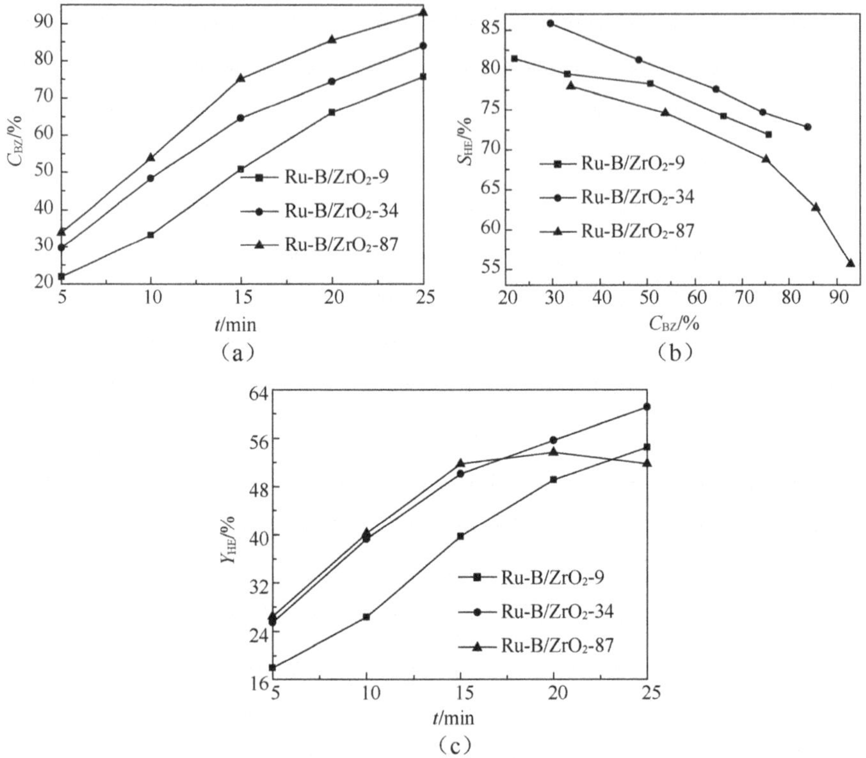

Fig. 3.4 Activity and selectivity of Ru–B/ZrO_2 loading on different specific surface area ZrO_2. **a** Change of benzene conversion with time; **b** Change of cyclohexene selectivity with benzene conversion; **c** Change of cyclohexene yield with time; reaction conditions: 423 K, 5 MPa H_2, 1000 r/min, 2 g catalyst, 45.7 g $ZnSO_4 \cdot 7H_2O$, 140 mL benzene, 280 mL H_2O; ZrO_2-9, ZrO_2-34 and ZrO_2-87 mean 9 m^2/g, 34 m^2/g and 87 m^2/g specific surface area of ZrO_2, respectively

Figure 3.4 shows the influences of loading ZrO_2 specific surface area on Ru–B/ZrO_2 activity and selectivity.

Figure 3.4 shows that with the increasing of ZrO_2 specific surface area, the conversion of benzene on the Ru–B/ZrO_2-*x* catalyst gradually increases, and the selectivity of cyclohexene is highest using the Ru–B/ZrO_2-34 catalyst at the same conversion rate. In response time, the maximum yield of cyclohexene in Ru–B/ZrO_2-34 catalyst can reach 61.1%, so ZrO_2-34 is the ideal carrier for the catalyst.

3.2 The Role of Components in Amorphous Alloy Catalysts Ru–B/ZrO_2

3.2.1 The Role of B in Amorphous Alloy Catalyst Ru–B/ZrO_2

In Ru–M–B/ZrO_2, Ru is the active component and provides the benzene hydrogenation activity center, M is the promoter, B is the nonmetallic element introduced by the reducing agent $NaBH_4$, which forms the amorphous alloy with Ru and M. ZrO_2 is the carrier. X-ray photoelectron spectroscopy (XPS) shows that elements B in Ru–M–B/ZrO_2 catalyst are in two forms of free and oxidation state, and enriches on the catalyst surface, part of free B electronics transfer to Ru and makes Ru a rich electronic active center. The Ru activity center of the rich electron can reduce the attraction of the double bond electron of cyclohexene, which is beneficial to the desorption of the middle product cyclohexene, so as to avoid further hydrogenation of cyclohexane. The oxidation state B is dispersed on the surface of the catalyst and can form intermolecular hydrogen bonds with water molecules. The free state B transfer partial electrons to Ru, electron-defect B is easy to be bonded with the lone pair of electrons in the water, so the surface hydrophilicity of the catalyst is increased. Catalyst owning Hydrophilic surface is easy to form a stable hysteretic membrane around it, restricting the diffusion of hydrogen, reducing the coverage of hydrogen, is helpful to reduce the chance of benzene directly adding hydrogen to generate cyclohexane.

Table 3.3 shows the effect of B content in Ru–B catalyst on selectivity and yield of cyclohexene [8].

It can be seen from Table 3.3, B content increased from 0 to 10.1%, the yield of cyclohexene increased from 6.3 to 30.4%, and the selectivity of cyclohexene increased from 19.4 to 50.8%. This indicates that B has an important effect on the

Table 3.3 Effect of B content in Ru–B catalyst on selectivity and yield of cyclohexene[a]

Ru–B	t/min	C_{BZ}/%	S_{HE}/%	Y_{HE}/%[b]
$Ru_{100}B_0$	90	32.5	19.4	6.3
$Ru_{98.1}B_{1.9}$	55	28.1	25.3	7.1
$Ru_{95.8}B_{4.2}$	65	40.8	33.6	13.7
$Ru_{93.3}B_{6.7}$	85	55.7	39.0	21.7
$Ru_{91.6}B_{8.4}$	85	60.4	45.4	27.4
$Ru_{90.1}B_{9.9}$	70	61.1	51.1	31.2
$Ru_{89.9}B_{10.1}$	75	59.9	50.8	30.4

[a]reaction condition: 423 K, 4 MPa, 1000 r/min, 35 mL benzene, 100 mL water, 0.15 g Ru–B, 1.5 g $ZnSO_4{\cdot}7H_2O$

[b]the maximum yield of cyclohexene

Note t is reaction time, C_{BZ} is the conversion of benzene, S_{HE} is the selectivity of cyclohexene, Y_{HE} is the yield of cyclohexene, following are the same

Table 3.4 Influence of different M on the properties of Ru–M–B amorphous alloy

Sample	*t*/min	C_{BZ}/%	S_{HE}/%	References
Ru–La–B/ZrO_2	30	58.7	75.1	[16]
Ru–Fe–B/ZrO_2	25	54.8	80.4	[10]
Ru–Zn–B	5	29.9	72.1	[9]
Ru–Zn–B/$Zr(OH)_4 \cdot H_2O$	55	73.1	62.3	[5]
Ru–Co–B/载体	20	76.3	68.3	[17]
Ru–Zn–B/ZrO_2	15	58.2	81.8	[18]

Note Reaction condition: 423 K, 4 MPa, 1000 r/min, 35 mL benzene, 100 mL water, 0.15 g Ru–B, 1.5 g $ZnSO_4 \cdot 7H_2O$

selectivity and yield of cyclohexene, so it can significantly influence the activity and selectivity of the catalyst by changing the B content in the Ru–B amorphous alloy.

3.2.2 The Role of M in Amorphous Alloy Catalyst Ru–B/ZrO_2

Table 3.4 gives the influence of different M on the properties of Ru–M–B amorphous alloy.

It can be seen from Table 3.4 that the catalyst has good activity and cyclohexene selectivity when M is La, Fe, Zn, and Co.

Amorphous alloy is a steady state structure in thermodynamics, which can be spontaneously transformed into crystalline structure under certain conditions. In the Ru–M–B/ZrO_2 amorphous alloy, M as the structural assistant can improve the dispersity of the active components, inhibit the accumulation of the active components of Ru, improve the stability of Ru–La–B amorphous alloy structure, and delay the crystallization. M affects the Ru activity center by the electron effect, the microcrystalline size and geometric arrangement of active components by geometrical effects, the catalyst surface properties by improving hydrophilicity, and thus improves the activity of the catalyst and the selectivity of cyclohexene [9, 19–23].

3.2.3 The Role of ZrO_2 in Amorphous Alloy Catalyst Ru–B/ZrO_2

ZrO_2 has two types of crystal: monocline (*m*-ZrO_2) and tetragonal (*t*-ZrO_2). *m*-ZrO_2 can obtain better selectivity and yield of cyclohexene than *t*-ZrO_2 due to its more surface hydroxyl groups. Calcinated under 700 °C, ZrO_2 mostly exists in monocline. Calcinated at above 1000 °C, ZrO_2 mostly exists in tetragonal [24, 25]. Adding a certain amount of ZrO_2 to the catalyst can increase the dispersion of the active component Ru and improve the utilization ratio of the active component. Hydrophilic ZrO_2

has the ability to form a hysteretic membrane on the surface of the catalyst, which is beneficial to cyclohexene desorption and to prevent the continued hydrogenation of cyclohexene. Lewis acid center in ZrO_2 has the adsorption and active effect on benzene, forming additional active center and having adsorption and active effect for benzene molecule, produces cyclohexene by stepwise hydrogenation with the overflow hydrogen atomic, and improves the selectivity and yield of cyclohexene.

Ru–M–B/ZrO_2, which were added to ZrO_2 before the catalyst was reduced, had the highest selectivity, and the best quality of Ru/ZrO_2 was 0.13–0.16. When the ratio was below 0.13, the catalyst activity and selectivity were lower. If it was above 0.16, the activity of the catalyst increased and the selectivity decreased [18, 26].

3.3 The Operating Conditions for Ru–M–B/ZrO_2 Amorphous Alloy Catalyst

3.3.1 Effect of Temperature

X-ray diffraction (XRD) shows that the diffraction peak of Ru–M–B/ZrO_2 amorphous alloy catalyst becomes sharp after hydrogenation, indicating that the grain size is larger and the higher the temperature is, the larger the Ru grain size is.

Under 145 °C, 153 °C, and 179 °C, the micrograin sizes of the catalyst after hydrogenation were 4.49 nm, 5.08 nm, and 5.59 nm, respectively. At the same time, a clear $NaBO_3$ diffraction peak was found in the XRD spectrum of the catalyst after hydrogenation, indicating that the amorphous alloy was decomposed during the hydrogenation process. The experimental results show that the amorphous alloy catalyst is suitable for the temperature of 135–145 °C [17, 18, 27–29].

3.3.2 Effect of Hydrogen Pressure

Under 4–6 MPa hydrogen pressure, the selectivity and yield of cyclohexene decrease with the increase of hydrogen pressure. Hydrogen pressure affects the surface hydrogen coverage of the catalyst, so the reaction process of benzene hydrogenation is changed. When the hydrogen coverage is low around the activated benzene molecules, the stepwise hydrogenation mechanism of benzene is dominant. When the hydrogen coverage is high around benzene molecules, the main effect of benzene is direct hydrogenation of cyclohexane. The maximum selectivity and yield of cyclohexene can be obtained only if the appropriate hydrogen cover and hydrogen/benzene molar ratio are obtained. Experiments show that the suitable hydrogen pressure for Ru–M–B/ZrO_2 amorphous alloy catalyst is 4.5–5 MPa [17, 27–29].

3.3.3 *Effect of Mixing Rate*

In heterogeneous catalysis system for selective hydrogenation of benzene to cyclohexene, there exists gas/liquid, liquid/liquid, and liquid/solid interface. A properly high stirring speed can eliminate the diffusion limitation from gas/liquid, liquid/liquid to reactants, and improves the benzene hydrogenation rate. Reactors, different catalysts, and catalytic systems are determined by the experiment. The experiment shows that in the 1 L reactor, the Ru–M–B/ZrO_2 catalyst can eliminate the external diffusion restriction when the stirring speed is higher than 900 r/min, and the suitable stirring speed is around 1000 r/min [17, 27–29].

3.3.4 *Effect of Additives*

Adding inorganic or organic additives to the reaction system is one of the most important ways to improve the selectivity of cyclohexene, with the best result by adding inorganic additives, especially using $ZnSO_4$.

Table 3.5 shows the influence of different Zn^{2+} concentrations on the performance of Ru–Fe–B/ZrO_2 catalyst.

It can be seen from Table 3.5 that as the $[Zn^{2+}] = 0.5–0.6$ mol/L and under the high benzene conversion, cyclohexene shows better selectivity. The effect of $ZnSO_4$ can be concluded as follows: the chemical adsorption on the surface of the catalyst through Zn^{2+} increases the surface hydrophilicity, and the $ZnSO_4$ in reaction slurry can promote the desorption of cyclohexene through a physical-chemical force [30]. The form of the element B depends on the addition amount of $ZnSO_4{\cdot}7\ H_2O$ and the reaction slurry pH. For Ru–Fe–B/ZrO_2, Zn^{2+} concentration of 0.5–0.6 mol/L, pH of 5.4–5.5, the catalyst showed better activity and cyclohexene selectivity.

Inorganic additives, especially the use of $ZnSO_4$ for getting a high yield of cyclohexene, has been applied to industrial production, but $ZnSO_4$ solution at high temperature is corrosive, where the hydrogenation is admitted in Hartz alloy kettle. The

Table 3.5 Influence of different Zn^{2+} concentration on the catalytic properties of Ru–Fe–B/ZrO_2

$[Zn^{2+}]$/(mol/L)	pH	C_{BZ}/%	S_{HE}/%	Y_{HE}/%
0	7.2	50.9	35.2	17.6
0.10	6.1	25.0	71.9	18.0
0.30	5.8	26.1	76.7	20.0
0.50	5.5	37.6	74.0	27.9
0.60	5.4	39.9	70.6	28.2
0.80	5.2	44.6	62.3	27.8

Note Reaction condition: 413 K, 5 MPa, 1000 r/min, 4 g Ru–Fe–B/ZrO_2 (0.64 g Ru), 140 mL benzene, 280 mL water, without pretreatment, at 5 min

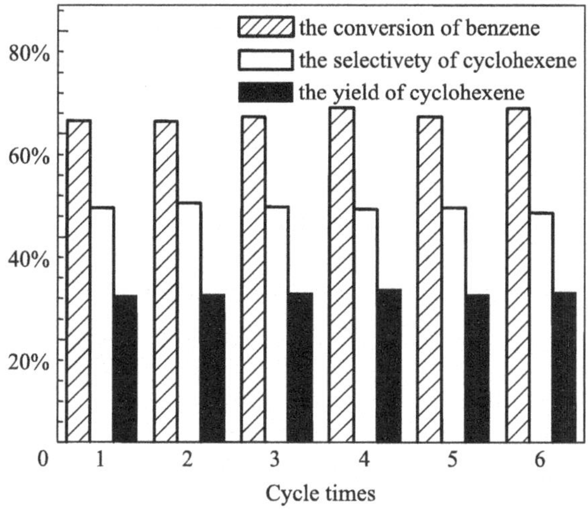

Fig. 3.5 Cycle performance of 3.8% Ru–1.2% Co–B/Al_2O_3 catalyst Reaction condition: 413 K, 5.0 MPa, 1000 r/min, 40 mg 3.8% Ru–1.2% Co–B/Al_2O_3, 3 mL benzene, 4 mL water

development of nonadditive catalytic system is an important research direction of benzene selective hydrogenation cyclohexene. At the Ru–Co–B/γ–Al_2O_3 catalyst, the yield of cyclohexene can reach 29% without additives under optimal conditions.

Figure 3.5 shows the cycle performance of 3.8% Ru–1.2% Co–B/Al_2O_3 catalyst [19].

It can be seen from Fig. 3.5 that the activity and selectivity of the catalysts after six cycles are almost invariable, indicating that the catalyst has good stability.

Table 3.6 shows the effect of different organic additives adding to Ru–Co–B/γ–Al_2O_3 catalyst on the hydrogenation of benzene [31].

Table 3.6 shows that the organic additive of alcamines has the effect of lowering activity and enhancing selectivity. Ethylenediamine and polyethylene glycol-6000 have the effect of increasing activity and selectivity, and the yield of cyclohexene is more than 32%.

Studies have shown that diethylamine hydrolysis interacts with $ZnSO_4$ can generate $Zn(OH)_2$. The mechanism of Diethanolamine and $Zn(OH)_2$ work together to improve the selectivity of cyclohexene is as follows: the diethanolamine adsorbing on the surface of the catalyst can form hydrogen bonding with cyclohexene molecules, which weakens the overlapping degree of C=C bond between cyclohexene and Ru d orbitals, promotes the desorption of cyclohexene molecules from Ru surface. Diethanolamine can also enhance the interaction between Zn and Ru and form a water layer on the surface of the catalyst. The electron from the N in diethanolamine is transferred to Ru, which makes Ru the center of the high-electron activity. The effects of other alcohols and amines on the active selectivity of Ru catalysts have a similar mechanism.

Table 3.6 Effect of different organic additives adding to Ru–Co–B/γ–Al_2O_3 catalyst on the hydrogenation of benzene

Additive	C_{BZ}/%	S_{BZ}/%	Y_{HE}/%
None[a]	51.9	39.2	20.4
Cholamine	45.9	66.9	30.7
Diethanolamine	46.8	68.8	32.2
Glycine	49.4	37.5	18.5
β-Alanine	19.1	63.9	12.2
Diphenylamine	59.2	45.2	26.7
Triethylamine	52.8	44.4	23.4
Ethanediamine	72.6	48.0	34.8
Diethylamine	44.2	62.3	27.5
1, 3-Propanediol	66.3	45.4	30.1
Glycol	60.9	43.1	26.3
Glycerol	51.5	47.2	24.3
Polyethylene glycol-6000	79.3	43.4	34.4
Polyethylene glycol-400	72.2	31.5	22.8

[a]reaction time 0.5 h

Note Reaction condition: 423 K, 5.0 MPa, 1500 r/min, 3 mL benzene, 4 mL water, 0.7 mmol $ZnSO_4$, 0.2 mmol additives, reaction time of 1 h

In addition, after the catalyst is pretreated by putting it in the reaction slurry containing $ZnSO_4$ and running for a period of time under hydrogenation reaction conditions without benzene, the activity of catalyst decreased obviously, and the selectivity of cyclohexene showed a significant rise [32]. This is because the pretreatment can make hydrogen adsorb on the strong active center in priority, benzene adsorb on the moderate intensity active center, the benzene can be activated at this time, and the intermediate cyclohexene becomes easily desorbed [29]. Pretreatment can also make the catalyst passivation and structural stabilization, prolong the service life of catalyst, and transform the catalyst size to the big hole direction at the same time, which is advantageous to the cyclohexene desorption and diffusion [33].

The study shows that adding acidic or alkaline substances to the reaction slurry can change the activity and selectivity of the catalyst. Without adding a new catalyst, the activity and selectivity deviation caused by the going time, fluctuating reaction conditions, and other factors maintain the normal operation of the catalyst [10, 34]. For the Ru–M–B/ZrO_2 catalyst, ZnO and H_2SO_4 can be used for adjusting activity and selectivity. With the increase of ZnO, the conversion rate of benzene decreased, and cyclohexene selectivity raised up. With the increase of H_2SO_4 amount, the conversion rate of benzene increased, and the selectivity of cyclohexene decreased. The mechanism is reversibly changing the active center and surface properties of the catalyst [14].

3.4 Pilot-Scale Study of Amorphous Alloy Catalyst Ru–M–B/ZrO_2

3.4.1 *Intermittent Pilot*

The pilot-scale study of Ru–M–B/ZrO_2 amorphous alloy catalyst was completed in a 50 L batch reactor. The influence of temperature, pressure, zinc sulfate concentration in the slurry, dispersant ZrO_2 and other factors on cyclohexene selectivity and yield by using Ru–Fe–B/ZrO_2 catalyst were investigated. The results showed that when the concentration of Zn^{2+} increased from 0.5 to 0.6 mol/L, not only the catalyst activity increased, but also cyclohexene selectivity increased. At Zn^{2+} concentration of 0.5 mol/L, adding dispersant ZrO_2 to the reaction slurry could improve not only the activity of the catalyst, but also the selectivity of cyclohexene. The optimal amount of ZrO_2 was obtained when $m_{Cat}/m_{ZrO2} = 1/5$. Once exceeded this ratio, the catalyst activity only increased slightly, but cyclohexene was significantly reduced. Under 403–408 K, 4.5–5.0 MPa of hydrogen pressure, water/benzene volume ratio of 2:1, Zn^{2+} concentration of 0.5 mol/L, $m_{Cat}/m_{ZrO2} = 1/5$, pretreatment of 12 h, the benzene conversion rate was 40%, cyclohexene selectivity was 80%, up to the level in the laboratory [35].

3.4.2 *Continuous Pilot*

In the 100 L stainless steel continuous device, the catalyst activity, selectivity, adjustability, regeneration, and life span of Ru–La–B/ZrO_2 were investigated in terms of the foreign catalysts [16, 20, 33].

Figure 3.6 shows the process flow chart of 100 L stainless steel continuous unit.

Figure 3.7 shows the activity and selectivity of Ru–La–B/ZrO_2 catalyst and import catalyst within 170 h.

In Fig. 3.7, Ru–La–B/ZrO_2 effective ruthenium content 0.18 kg, was only 40% of the 0.45 kg for foreign catalysts. It can be seen that before 100 h, the rate of benzene conversion with Ru–La–B/ZrO_2 catalyst was much higher than that with foreign catalysts (nearly 10% points). During 100–170 h, the conversion rate of the two is close. The selectivity of Ru–La–B/ZrO_2 gradually increased steadily with time. At 40–170 h, The catalyst selectivity was stable and significantly higher than that of foreign catalysts. The results show that comparing to the foreign catalyst, the Ru–La–B/ZrO_2 not only has good activity and selectivity, but also improves the Ru utilization ratio under the same conditions.

Figure 3.8 shows the TEM and SAED photos of Ru–La–B/ZrO_2 catalyst in the pilot plant at different times [16, 33].

As can be seen from Fig. 3.8, with the extension of running time, Ru–La–B/ZrO_2 amorphous alloys began to reunite and crystallize, which may be one of the reasons for the decreasing activity of the catalyst [16].

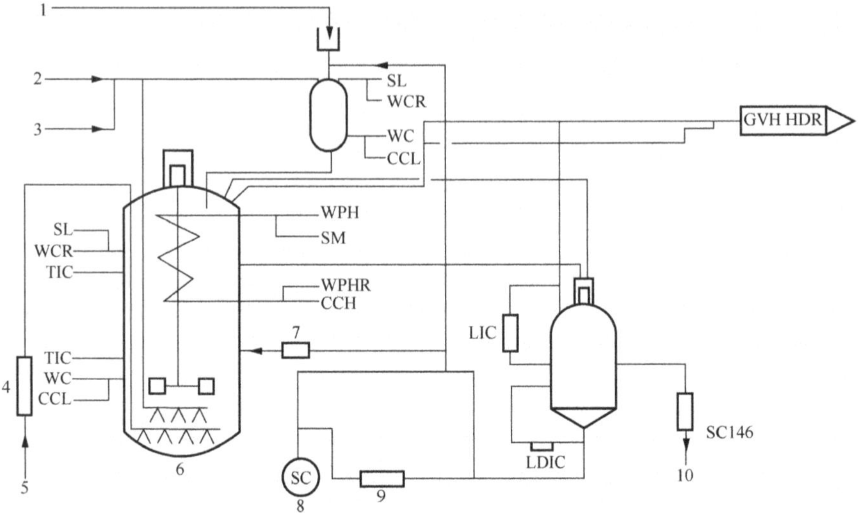

Fig. 3.6 Process flow chart of 100 L stainless steel continuous unit. 1. Catalyst entrance; 2. Hydrogen inlet; 3. Nitrogen inlet; 4. Benzene flowmeter; 5. Benzene feed inlet; 6. Discharge port; 7. Slurry flowmeter; 8. Catalyst sampling port; 9. Slurry circulating pump: 10. Product outlet; *SL* low-pressure steam; *SM* medium pressure steam; *WC* cooling water (to); *WCR* cooling water (return); *CCL* low-temperature condensation; *CCH* high-temperature condensation; *WPH* high purity water (to); *WPHR* high purity water (return); *TIC* temperature indication adjustment; *LIC* liquid level indicator regulation; *LDIC* liquid level meter; *SC* velocity control; *GVH HDR* general valve manual adjustment

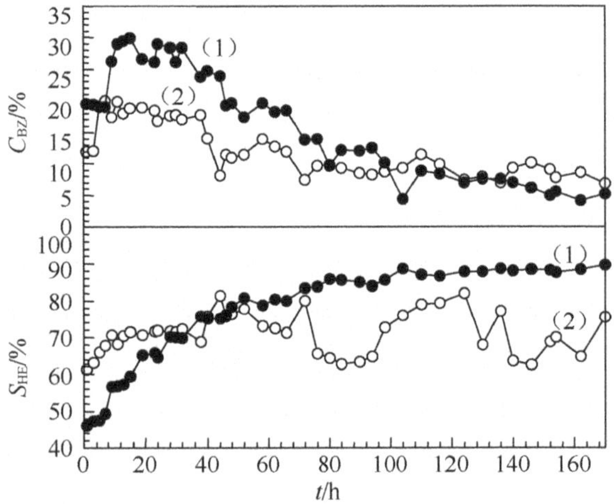

Fig. 3.7 Performance of two kinds of catalysts in the pilot plant. (1) Ru–La–B/ZrO_2; (2) the foreign catalysts. Reaction condition: 135 °C, 4.4–4.5 MPa, 900 r/min, 1.35 kg catalyst, 170 L water, 6.75 kg ZrO_2, 17 kg $ZnSO_4 \cdot 7H_2O$, pretreatment of 10–22 h, slurry circulation flow rate of 130–150 kg/h, benzene flow rate of 65–75 kg/h

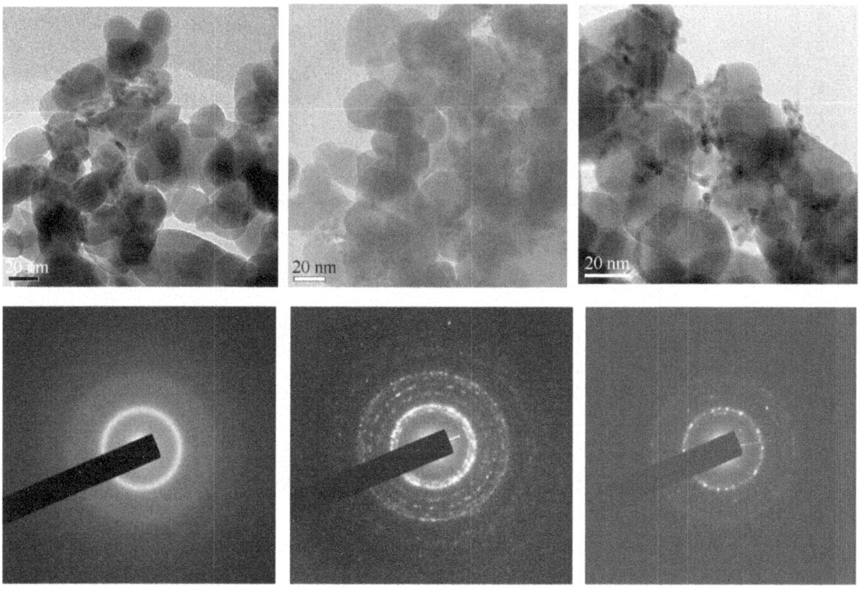

(a) new catalyst (b) running 100 h (c) running 170 h

Fig. 3.8 TEM and SAED photos of Ru–La–B/ZrO_2 catalyst in the pilot plant at different times

The iron ions and zinc ions in Slurry and catalyst before and after the reaction were tested, and it was found that the adsorption of excessive Zn^{2+} in slurry and the Fe^{2+} down by the way corrosion during the process of long-term operation, that is why catalyst activity decrease gradually. The catalyst was regenerated by acid washing with water and acid, respectively, and the activity and selectivity of the catalyst before and after water washing and acid washing were given in Table 3.7.

Table 3.7 shows that the water washing has no effect on the reactivation of the catalyst, while the activity and selectivity of the catalyst can be basically restored after acid washing. According to the analysis of the content of Zn^{2+} and Fe^{2+} in the

Table 3.7 Activity and selectivity of the catalyst before and after water washing and acid washing

	5 min		10 min		15 min	
	C_{BZ}/%	S_{HE}/%	C_{BZ}/%	S_{HE}/%	C_{BZ}/%	S_{HE}/%
Before washing	6.5	60.4	11.4	50.2	16.5	31.4
Water washing	6.1	65.4	9.5	61.5	13.4	48.5
Acid washing with 0.5 mol/L HCl	10.8	65.8	18.8	61.2	27.6	49.6
Acid washing with 2.0 mol/L HCl	14.5	59.6	23.2	57.8	32.5	51.4
Acid washing with 1.0 mol/L H_2SO_4	11.6	68.0	20.2	57.0	29.3	48.9

Note Reaction condition: 423 K, 5.0 MPa, 900 r/min, 58 mL water, 29 mL benzene, 0.45 g catalyst, 7.2 g $ZnSO_4 \cdot 7H_2O$

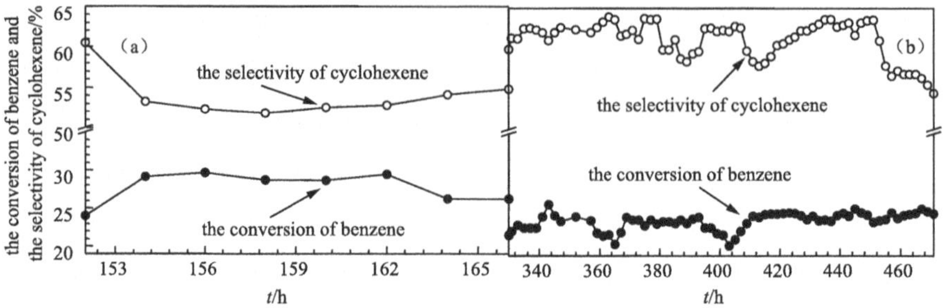

Fig. 3.9 Activity and selectivity after adding the regenerated catalyst into the pilot system (**a**) in 340–466 h (**b**) [35]. Reaction condition: 135°, 4.4–4.5 MPa, 900r/min, 1.35 kg catalyst, 170 L water, 6.75 kg ZrO_2, 17 kg $ZnSO_4{\cdot}7H_2O$, pretreatment 10–22 h, slurry circulation flow rate 130–150 kg/h, benzene flow rate 65–75 kg/h

washing solution, it was found that the 2.0 mol/L HCl could wash out more Fe^{2+}, and the catalyst showed the best effect after regeneration.

Figure 3.9 gives the activity and selectivity of the catalyst after adding the regenerated catalyst to the pilot system.

From Fig. 3.9a, it can be seen that after replenishing the regeneration catalyst, the conversion rate of benzene is up to 30.0% and maintains over 24% for 14 h. The selectivity decreased, but increased slightly over time. As can be seen from Fig. 3.9b, the activity of the catalyst in 466 h has remained above 20% through regeneration and modulation, and the selectivity of cyclohexene has remained above 55%. The catalytic system, process flow, and operating conditions of the selective hydrogenation of benzene by using the amorphous alloy catalyst were established through the pilot test.

References

1. Zong, B.N.: Applications of the amorphous alloy catalyst and magnetically stabilized bed technology in petrochemical process. Chin. J. Catal. **29**, 873–877 (2008)
2. Yokoyama, A., Komiyama, H., Inouet, H., et al.: The hydrogenation of carbon monoxide by amorphous ribbons. J. Catal. **68**, 355–361 (1981)
3. Liu, S.C., Li, L.M., Wang, X.Y.: Benzene selective hydrogenation cyclohexene catalyst and its manufacturing method. Chinese Patent, CN 0112208.5 (2011)
4. Liu, S.C., Li, L.M., Wang, X.Y.: The preparation, modulation and regeneration method of catalyst for benzene selective hydrogenation preparing cyclohexene. Chinese Patent, CN 10060451.0 (2004)
5. Qiao, M.H., Wang, J.Q., Xie, S.H., et al.: Amorphous catalyst containing ruthenium boron for benzene selective hydrogenation and its preparing method. Chinese patent, CN 031156665 (2003)
6. Wu, M., Chen, Z.X., Sun Z.L., et al.: The preparation and application of catalyst for benzene selective hydrogenation preparing cyclohexene. Chinese patent, CN 101260628 (2005)

7. Chen, Z.X., Wu, M., Sun, Z.L., Gu, W.: The preparation and application of a kind of catalyst for benzene selective hydrogenation preparing cyclohexene. Chinese patent, CN 101260628 (2005)
8. Liu, Z., Dai, W.L., Liu, B., et al.: The effect of boron on selective benzene hydrogenation to cyclohexene over ruthenium boride powders. J. Catal. **187**, 253–256 (1999)
9. Qin, H.A., Huang, Z.X., Liu, S.C.: Study on the unsupported Ru–Zn–B catalysts for selective hydrogenation of benzene to cyclohexene. J. Henan Normal Univer. (Natural Sci.) **20**, 350–352 (2007)
10. Sun, H.J., Zhang, C., Yuan, P., et al.: Preparation and modifiable character of a novel amorphous catalyst Ru–Fe–B/ZrO_2 for the selective hydrogenation of benzene to cyclohexene. J. Catal. **29**, 441–446 (2008)
11. Xie H.F., Nan J., Liu Y.Q., et al.: Determination of reduction reaction between Ru^{3+} and BH_4^- by Use of pH value. In: 230th ACS National Meeting. American Chemical Society, Washington (2005)
12. Shen, J.Y., Li, Z.Y., Chen, Y.: Study on chemical preparation, reaction mechanism and properties of Ni–B superfine amorphous alloy. Chin. J. Inorg. Chem. **11**, 1–7 (1995)
13. Li, H.X., Li, H., Dai, W.L., et al.: Preparation of the Ni–B amorphous alloys with variable boron content and its correlation to the hydrogenation activity. Appl. Catal. A **238**, 119–130 (2003)
14. Liu, Z.Y., Sun, H.J., Wang, D.B., et al.: The modifiable character of a novel Ru–Fe–B/Zr0 = catalyst for benzene selective hydrogenation to cyclohexene. Chin. J. Chem. **28**, 1927–1934 (2010)
15. Ji, Y.L., Wang, Z.W., Liu, S.C., et al.: A study on Ru catalyst for selective hydrogenation of benzene to cyclohexene. Precious. Met. **24**, 26–30 (2003)
16. Liu, S.C., Liu, Z.Y., Wang, Z., et al.: Characterization and study on performance of the Ru–La–B/ZrO_2 amorphous alloy catalysts for benzene selective hydrogenation to cyclohexene under pilot conditions. Chem. Eng. J. **139**, 157–164 (2008)
17. Wu, M., Sun, Z.L., Chen, Z.X., et al.: Reaction conditions affecting liquid phase selective hydrogenation of benzene to cyclohexene. Chin. Synth. Fiber Ind. **28**, 4–9 (2005)
18. Liu, S.C., Wang, H.R., Han, M.L., et al.: Study on the catalyst with high activity and high selectivity for partial hydrogenation of benzene to cyclohexene. J. Fuel Chem. Tech. **29**, 126–129 (2001)
19. Fan, G.Y., Jiang, W.D., Wang, J.B., et al.: Selective hydrogenation of benzene to cyclohexene over RuCoB/γ–$A1_2O_3$ without additive. Catal. Commun. **10**, 98–102 (2008)
20. Liu, S.C., Liu, Z.Y., Wang, Z., et al.: A novel amorphous alloy Ru–La–B/ZrO_2 catalyst with high activity and selectivity for benzene selective hydrogenation. Appl. Catal. A **313**, 49–57 (2006)
21. Yang, X.L., Guo, Y.Q., Zhang, Z.J., et al.: Study on the Ru–Co–B/ZrO_2 catalyst for selective hydrogenation of benzene to cyclohexene.]. Chem. Res. Appl. **15**, 664–665 (2003)
22. Yang, X.L., Guo, Y.Q., Liu, S.C.: Catalytic hydrogenation of benzene to cyclohexene over amorphous Ru–Co/ZrO_2. Chin. J. Appl. Chem. **20**, 379–381 (2003)
23. Han, M.L., Liu, S.C., Yang, X.D., et al.: The effects of promoters on the catalytic properties and surface character of Ru–B/ZrO_2 amorphous alloy catalysts. J. Mol. Catal. (Chin.) **18**, 47–50 (2004)
24. Liu, Z.Y., Sun, H.J., Li, X.Y., et al.: Effect of support calcination temperature on the Ru–Fe–B/ZrO_2 catalytic performance for benzene selective hydrogenation to cyclohexene. J. Henan Normal Univer. (Natural Sci.) **23**, 423–425 (2010)
25. Pang, X.S., Ye, D.Q., Chen, F.L., et al.: A study on liquid phase partial hydrogenation of benzene to cyclohexene over supported Ru catalysts. Petrochem. Ind. **23**, 566–572 (1994)
26. Liu, S.C., Zhu, B.Z., Luo, G., et al.: Characterization of amorphous Ru–M–B/ZrO_2 catalysts for partial hydrogenation of benzene to cyclohexene. J. Mol. Catal. (Chin.) **16**, 217–222 (2002)
27. Ning, J.B., Xu, J., Liu, J., et al.: Selective hydrogenation of benzene to cyclohexene over colloidal ruthenium catalyst stabilized by silica. Catal. Lett. **109**, 175–180 (2006)

28. Ning, J.B.: Selective hydrogenation of benzene to cyclohexene over Ru catalysts. Ph.D. thesis. Dalian Institute of Chemical Physics, Chinese Academy of Sciences, Dalian (2007)
29. Liu, S.C., Luo, G., Wang, H.R., et al.: Study on operation conditions for liquid phase selective hydrogenation of benzene to cyclohexene over Ru–M–B. ZrO_2 Catalyst. Chin. J. Catal. **23**, 317–320 (2002)
30. Wang, S.Z., Li, J.P., Luo, G., et al.: Liquid phase selective hydrogenation of benzene to cyclohexene over Ru–M–B/ZrO_2 Catalysts. J. Zheng Zhou Univer. **34**, 78–81 (2002)
31. Fan, G.Y., Li, R.X., Li, X.J., et al.: Effect of organic additives on partial hydrogenation of benzene. Catal. Commun. **9**, 1394–1397 (2008)
32. Ashida, K., Iwasaki, M.: Preparation methods of cyclohexene and cyclohexane. Japanese patent, JP 981087353 (1998)
33. Wang, H., Liu, Z.Y., Shi, R.J., et al.: Deactivation and regeneration of amorphous Ru–La–B/ZrO_2 catalyst for selective hydrogenation of benzene to cyclohexene. Chin. J. Catal. **26**, 407–411 (2005)
34. Ashida, K., Iwasaki, M.: The method of producing cycloolefins. Japanese patent, JP 981241395 (1998)
35. Liu, S.C., Liu, Z.Y., Zhao, S.H., et al.: Study on the nanosized amorphous Ru–Fe–B/ZrO_2 alloy catalyst for benzene selective hydrogenation to cyclohexene. J. Nat. Gas Chem. **15**, 319–326 (2006)

Chapter 4
The Second-Generation Catalyst for Selective Hydrogenation of Benzene to Cyclohexene-Ru-Zn-Na_2SiO_3-PEG-10000

The main technical indexes of the second-generation of catalysts, Ru-Zn-Na_2SiO_3-PEG-10000, are as follows: conversion of benzene was 50%, selectivity of cyclohexene was over 80%, yield of cyclohexene was more than 42%, reached the second-generation of similar catalyst level abroad. Compared with the first generation amorphous alloy Ru-M-B/ZrO_2, the second-generation catalyst belongs to the crystalline structure with better thermal stability and longer service life. Because of the modification of Na_2SiO_3-PEG-10000 on the surface of the catalyst, it has better catalytic activity and cyclohexene selectivity, the conversion rate of benzene increased by 10 percentage points, and the cyclohexene yield increased by 8 percentage points.

On the basis of extensive research on inorganic and organic modifiers, Na_2SiO_3, alcohols, and amines can significantly improve the selectivity and yield of cyclohexene, but the mechanism are different. Na_2SiO_3 mainly exists on the surface of the Ru-Zn catalyst, while alcohols and amines are mainly present in the slurry. The role of amines can be substituted by Na_2SiO_3 and PEG, and therefore, developed the Ru-Zn-Na_2SiO_3-PEG-10000 catalytic system. Zn in Zn-Ru catalyst exists in the form of ZnO. Through pretreatment on the surface, ZnO interacts with $ZnSO_4$ in the slurry, and generate insoluble alkali salt, which forms mesh structure covering on the surface of the catalyst with Na_2SiO_3-PEG-10000, at the same time, in the slurry by PEG-10000 and cyclohexene form intermolecular hydrogen bond to stabilize the intermediate cyclohexene, jointly improve the cyclohexene selectivity.

The second-generation catalyst, which realized industrialization in 2010, was used for 100,000 t/a benzene selective hydrogenated to produce cyclohexene and its downstream industrial devices.

The main contents of this chapter include the influence of Na_2SiO_3 on the performance of Ru-Zn catalyst, the influence of PEG-10000 on the performance of Ru-Zn catalyst, and on the second-generation of the Ru-Zn-Na_2SiO_3-PEG-10 catalytic system for benzene selective hydrogenation.

Z. Liu et al., *Catalytic Technology for Selective Hydrogenation of Benzene to Cyclohexene*,
https://doi.org/10.1007/978-981-15-6411-6_4

4.1 Influence of Na_2SiO_3 on the Catalyst Performance of Ru-Zn

It's easy to hydrogenate the whole benzene ring when its π-bond is opened, so it's difficult to get cyclohexene as the partial hydrogenated product. However, the properties can be changed by adding a modifying agent to catalyst surface, which improves the catalytic activity and cyclohexene selectivity [1]. Using inorganic salt or organic materials containing polar groups as additives to modify the catalyst can greatly improve the selectivity and yield of cyclohexene [2–16]. On the basis of further study of the inorganic modifier, Na_2SiO_3 is a good modifier for Ru-Zn catalyst.

4.1.1 Catalyst Activity and Selectivity of Premodified Ru-Zn

Table 4.1 gives the catalyst activity and selectivity of premodified Ru-Zn.

According to the activity and selectivity date, C_{BZ}-t curve was obtained by plotting the benzene conversion rate C_{BZ} to the time t, S_{HE}-C_{BZ} curve was obtained by plotting the cyclohexene selectivity S_{HE} against benzene conversion rate C_{BZ}. Then the time $t40$ and $t50$, which mean that benzene has a conversion rate of 40% and 50%, respectively, can be obtained by applying the interpolation method based on the above C_{BZ}-t curve. Similarly, the time $S40$ and $S50$ of cyclohexene selectivity 40 and 50%, can be obtained based on S_{HE}-C_{BZ} curve. Thus, the main performance index of the benzene selective hydrogenation catalyst can be determined.

Figure 4.1 shows the C_{BZ}-t curve and S_{HE}-C_{BZ} of premodified Ru-Zn catalyst.

As shown in Fig. 4.1, at 11 min, the conversion of benzene was 40% (mole fraction, following are the same) and the selectivity of cyclohexene was 74%. At 15 min, the conversion of benzene was 50% (mole fraction, following are the same) and the selectivity of cyclohexene was 71%. The activity and selectivity of premodified

Table 4.1 Catalyst activity and selectivity of premodified Ru-Zn

t/min	C_{BZ}/%	S_{HE}/%	Y_{HE}/%
5	19	76	15
10	38	74	28
15	50	71	36
20	65	68	44
25	72	65	46

Note Evaluation conditions: 150 °C, 5.0 MPa, 1200 r/min, 140 mL benzene, 280 mL water, 1.96 g Ru-Zn catalyst, 9.8 g ZrO_2, without pretreatment; t is sampling time, C_{BZ} is benzene conversion rate (mole fraction), S_{HE} is cyclohexene selectivity (mole fraction), Y_{HE} is cyclohexene yield, following are the same

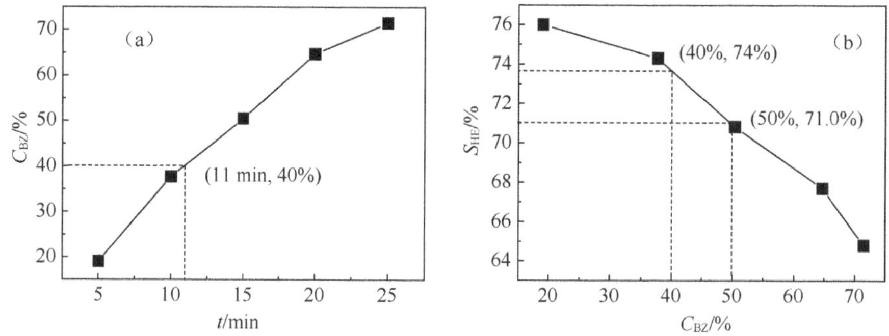

Fig. 4.1 **a** C_{BZ}-t curve and **b** S_{HE}-C_{BZ} of premodified Ru-Zn catalyst

Ru-Zn catalyst cannot reach the level of the first generation amorphous alloy Ru-M-B/ZrO_2.

In the preparation process of Ru-Zn catalyst, the materials RuC13·3H2O and $ZnSO_4 \cdot 7H_2O$ were used and reduced in $ZnSO_4$ solution, thus the mother solution contained a large number of Cl^- and SO_4^{2-}. Washing the catalyst with solutions containing different materials can remove pare of the residual ions, and make the ions in the solutions absorbed on the catalyst surface. This is a kind of method of surface modification of Ru-Zn catalyst. This method was used to investigate the modification effect of the surface of Ru-Zn catalyst using pure water, NaOH, and Na_2SiO_3.

4.1.2 Washing Catalyst with Pure Water

Table 4.2 shows the activity and selectivity of Ru-Zn catalyst after washing with pure water until there's no chloride ion and sulfate ion.

Figure 4.2 shows the Ru-Zn catalyst C_{BZ}-t curve and S_{HE}-C_{BZ} curve after washing by pure water.

Table 4.2 Activity and selectivity of Ru-Zn catalyst after washing with pure water

t/min	C_{BZ}/%	S_{HE}/%	Y_{HE}/%
5	27	85	23
10	41	83	34
15	54	80	44
20	65	78	51
25	73	75	55

Evaluation conditions: 150 °C, 5.0 MPa, 1200 r/min, 140 mL benzene, 280 mL water, 1.96 g Ru-Zn catalyst, 9.8 g ZrO_2, without pretreatment

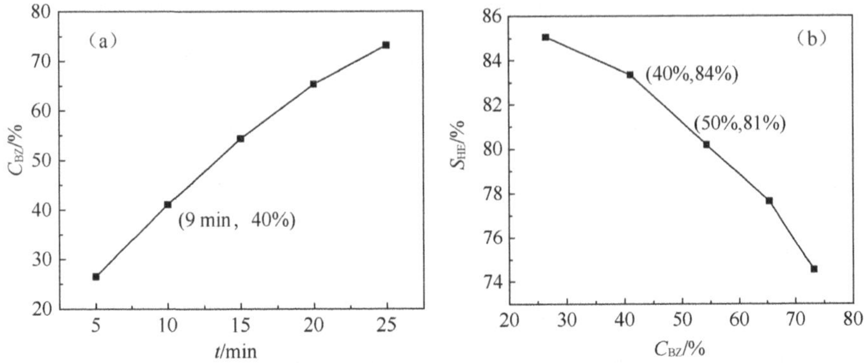

Fig. 4.2 Ru-Zn catalyst **a** C_{BZ}-t curve and **b** S_{HE}-C_{BZ} curve after washing by pure water

As shown in Fig. 4.2, when the catalyst was washed by pure water, the conversion of benzene was 40% and the selectivity of cyclohexene was 84% at 9 min. At 15 min, the benzene conversion was 54% and cyclohexene selectivity was 80%.

The SO_4^{2-} and $C1^-$ adsorbed on the catalyst occupied the surface active center of Ru catalyst, affected the adsorption, activation, and desorption of cyclohexene, thus affected the activity selectivity of the catalyst. The catalyst was washed by NaOH and Na_2SiO_3 solution, and the effect was compared with that washed by pure water.

Figure 4.3 shows the C_{BZ}-t curve and the S_{HE}-C_{BZ} curve of the catalyst after washing by H2O, NaOH, and Na_2SiO_3 solution, respectively.

As can be seen from Fig. 4.3a, the activity of Ru-Zn catalyst washed by NaOH solution and water is basically the same, however, the activity of catalyst washed by Na_2SiO_3 solution is significantly reduced. As shown in Fig. 4.3b, the selectivity of cyclohexene washed by Na_2SiO_3 was highest, the selectivity of cyclohexene washed by NaOH solution was higher than that of washed by pure water.

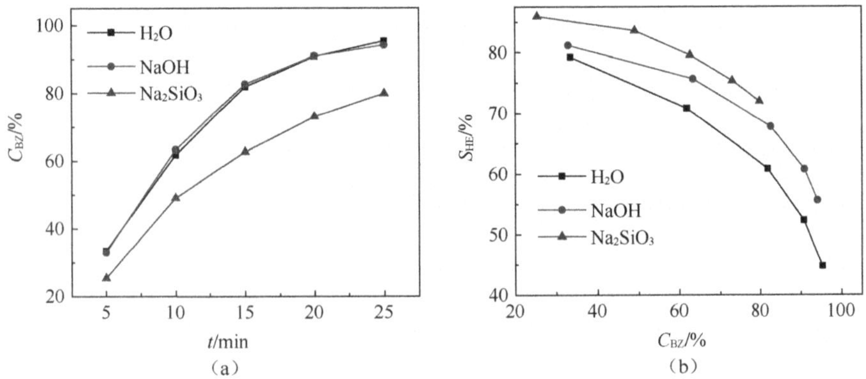

Fig. 4.3 **a** C_{BZ}-t curve and **b** S_{HE}-C_{BE} curve of the catalyst washed by H_2O, NaOH, and Na_2SiO_3 solution, respectively

Table 4.3 The main performance indicators of the catalyst washed by pure water, NaOH and Na_2SiO_3 solution, respectively

Solution	$t40$/min	$S40$/%	$Y40$/%	$t50$/min	$S50$/%	$Y50$/%	$t60$/min	$S60$/%	$Y60$/%
H_2O	6	77	31	8	74	37	10	71	42
NaOH	6	80	32	8	78	39	10	76	46
Na_2SiO_3	8	84	34	10	83	42	14	80	48

Notes Y_{40}, Y_{50}, Y_{60} are the cyclohexene yield corresponding to 40, 50, 60% benzene conversion rate. Evaluation conditions: 150 °C, 5.0 MPa, 1200 r/min, 140 mL benzene, 280 mL water, 1.96 g Ru-Zn catalyst, 9.8 g ZrO_2, 49.2 g $ZnSO_4 \cdot 7H_2O$, without pretreatment

Table 4.3 gives the main performance indicators of the catalyst washed by pure water, NaOH, and Na_2SiO_3 solution, respectively.

As can be seen from Table 4.3, when washed by Na_2SiO_3 solution, the conversion of benzene at 14 min was 60%, the selectivity of cyclohexene was 80%, the yield was 48%, the catalyst had good activity and high selectivity of cyclohexene. When washed by Na_2SiO_3, the residue of Na_2SiO_3 on the catalyst surface hydrolyzed to NaOH and H_2SiO_3 in the hydrogenation reaction conditions, chemisorbed on the surface of the catalyst, interacting with $ZnSO_4$ in the slurry. Thus modified the catalyst surface. In contrast, washed by NaOH solution, the adsorpted NaOH on the catalyst surface in the hydrogenation reaction conditions, was lack of hydrolysis-generated H_2SiO_3 comparing with Na_2SiO_3. When washed with pure water, it was difficult to remove the residual chloride. The data comparison showed that different surface modifications had different effects on the selectivity of cyclohexene.

Table 4.4 shows the Zn content in catalysts determined by atomic absorption spectroscopy (AAS) and inductively coupled plasma atomic emission spectrometry (ICP-AES) after washing with different solutions.

As can be seen from Table 4.4, the Zn content in the catalyst decreased after washing with NaOH solution and Na_2SiO_3 solution. This is because the ZnO in the catalyst belongs to amphoteric oxide, both soluble in acid and in alkali. In contrast, Na_2SiO_3 was weaker than NaOH, so the content of Zn decreased after washing with Na_2SiO_3, while the content of Zn decreased more after washing with NaOH solution, and the content of Zn was the highest after washing with pure water.

The effect of Zn on the activity and selectivity of Ru-Zn catalyst is that the activity decreases and selectivity of cyclohexene increases with the increase of Zn content. However, compared to the catalyst treated with NaOH and Na_2SiO_3 solution with pure water washing, the selectivity of cyclohexene increases, which means that the

Table 4.4 Zn content in catalysts after washing with different solutions

Cleaning solution	Zn/%(wt)	
	ICP-AES	AAS
NaOH	2.90	2.80
Na_2SiO_3	4.65	4.35
H_2O	5.52	5.45

activity and selectivity of the catalyst is not only determined by Zn content, but also closely related to the surface properties of the catalyst.

The Zn content in catalyst after NaOH washing decreased more, but the activity of the catalyst was not changed compared with water washing (the same time is needed for the same benzene conversion), and the selectivity increased. This is because NaOH improves the hydrophilicity of the catalyst surface. However, the concentration of NaOH cannot be too high, otherwise, it will lead to more loss of catalyst of Zn and destruction of surface structure, only proper control of the concentration and washing times of the washing liquid can achieve the best effect.

The Zn content in catalyst after Na_2SiO_3 solution washing was reduced slightly, although the activity of the catalyst decreased and the selectivity of cyclohexene was greatly improved, so Na_2SiO_3 was a good surface modifier for the Ru-Zn catalyst. With the appropriate concentration of Na_2SiO_3 solution, high cyclohexene selectivity and yield can be obtained.

Figures 4.4, 4.5, and 4.6 show the activity, selectivity, and yield of cyclohexene at 15 min, respectively, with following treatments: the catalyst washed by different solution, after hydrocracking pretreatment for 22 h, removing the upper organic phase and reusing 7 times.

From Figs. 4.4 4.5, and 4.6, the catalyst stability of catalyst after water or NaOH solution washing was poor, while the activity and selectivity showed large fluctuation. Washed by Na_2SiO_3 solution, the catalyst activity and selectivity showed better and stable performance. At the seventh time, the conversion rate of benzene was stable at about 70%, and the selectivity of cyclohexene was still more than 70%.

In conclusion, Na_2SiO_3 mainly exists on the surface of the catalyst, and the Ru-Zn catalyst modified by Na_2SiO_3 has not only higher activity and selectivity, but also good stability.

Considering the interaction of catalyst and catalyst system in the process of benzene selective hydrogenation can bring into in situ modification of the surface

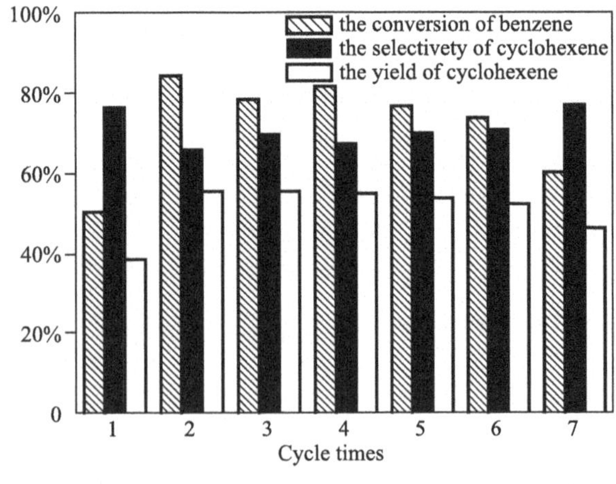

Fig. 4.4 Ru-Zn catalyst after water washing. Reaction conditions: 150 °C, 5.0 MPa, 1200 r/min, 140 mL benzene, 280 mL water, 1.96 g Ru-Zn catalyst, 9.8 g ZrO_2, 49.2 g $ZnSO_4 \cdot 7H_2O$, without pretreatment

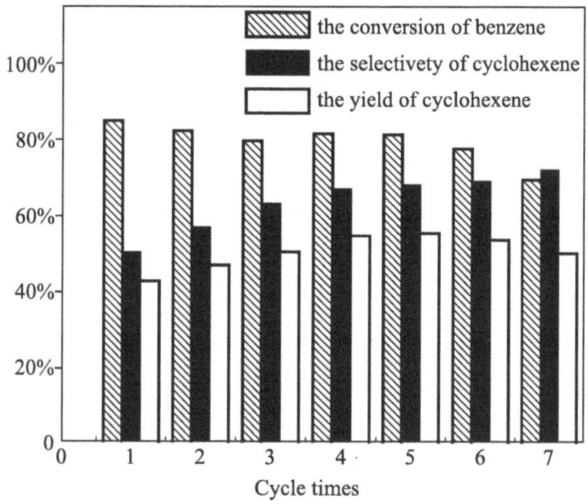

Fig. 4.5 Ru-Zn catalyst after NaOH washing. (reaction conditions were the same with Fig. 4.4)

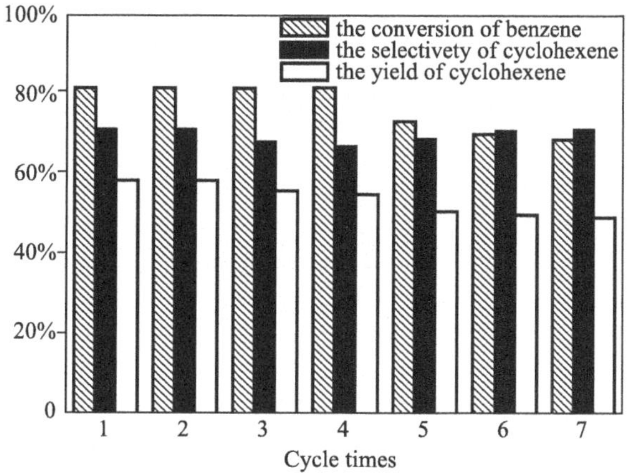

Fig. 4.6 Ru-Zn catalyst after Na_2SiO_3 washing. (reaction conditions was the same as Fig. 4.4)

properties of the catalyst, thus the activity and selectivity of catalyst after adding NaOH and Na_2SiO_3 in reaction slurry were further investigated.

4.1.3 Adding NaOH and Na_2SiO_3 in Reaction Slurry

Figure 4.7 shows the activity of Ru-Zn catalyst and selectivity of cyclohexene after adding NaOH and Na_2SiO_3 in reaction slurry.

As shown in Fig. 4.7a and b, with pure water hydrogenation, the highest activity of the catalyst and the lowest selectivity of cyclohexene can be obtained. Adopting

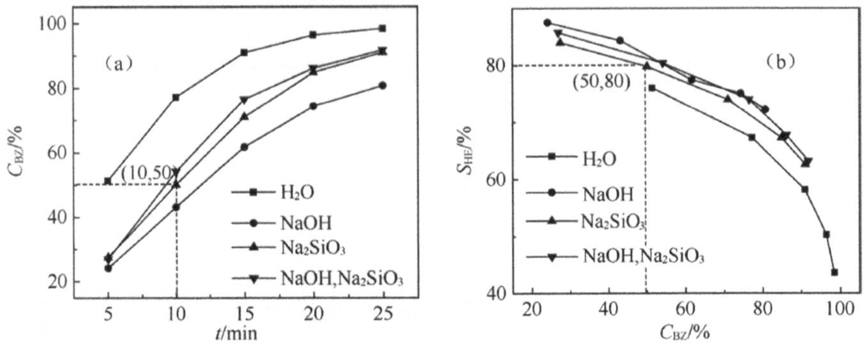

Fig. 4.7 **a** C_{BZ}-t curve and **b** S_{HE}-C_{BZ} curve of Ru-Zn catalyst after adding NaOH and Na_2SiO_3 in the reaction slurry, respectively. Reaction conditions: 150 °C, 5.0 MPa, 1200 r/min, 140 mL benzene, 280 mL water, 1.96 g Ru-Zn catalyst, 9.8 g ZrO2, 49.2 g $ZnSO_4 \cdot 7H_2O$ without pretreatment, adopting pure water, NaOH, Na_2SiO_3, NaOH + Na_2SiO_3 aqueous solution as Hydrogenation solution

NaOH solution hydrogenation, the lowest activity and the highest selectivity of cyclohexene can be obtained. The hydrogenation by using Na_2SiO_3 solution or NaOH + Na_2SiO_3 solution can not only keep the high activity of the catalyst, but also obtain high selectivity of cyclohexene, and thus get the higher yield of cyclohexene. When the Na_2SiO_3 solution was added to the solution, the benzene conversion was 50%, the selectivity of cyclohexene was 80%, the yield of cyclohexene reached 40% at 10 min, showing high activity and high selectivity.

Figure 4.8 shows the activity of Ru-Zn catalyst and the selectivity of cyclohexene after the pretreatment of 22 h by adding NaOH and Na_2SiO_3 in the reaction slurry.

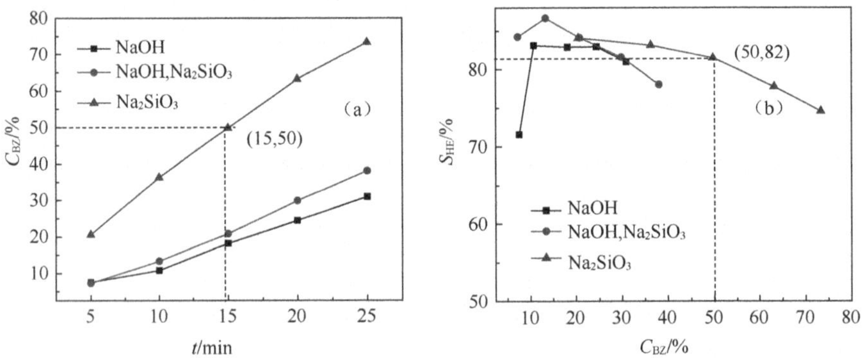

Fig. 4.8 Ru-Zn catalyst **a** C_{BZ}-t curve and **b** S_{HE}-C_{BZ} curve after pretreatment with different slurry for 22 h. Reaction conditions: 150 °C, 5.0 MPa, 1200 r/min, 140 mL benzene, 280 mL water, 1.96 g Ru-Zn catalyst, 9.8 g ZrO_2, 49.2 g $ZnSO_4 \cdot 7H_2O$, 0.3 g NaOH, or 0.3 g Na_2SiO_3, or 0.3 g (NaOH + Na_2SiO_3), pretreatment for 22 h

From Fig. 4.8, after the pretreatment by adding Na_2SiO_3 to the reaction system, 50% direct hydrogenation of benzene was extended from 10 min to 15 min, the selectivity of cyclohexene increased from 80 to 82%, and the selectivity of cyclohexene decreased slowly with stable performance. It showed that in the pretreatment process, the interaction between catalyst and reaction system could modify the catalyst surface in situ, improve the selectivity and yield of cyclohexene, and stabilize the structure and surface properties of the catalyst.

In conclusion, whether in the Ru-Zn catalyst washing process or in the catalytic system, the use of Na_2SiO_3 to modify the surface of the catalyst is an important way to obtain high selectivity and yield of cyclohexene.

4.2 Effect of PEG-10000 on the Performance of Ru-Zn Catalyst

Both amines and alcohols containing polar groups can significantly improve the selectivity and yield of cyclohexene, while alcohols show better performance but the effect depends on the degree of polymerization of alcohols. On the basis of extensive research, PEG-10000 is the best modifier for Ru-Zn catalyst.

4.2.1 Alcohol Additives

Table 4.5 shows the composition and pH of Ru-Zn (2.8%) catalyst determined by XRF after adding different alcohols to the reaction system.

As can be seen from Table 4.5, in the benzene selective hydrogenation reaction system, the effects of adding ethanol, glycerol, mannitol, ethylene glycol, and polyvinyl alcohol on the catalyst Zn/Ru atom ratio are very slight. After adding 0.02 g polyvinyl alcohol PVA-1750, the ratio of Zn/Ru atom in the catalyst increased slightly, which indicated that PVA-1750 could promote the chemical adsorption of Zn on Ru surface. After adding the same amount of polyethylene glycol, with the increase of molecular weight, the ratio of Zn/Ru atom in the catalyst increased gradually, which indicated that the amount of Zn adsorbed on Ru surface increased gradually. However, with the addition of PEG-10000, the content of Zn in the catalyst changed little. In addition, the Zr/Ru atom ratio was fixed at about 5.00 because of the fixed amount of ZrO_2 and Ru. The pH of the aqueous phase was about 6.0 at room temperature.

Figure 4.9 shows the XRD pattern of Ru-Zn (2.8%) catalyst after the hydrogenation with alcohol additive.

As can be seen from Fig. 4.9, the Ru-Zn (2.8%) catalyst showed only the characteristic peak (JCPDS 01-070-0247) of Ru after adding alcohols to the reaction system. The size of Ru crystallites was 4.2 nm obtained by using the maximum peak

Table 4.5 Different Ru-Zn (2.8%) catalyst composition and pH of water phase after hydrogenation of different alcohols

Sample	Zn/Ru (mol/mol)[a]	Zr/Ru (mol/mol)[a]	pH[b]
Without additive	0.26	4.80	5.8
0.2 g Ethanol	0.23	5.09	5.9
0.6 g Ethanol	0.22	4.89	6.0
0.2 g PVA-26	0.25	5.30	5.3
0.02 g PVA-1750	0.28	4.98	5.6
0.2 g Glycerol	0.25	4.94	6.1
0.6 g Glycerol	0.24	4.95	5.7
0.2 g Mannitol	0.24	5.19	5.6
0.6 g Mannitol	0.24	5.34	5.5
0.2 g ethylene Glycol	0.25	5.21	5.9
0.6 g ethylene Glycol	0.24	4.66	6.0
0.2 g PEG-400	0.28	5.19	5.7
0.2 g PEG-600	0.27	4.97	5.6
0.2 g PEG-6000	0.32	5.23	5.8
0.2 g PEG-10000	0.34	4.68	5.9
0.4 g PEG-10000	0.30	5.00	6.2
0.6 g PEG-10000	0.32	5.00	5.7
1.0 g PEG-10000	0.31	4.97	5.6
0.2 g PEG-20000	0.33	5.33	5.3

a XRF determination; **b** Determination of pH meter at room temperature; Hydrogenation conditions: 150 °C, 4.0 MPa, 1400 r/min, 2 g Ru-Zn (2.8%) catalyst, 9.8 g ZrO_2, 49.2 g $ZnSO_4 \cdot 7H_2O$, 140 mL benzene, 280 mL water

width at 44.0° and schemer formula. Except for 44.0° and the weak diffraction peak near Ru, the other diffraction peaks belonged to the dispersant monoclinic phase ZrO2 (JCPDS 01-070-0274).

Figure 4.10 shows the FTIR spectra of Ru-Zn (2.8%) catalyst after adding different amounts of PEG-10000 to the reaction system.

As can be seen from Fig. 4.10, with the increase of PEG-10000, the absorption peak of PEG-10000 did not appear on the hydrogenation catalyst, which indicated

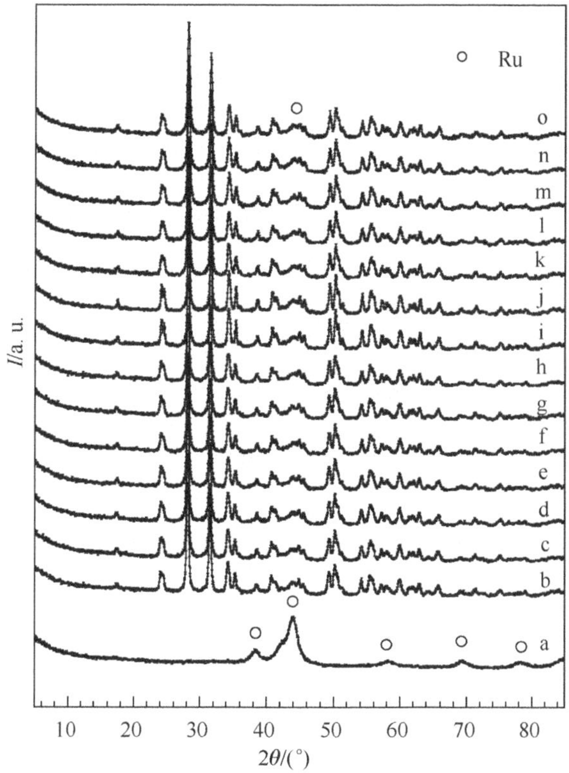

Fig. 4.9 XRD pattern of Ru-Zn (2.8%) catalyst after the hydrogenation with alcohol additive
a Ru-Zn catalyst; **b** without additives; **c** 0.2 g PVA-26; **d** 0.02 g PVA-1750; **e** 0.2 g mannitol; **f** 0.6 g mannitol; **g** 0.2 g propanetriol; **h** 0.6 g propanetriol; **i** 0.2 g ethylene glycol; **j** 0.6 g ethylene glycol; **k** 0.2 g PEG-400; **l** 0.2 g PEG-600; **m** 0.2 g PEG-6000; **n** 0.2 g PEG-10000; **o** 0.2 g PEG-20000

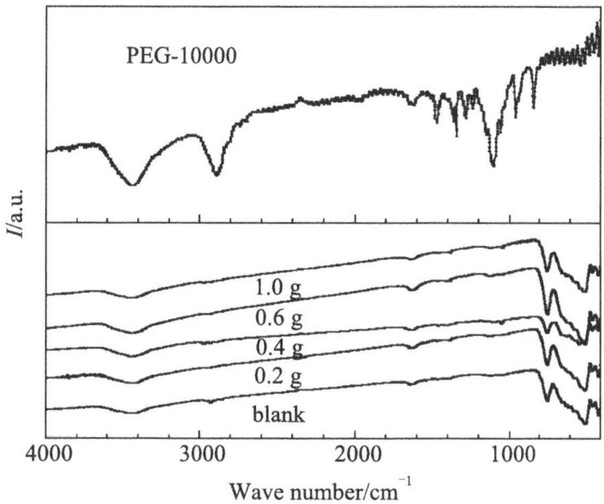

Fig. 4.10 FTIR spectra of Ru-Zn (2.8%) catalyst after adding different amount of PEG-10000 to the reaction system

that PEG mainly existed in the reaction system, and the amount of PEG adsorbed on the surface of the catalyst was so little that cannot be detected.

Table 4.6 shows the structural formula and the optimized hydrogen bond length of the intermolecular hydrogen bond formed by the alcohols and cyclohexene molecules in the liquid phase.

It can be inferred from Table 4.6 that alcohols mainly exist in the slurry as additives. They stabilize the intermediate product cyclohexene, effectively inhibit its re-adsorption, avoid further hydrogenation, and improve the selectivity of cyclohexene through the interaction with cyclohexene molecules.

Table 4.7 shows the highest yield of cyclohexene on Ru-Zn (2.8%) catalyst and the corresponding benzene conversion and cyclohexene selectivity when making alcohols as additive.

As can be seen from Table 4.7, Using ethanol, polyvinyl alcohol, and mannitol as additives, the highest yield of cyclohexene in Ru-Zn (2.8%) catalyst has little change, and the effect is very slight. Using glycerol and ethylene glycol as additives,

Table 4.6 Intermolecular hydrogen bond structural formula and optimized bond length between alcohols and cyclohexene in liquid phase

Addtive	Structure	Hydrogen bond /Å
Ethanol	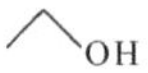	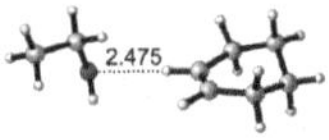
PVA	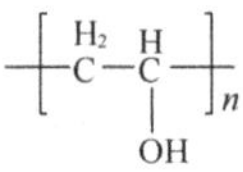	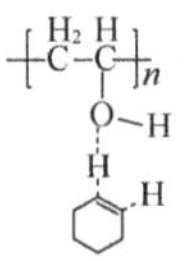
Glycerol	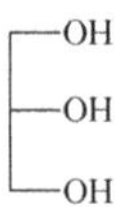	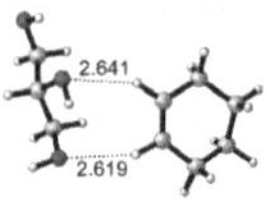
Mannitol	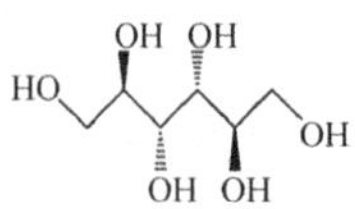	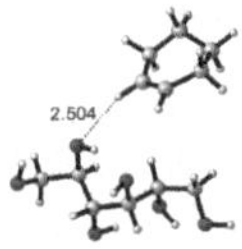
Glycol		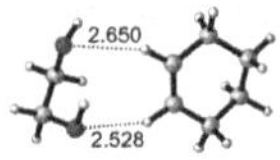
PEG	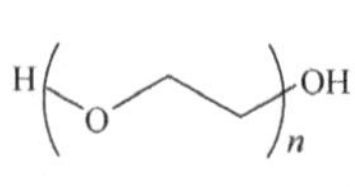	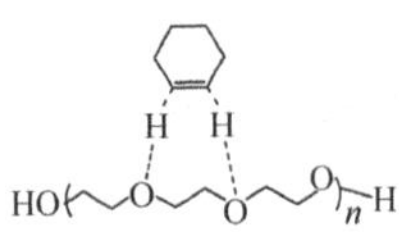

Note: All theoretical calculations were completed on the Gaussian 09 program. All structures were optimized with the B3LYP method at 61-31FG (d, p) level. The vibration frequency was calculated on the optimization structure to ensure that all structures did not have a fictitious frequency

Table 4.7 Using alcohols as an additive, the highest yield of cyclohexene on Ru-Zn (2.8%) catalyst and the corresponding conversion of benzene and selectivity of cyclohexene[a]

Sample	C_{BZ}/%[b]	S_{BZ}/%[b]	Y_{HE}/%[b]	t/min[b]
Without additive	92.9	56.5	52.5	15
0.2 g Ethanol	81.3	64.1	52.1	10
0.6 g Ethanol	92.6	54.6	50.6	15
0.2 g PVA-26	79.2	64.6	51.2	5
0.02 g PVA-1750	89.2	60.7	54.1	15
0.2 g Glycerol	84.5	64.5	54.5	10
0.6 g Glycerol	83.3	67.1	55.9	10
0.2 g Mannitol	87.0	58.9	51.3	10
0.6 g Mannitol	83.4	60.5	50.5	10
0.2 g Glycol	85.9	66.4	57.0	10
0.6 g Glycol	86.6	65.6	56.8	15
0.2 g PEG-400	83.6	68.8	57.5	10
0.2 g PEG-600	88.2	63.5	56.0	10
0.2 g PEG-6000	92.6	65.2	60.4	10
0.2 g PEG-10000	86.8	71.8	62.3	10
0.4 g PEG-10000	79.7	74.8	59.6	10
0.6 g PEG-10000	89.0	67.3	59.9	15
1.0 g PEG-10000	91.5	67.0	61.3	15
0.2 g PEG-20000	85.5	71.8	61.4	15

a Reaction conditions: 150 °C, 4.0 MPa, 1400 r/min, 2 g Ru-Zn (2.8%) catalyst, 9.8 g ZrO_2, 49.2 g $ZnSO_4 \cdot 7H_2O$, 140 mL benzene, 280 mL water, 0.2 g alcohol; **b** the highest yield of cyclohexene

the highest yield of cyclohexene increases slightly. With the addition of PEG as an additive, the highest yield of cyclohexene increases with the increase of molecular weight. When PEG-10000 and PEG-20000 was, respectively, selected as the additive, the yield of cyclohexene could reach up to 62.3 and 61.4%. With the further increasment of PEG-10000, the yield will be stabilized by 60%.

Figure 4.11 displays the C_{BZ}-t and S_{HE}-C_{BZ} curves of Ru-Zn catalyst in the presence of PEG with different molecular weights.

As the comparative results with blank experiments listed in Fig. 4.11, the activation of Ru-Zn catalyst stays the same in the presence PEG-400, increases in combination with PEG-600 and PEG-6000, and declines when PEG-10000 and PEG-20000 was used. The selectivity of cyclohexene also climbs with the increasing molecular weight of polyethylene glycol. In overall consideration, PEG-10000 can promote not only the activity of catalyst, but also the selectivity of cyclohexene, leading to the highest yield of cyclohexene [17].

Figure 4.12 presents the C_{BZ}-t and S_{HE}-C_{BZ} curves of Ru-Zn catalyst in the presence of PEG-10000.

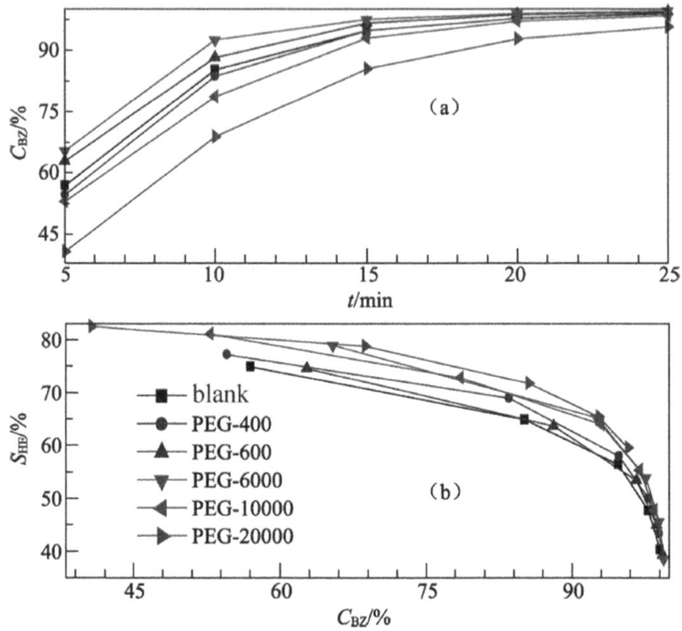

Fig. 4.11 Activation of Ru-Zn catalyst in the presence of PEG. **a** C_{BZ}-t curve; **b** S_{HE}-C_{BZ} curve; the number after PEG means the molecular weight

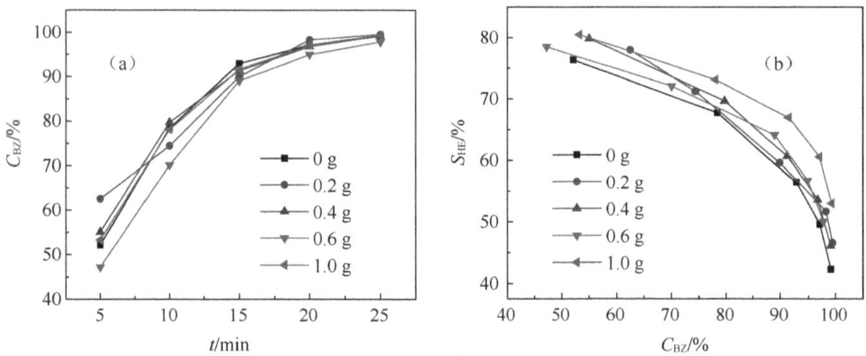

Fig. 4.12 The reaction system of Ru-Zn catalyzed hydrogenation in the presence of the different amount of PEG-10000. **a** C_{BZ}-t; **b** S_{HE}-C_{BZ}; Other reaction conditions: 150 °C, 5.0 MPa, 1200 r/min, 140 mL benzene, 280 mL H_2O, 1.96 g Ru-Zn catalyst, 49.2 g $ZnSO_4 \cdot 7H_2O$, 9.8 g ZrO_2

It could be proposed that even a little amount of PEG is chemically adsorbed on the catalyst surface, the hydroxyl group could enhance the hydrophily of the catalyst surface. If the lone-pair electron partially transfers toward Ru, the Ru will become the electron-rich center, which could benefit for the desorption of cyclohexene [17].

Figure 4.12 presents the C_{BZ}-t and S_{HE}-C_{BZ} curves of Ru-Zn catalyst in the presence of PEG-10000.

It could be seen from Fig. 4.12a that the activity of Ru-Zn catalyst gradually decreases with the increasing amount of PEG-10000 and reaches the bottom when 0.6 g PEG-10000 is used. While the activity increases when the amount of PEG-10000 is 1.0 g. Figure 4.12b shows that the selectivity of cyclohexene reaches the highest when 1.0 g PEG-10000 is used. The conversion is 80 with 72% selectivity after 10 min of reaction.

Figure 4.13 displays the activity of pretreated Ru-Zn catalyst, yield and selectivity of cyclohexene after PEG-10000 is added into the reaction system.

It shows that when the pretreated Ru-Zn (2.8%) catalyst is used in combination with PEG-10000, the conversion of benzene can be 82.0% with the yield and selectivity of 78.6 and 64.5% after 15 min. In addition, the yield of cyclohexene can, respectively, reach up to 63.1, 62.8, and 61.3% after 20, 25, and 30 min. This indicates that the yield and selectivity of cyclohexene could be very high in such hydrogenation reaction system.

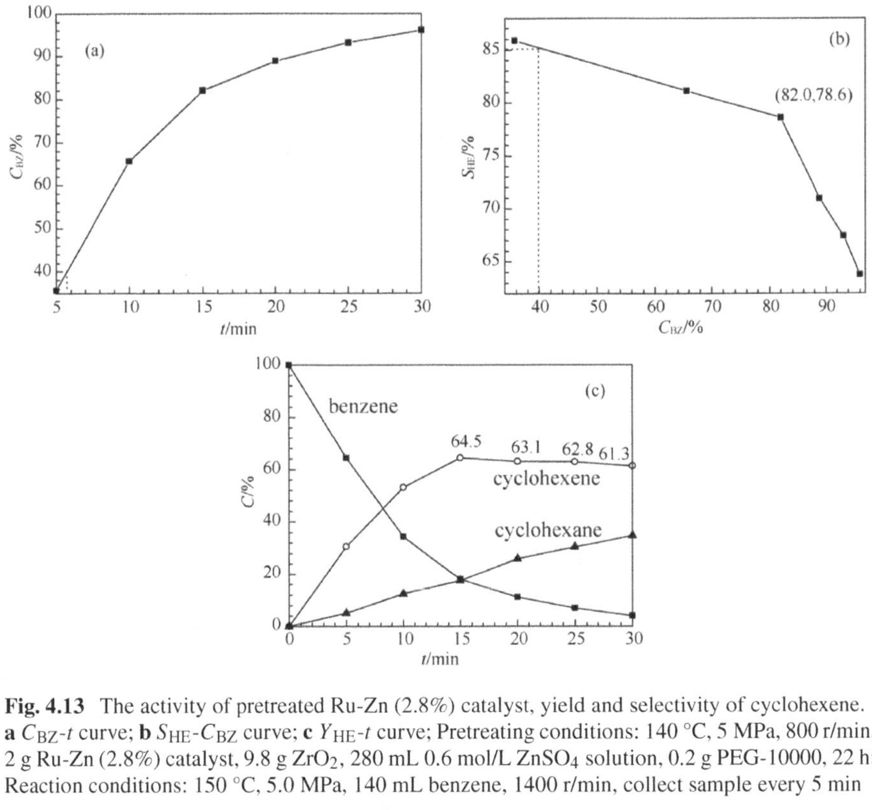

Fig. 4.13 The activity of pretreated Ru-Zn (2.8%) catalyst, yield and selectivity of cyclohexene. **a** C_{BZ}-t curve; **b** S_{HE}-C_{BZ} curve; **c** Y_{HE}-t curve; Pretreating conditions: 140 °C, 5 MPa, 800 r/min, 2 g Ru-Zn (2.8%) catalyst, 9.8 g ZrO_2, 280 mL 0.6 mol/L $ZnSO_4$ solution, 0.2 g PEG-10000, 22 h; Reaction conditions: 150 °C, 5.0 MPa, 140 mL benzene, 1400 r/min, collect sample every 5 min

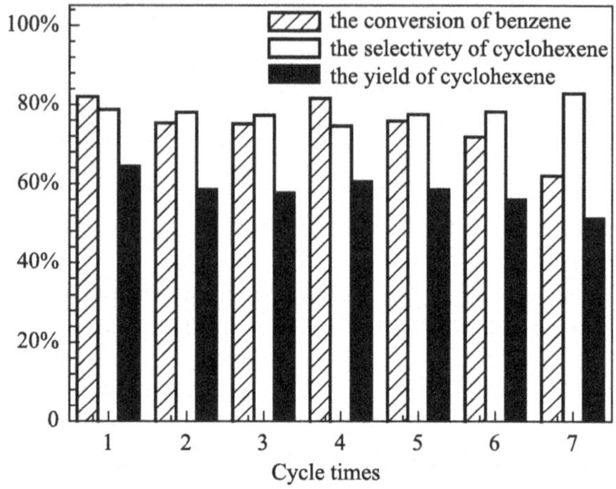

Fig. 4.14 The conversion of benzene, yield and selectivity of cyclohexene during the seven times repeating use of Ru-Zn (2.8%) catalyst. Pretreating conditions: 140 °C, 5 MPa, 800 r/min, 2 g Ru-Zn (2.8%) catalyst, 9.8 g ZrO_2, 280 mL 0.6 mol/L $ZnSO_4$ solution, 0.2 g PEG-10000, 22 h; Reaction conditions: 150 °C, 5.0 MPa, 140 mL benzene, 1400 r/min, collect sample every 15 min

Figure 4.14 shows the reused results after seven times. The catalyst is firstly pretreated with 0.6 mol/L $ZnSO_4$ solution for about 22 h, then it is added in the reaction system with PEG-10000.

It can be seen from Fig. 4.14 that before the seventh recycling use, the conversion of benzene will be stabilized in above 71.8% with the yield and selectivity of 78.2 and 56.2%. The conversion decreases to 62.1% after using for six times, while the yield and selectivity are still as high as 51.6 and 82.8%. So it indicates that when PEG-10000 is used as an additive, Ru-Zn (2.8%) catalyst has good stability and industrial perspective.

In conclusion, the presence of PEG and PVA-1750 can promote the chemical adsorption of Zn^{2+} on the catalyst surface, which is a benefit for the selectivity of cyclohexene. With the increment in the molecular weight of PEG, the adsorption amount of Zn^{2+} on the catalyst surface and yield of cyclohexene can be enhanced. The hydrophily of the catalyst can be improved when a small amount of alcohol-type compounds adsorbs on the catalyst surface. The hydrogen bonding interaction between alcohol and cyclohexene can inhibit the adsorption of cyclohexene on the Ru surface and further hydrogenation into cyclohexane [18, 19]. Ethanol, PVA, and mannitol can only form one intermolecular hydrogen bond with low bond energy (about −4.1 kJ/mol), which has little effect on the yield and selectivity of cyclohexene. While glycerol, glycol, and PEG can form two hydrogen bonds in the octatomic ring, which is a benefit for the yield of cyclohexene. When PEG is used as an additive, it can not only promote the formation of two hydrogen bonds in the octatomic ring, but also lead to the increasing adsorption of Zn^{2+} on Ru-Zn catalyst surface. When PEG-10000 and PEG-20000 are used as additives, the highest yield of cyclohexene can, respectively, reach up to 62.3 and 61.4% with the catalysis of Ru-Zn (2.8%).

4.2.2 Amine Additives

Table 4.8 presents the surface area S_{BET}, pore diameter d, and average pore volume Vp of Ru-Zn (4.9%) catalyst modified with diethanolamine after hydrogenation.

As shown in Table 4.8, the surface area, pore diameter, and pore volume are 62 m^2/g, 5.74 nm, and 0.089 cm^3/g, respectively. After hydrogenation reactions and recycling use, the surface area, pore diameter, and pore volume of catalyst are all close to ZrO_2.

Figure 4.15 shows the TEM image and microcrystal size distribution of Ru-Zn (4.9%) catalyst.

It can be seen from Fig. 4.15 that the microcrystal size of Ru-Zn (4.9%) catalyst is almost concentrated on 3.5 nm and increases to 3.8 nm after five times of recycling use.

Figure 4.16 shows the XRD spectrum of Ru-Zn (4.9%) catalyst in combination with diethanolamine after hydrogenation.

As shown in Fig. 4.16 that when diethanolamine is added into the reaction system, the characteristic peaks of Ru appear at $2\theta = 38.4°$, $42.4°$, $43.9°$, $58.3°$, $69.3°$, and $78.3°$ (JCPDS 01-070-0274), and the remainings are attributed to the monoclinic ZrO2 (JCPDS 00-024-1165). When the amount of diethanolamine attains to 0.4 g, the characteristic peak of $(Zn(OH)_2)_3(ZnSO_4)(H_2O)_3$ appears at $9.8°$ (JCPDS 01-078-0247), which indicates that the transformation of diethanolamine and $ZnSO_4$ into $(Zn(OH)_2)_3(ZnSO_4)(H_2O)_3$. The reaction mechanism is shown as follows:

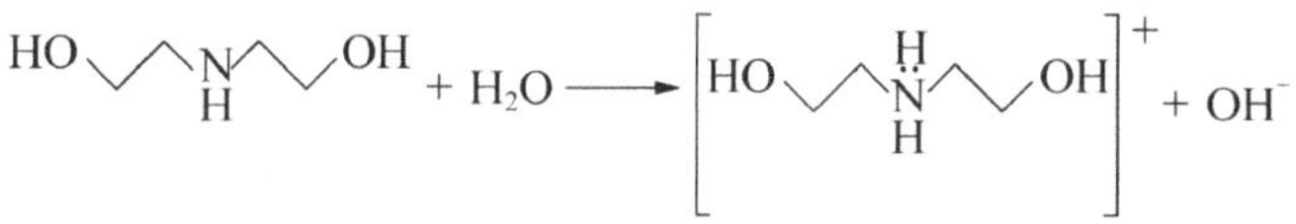

(4.2.1)

$$6OH^- + 4Zn^{2+} + SO_4^{2-} + 3H_2O \longrightarrow (Zn(OH)_2)_3(ZnSO_4)(H_2O)_3 \downarrow \quad (4.2.2)$$

Table 4.8 The surface area S_{BET}, pore diameter d, and average pore volume V_p of Ru-Zn (4.9%) catalyst

Sample	$S_{BET}/(m^2/g)$	d/nm	$V_p/(cm^3/g)$
ZrO_2	34	15.99	0.13
Ru-Zn(4.9%) BH	62	5.74	0.089
Ru-Zn(4.9%) AH	37	19.88	0.179
Ru-Zn(4.9%)-0.2 g diethanol amine AH	35	10.58	0.092
Ru-Zn(4.9%)-0.3 g diethanol amine AH	36	17.78	0.159
Ru-Zn(4.9%)-0.4 g diethanol amine AH	37	13.26	0.124
Ru-Zn(4.9%)-0.3 g diethanol amine AR	42	15.74	0.168

Note BH and AH mean the catalyst before and after hydrogenation respectively, AR means the catalyst after recycling use; Hydrogenation conditions: 150 °C, 5 MPa H_2, 1000 r/min, 1.96 g Ru-Zn (4.9%) catalyst, 9.8 g ZrO_2, 49.2 g $ZnSO_4 \cdot 7H_2O$, 140 mL benzene, 280 mL H_2O

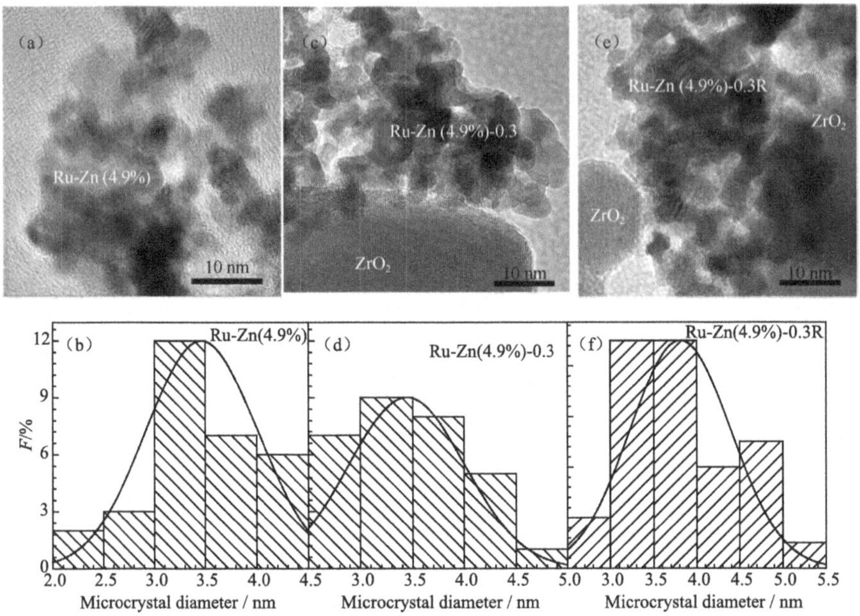

Fig. 4.15 TEM image and microcrystal size distribution of Ru-Zn catalyst. **a**, **b** Ru-Zn (4.9%) BH; **c**, **d** Ru-Zn (4.9%)-0.3 g diethanolamine AH; **e**, **f** Ru-Zn (4.9%)-0.3 g diethanolamine AR

Fig. 4.16 XRD spectrum of Ru-Zn (4.9%) catalyst in combination with diethanol amine after hydrogenation

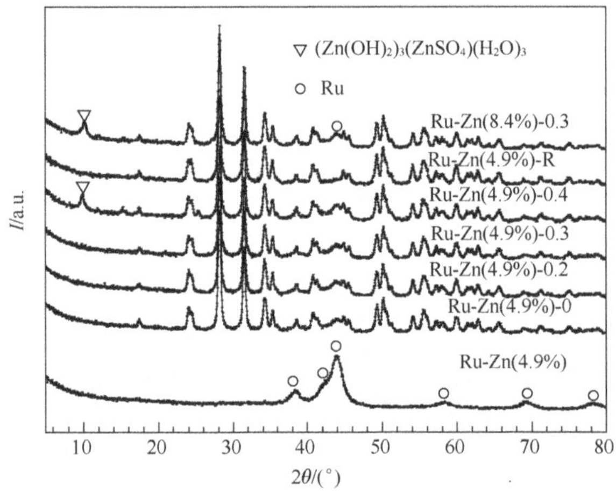

Diethanol amine is firstly hydrolyzed to produce OH^- followed by the conversion of $ZnSO_4$ into basic salt, which is chemically adsorbed on the catalyst surface.

Table 4.9 listed the metal content in Ru-Zn catalyst after hydrogenation in the presence of diethanolamine.

Table 4.9 Metal content in Ru-Zn catalyst after hydrogenation in the presence of diethanolamine

Sample	Ru/%	Zn/%	Zr/%	Hf/%
ZrO2	–	–	94.65	5.35
Ru-Zn(4.9%) BH	17.15	2.91	76.44	3.51
Ru-Zn(4.9%)-0.2 diethanol amine AH	16.67	3.60	76.47	3.25
Ru-Zn(4.9%)-0.3 diethanol amine AH	17.28	4.56	74.90	3.26
Ru-Zn(4.9%)-0.4 diethanol amine AH	16.62	6.15	74.25	2.98
Ru-Zn(4.9%)-0.3 diethanol amine AR	9.32	3.42	82.81	3.45
Ru-Zn(8.4%)-0.3 diethanol amine AH	17.32	6.23	73.98	2.47

It can be been from Table 4.9 that the sample after hydrogenation mainly contains Ru, Zn, Zr, and Hf, which comes from the HfO_2 existed in ZrO_2 dispersant. The $ZnSO_4$ in reaction slurry transforms into $(Zn(OH)_2)_3(ZnSO_4)(H_2O)_3$ after the hydrolysis of diethanolamine. Besides, the production of basic salt and Zn content in the catalyst is enhanced with the increasing amount of diethanolamine.

Figure 4.17 shows the FTIR and TPR spectra of Ru-Zn (4.9%) catalyst after hydrogenation in the presence of diethanolamine.

It can be obtained from Fig. 4.17a that with the increasing amount of diethanolamine, the adsorption peak of it does not appear in the FTIR spectrum of the catalyst after hydrogenation, indicating that diethanolamine only exists in the reaction system and only a small amount of it is adsorbed on the catalyst surface.

Figure 4.17b shows that only a reduction peak appears in the TPR spectra of all samples, which is attributed to the substep reduction of RuOx. With the increasement of diethanolamine, the reduction peak moves toward low temperature until the amount reaches to 0.4 g. The adsorbed diethanolamine on the catalyst surface can inhibit the aggregation of Ru-Zn catalyst due to the space effect. The more the amount of diethanol amine used, the smaller the partical catalyst will be after hydrogenation, so the reduction temperature is gradually decreased. While the amount of

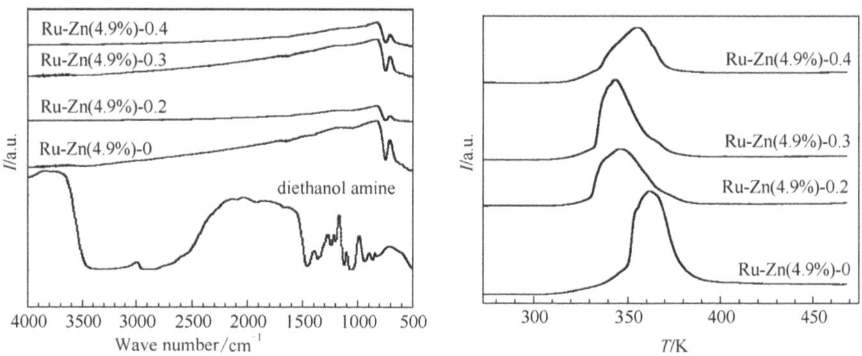

Fig. 4.17 FTIR **a** and TPR **b** spectra of Ru-Zn (4.9%) catalyst after hydrogenation in the presence of diethanolamine

diethanolamine is above 0.4 g, the reduction of RuOx can be delayed due to the increasement of $(Zn(OH)_2)_3(ZnSO_4)(H_2O)_3$ on the catalyst surface. But the reduction temperature of all samples is below 150 °C, indicating the existence of Ru under reaction condition.

Figure 4.18 shows the hydrogenation activity of Ru-Zn (4.9%), yield, and selectivity of cyclohexene in the presence of diethanolamine.

It can be seen from Fig. 4.18 that when the diethanolamine is increased to 0.3 g, the catalytic activity decreases and the selectivity of cyclohexene increases under the same benzene conversion. The highest yield of cyclohexene during reaction time can be enhanced from 55.2 to 59.4%. When the diethanolamine is increased to 0.4 g, the catalytic activity decreases obviously, but the selectivity of cyclohexene increases under the same benzene conversion. The highest yield of cyclohexene declines to 41.9%. When the amount of diethanolamine is 0.3 g, the basic salt is highly distributed on the surface of catalyst and ZrO_2. The Ru-Zn (4.9%) catalyst shows the highest performance on selective hydrogenation of benzene. The oxygen or nitrogen atom of diethanolamine in reaction slurry can form a hydrogen bond

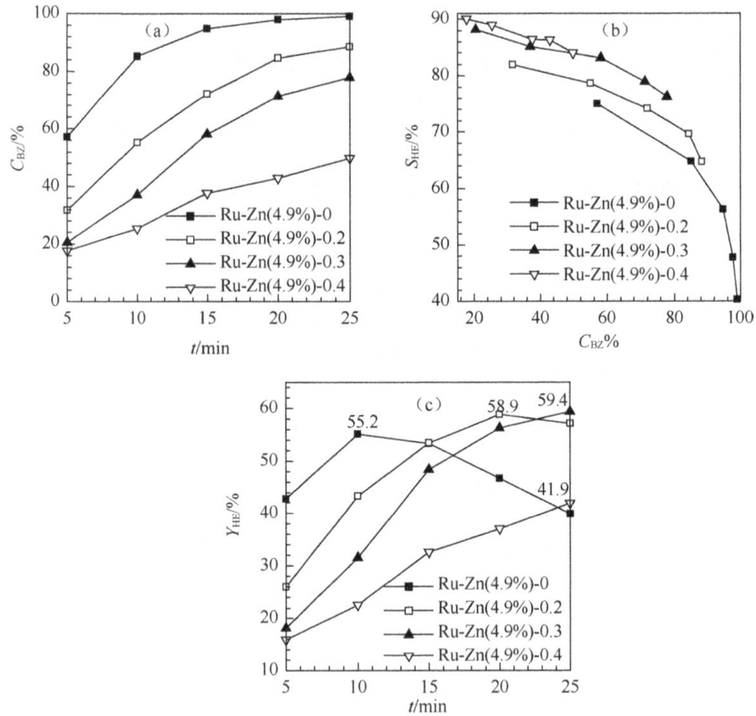

Fig. 4.18 The hydrogenation activity of Ru-Zn (4.9%), yield and selectivity of cyclohexene in the presence of diethanolamine. **a** C_{BZ}-t curve; **b** S_{HE}-C_{BZ} curve; **c** Y_{HE}-t curve; Reaction condition: 150 °C, 5 MPa H_2, 1000 r/min, 1.96 g Ru-Zn (4.9%) catalyst, 9.8 g ZrO2, 49.2 g $ZnSO_4 \cdot 7H_2O$, 140 mL benzene, 280 mL H_2O

with cyclohexene via lone electron pairs [20], which can stabilize the cyclohexene molecule and prevent the adsorption and further hydrogenation of it into cyclohexane. With the synergistic action of basic salt and diethanolamine, the catalyst shows good selectivity for cyclohexene.

Figure 4.19 shows the activity of Ru-Zn catalyst with different Zn content in the presence of diethanolamine.

Figure 4.20 presents the benzene conversion, yield, and selectivity of cyclohexene with the five repeating times of Ru-Zn (4.9%) catalyst in the presence of 0.3 g diethanolamine.

It can be seen from Fig. 4.20 that the conversion of benzene, yield, and selectivity of cyclohexene can reach up to 84.3, 75.5, and 63.6% when the catalyst is used for the third time. The benzene conversion and cyclohexene yield can be still kept above 75 and 58%. So it indicates that the Ru-Zn (4.9%) catalyst has good stability and industrial perspective when diethanolamine is used as an additive.

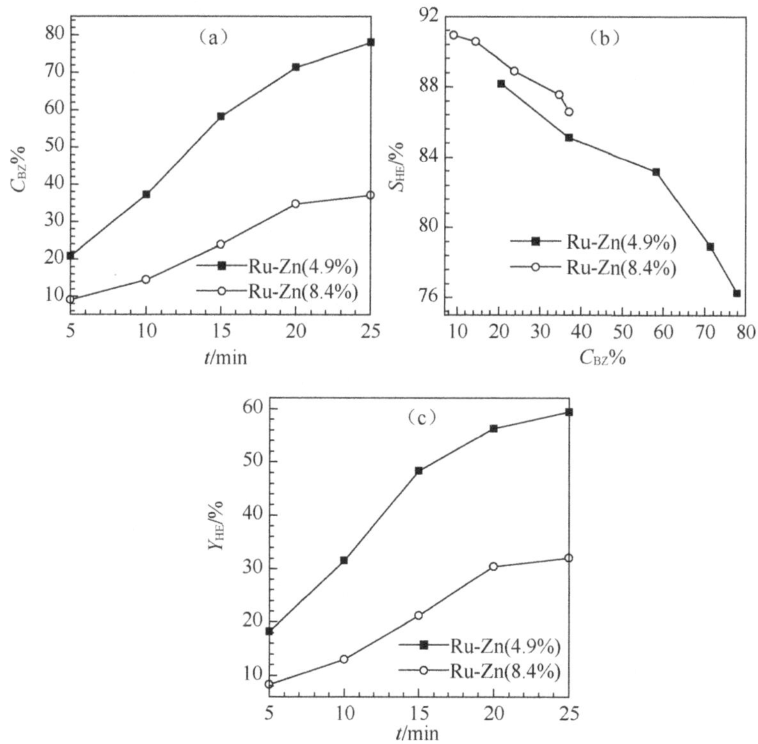

Fig. 4.19 The activity of Ru-Zn catalyst with different Zn content in the presence of diethanolamine **a** C_{BZ}-t curve, **b** S_{HE}-C_{BZ} curve, **c** Y_{HE}-t curve; Reaction condition: 150 °C, 5 MPa H2, 1000 r/min, 1.96 g Ru-Zn (4.9%) catalyst, 9.8 g ZrO_2, 49.2 g $ZnSO_4 \cdot 7H_2O$, 0.3 g diethanolamine, 140 mL benzene, 280 mL H_2O

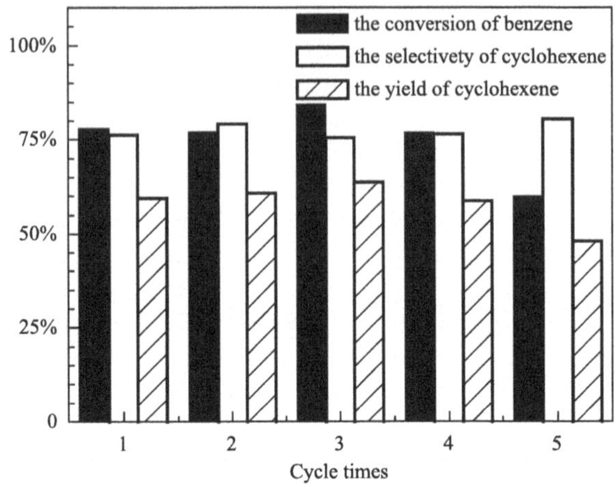

Fig. 4.20 The recycling performance of Ru-Zn (4.9%) catalyst in the presence of diethanolamine. Reaction condition: 150 °C, 5 MPa H_2, 1000 r/min, 1.96 g Ru-Zn (4.9%) catalyst, 9.8 g ZrO_2, 49.2 g $ZnSO_4 \cdot 7H_2O$, 0.3 g diethanol amine, 140 mL benzene, 280 mL H_2O

Although amine can obviously promote the yield and selectivity of cyclohexene, the basic salt from $ZnSO_4$ through hydrolysis plays an important role. This interaction mechanism is the same as Na_2SiO_3 but different from alcohol. The diethanolamine can modify Ru-Zn catalyst and be replaced with Na_2SiO_3 or PEG. While the Na_2SiO_3-PEG-10000 system can enhance the yield and selectivity of cyclohexene via the interaction with the slurry.

4.3 Main Performance Index the Second-Generation Catalytic System of Ru-Zn-Na_2SiO_3-PEG-10000 for Benzene Selective Hydrogenation

4.3.1 The Modification Mechanism of Na_2SiO_3-PEG-10000 on Ru-Zn Catalyst

Figure 4.21 shows the contact angle of Ru-Zn catalyst modified with Na_2SiO_3-PEG-10000 with water under the electron microscope.

Figure 4.21 shows that the contact angle of the catalyst surface is close to 90° before modification, while it changes to 45° after being modified with Na_2SiO_3-PEG-10000. So it indicates that the hydrophily of Ru-Zn catalyst surface is promoted after modification.

Figure 4.22 presents the net structure of Na_2SiO_3-PEG-10000 through TEM characterization.

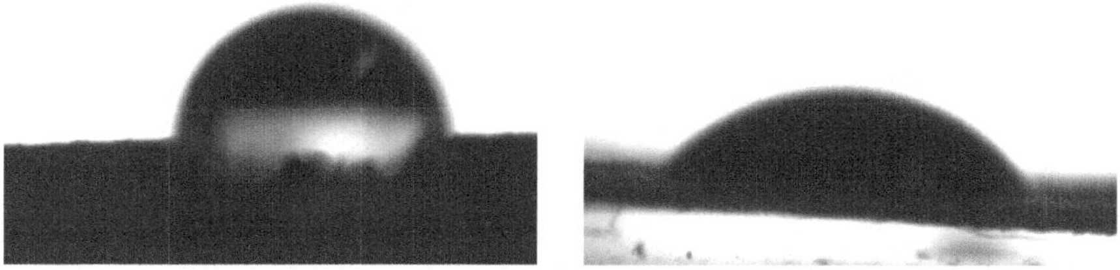

(a) Before modification (b) Modified with Na_2SiO_3-PEG-10000

Fig. 4.21 The contact angle of Ru-Zn catalyst with water before **a** and after **b** modification

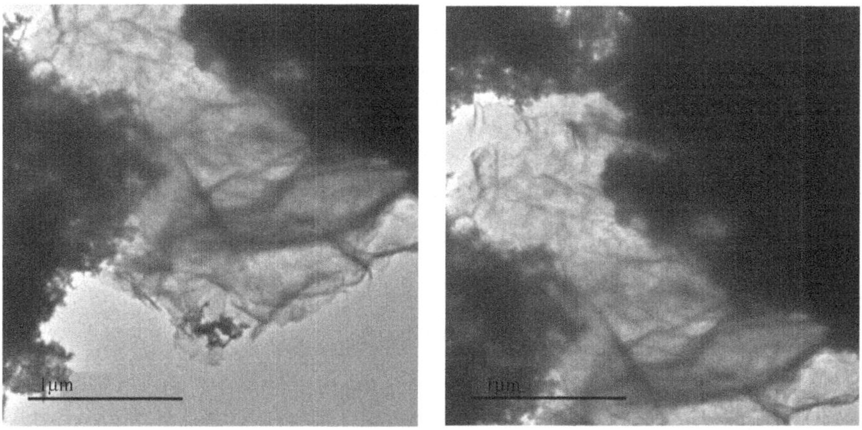

Fig. 4.22 The net structure of Na_2SiO_3-PEG-10000

It can be seen from Fig. 4.22 that Na_2SiO_3-PEG-10000 can form a net structure, which contains basic salt and chemically adsorbed on Ru-Zn catalyst surface. When the amount of net structure is suitable, it distributes on catalyst surface as single-layer distribution. While the content above the threshold of single-layer distribution, it will be randomly distributed onto the catalyst surface and even the slurry via intermolecular van der Waals force. The Ru-Zn catalyst shows the highest yield and selectivity of cyclohexene when the net structure of Na_2SiO_3-PEG-10000 is distributed on catalyst surface as a single layer.

Figure 4.23 shows the surface structure of Ru-Zn catalyst modified with Na_2SiO_3-PEG-10000 according to the characterization results of XRD, electron microscope, and TEM.

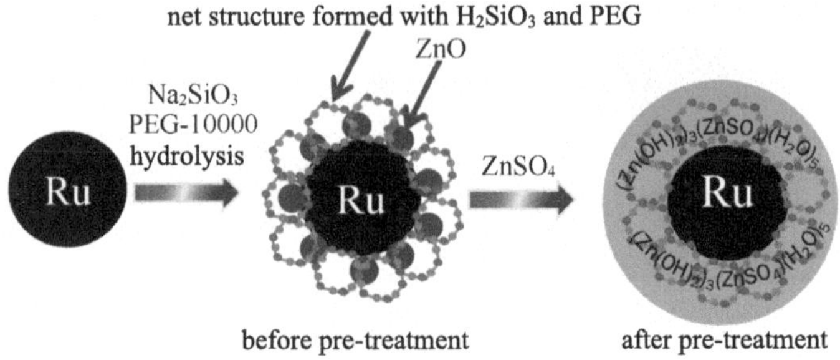

Fig. 4.23 The surface structure of Ru-Zn catalyst modified with Na_2SiO_3-PEG-10000

As shown in Fig. 4.23, the Zn exists in the form of ZnO. The ZnO on the surface of the pretreated catalyst can be transformed into basic zinc sulfate via the interaction with $ZnSO_4$, followed by the formation of a net structure with Na_2SiO_3-PEG-10000 to cover the catalyst surface and change the surface property. Such a surface structure can effectively prevent the crash and aggregation of active Ru component, prolonging the catalyst life and enhancing the yield and selectivity of cyclohexene.

4.3.2 Main Performance Index of Ru-Zn-Na_2SiO_3-PEG-1000 Catalytic System

The Ru-Zn-Na_2SiO_3-PEG-10000 catalyst system is the second-generation catalyst for benzene selective hydrogenation, and the main performance indexes include benzene conversion, cyclohexene selectivity, and yield.

Table 4.10 shows the activity and selectivity of direct hydrogenation over Ru-Zn-Na_2SiO_3-PEG-10000 catalyst.

Table 4.10 Activity and selectivity of direct hydrogenation over Ru-Zn-Na_2SiO_3-PEG-10000 catalyst

t/min	T/°C	C_{BZ}/%	S_{HE}/%	Y_{HE}/%
5	163	56.66	76.26	43.21
10	163	85.28	67.06	57.19
15	155	94.24	57.10	53.81
20	149	96.99	49.38	47.89
25	147	98.96	42.76	42.32

Note Slurry composition: 1.96 g Ru-Zn catalyst, 280 mL of 0.007887 mol/L Na_2SiO_3, 0.2 g PEG-10000, 9.8 g ZrO_2, 45.7 g $ZnSO_4 \cdot 7H_2O$; Hydrogenation conditions: 150 °C, 5 MPa, 1400 r/min, time is recorded after adding 140 mL of benzene

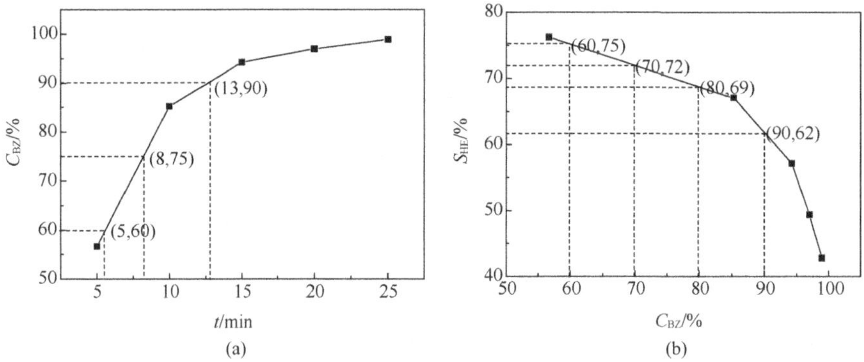

Fig. 4.24 Activity and selectivity of direct hydrogenation over Ru-Zn-Na_2SiO_3-PEG-10000 catalyst. **a** C_{BZ}-t curve; **b** S_{HE}-C_{BZ} curve

Figure 4.24 shows the C_{BZ}-t and S_{HE}-C_{BZ} curves for the selective hydrogenation of benzene over Ru-Zn-Na_2SiO_3-PEG-10000 catalyst.

From Fig. 4.24, it can be seen that the Ru-Zn-Na_2SiO_3-PEG-10000 is used for direct hydrogenation of benzene, the benzene conversion is 60% at 5 min, 75% at 8 min, and 90% at 13 min, indicating high catalytic activity. The selectivity of cyclohexene is 75% at benzene conversion of 60, 72% at benzene conversion of 70, 69% at benzene conversion of 80, and 62% at benzene conversion of 90%, showing a high selectivity of cyclohexene. Such a catalyst can obtain a high yield of cyclohexene after pretreatment.

Table 4.11 gives the hydrogenation activity and selectivity of Ru-Zn-Na_2SiO_3-PEG-10000 after pretreatment for 22 h.

Figure 4.25 shows the C_{BZ}-t and S_{HE}-C_{BZ} curves for the selective hydrogenation of benzene over Ru-Zn-Na_2SiO_3-PEG-10000 after pretreatment for 22 h.

From Fig. 4.25, it can be seen that the main performance indicators of the Ru-Zn-Na_2SiO_3-PEG-10000 catalyst after pretreatment for 22 h are as follows: $t40 =$

Table 4.11 Hydrogenation activity and selectivity of Ru-Zn-Na_2SiO_3-PEG-10000 after pretreatment for 22 h

t/min	T/°C	C_{BZ}/%	S_{HE}/%	Y_{HE}/%
5	149	16.18	90.42	14.63
10	152	34.82	86.82	30.23
15	153	50.10	84.91	42.54
20	153	62.74	82.15	51.54
25	151	72.44	78.24	56.68

Note Pretreatment conditions: 150 °C, 5 MPa, 800 r/min, 22 h; reaction system: 280 mL of 0.007887 mol/L Na_2SiO_3, 0.2 g PEG-10000, 9.8 g ZrO_2, 45.7 g of $ZnSO_4 \cdot 7H_2O$; Hydrogenation conditions: 150 °C, 5 MPa of H_2, 1400 r/min, 140 mL benzene

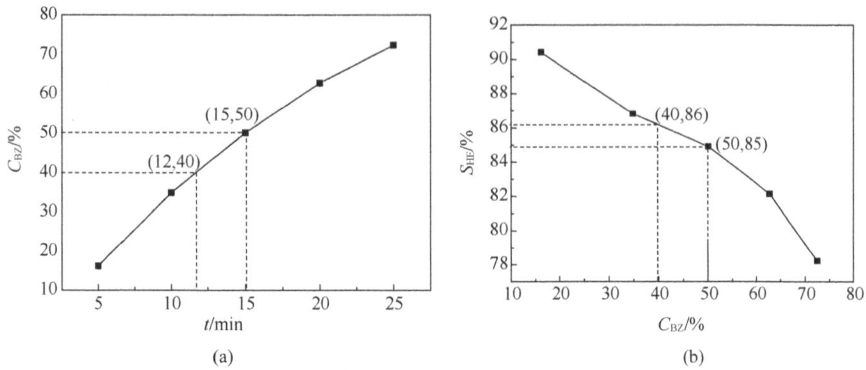

Fig. 4.25 Activity and selectivity of Ru-Zn-Na_2SiO_3-PEG-10000 after pretreatment for 22 h. **a** C_{BZ}-t curve; **b** S_{HE}-C_{BZ} curve

12 min, $\gamma40 = 126$, $S40 = 86\%$; $t50 = 15$ min, $\gamma50 = 136$, $S50 = 85\%$, reaching the level of the second-generation catalyst abroad.

After the pilot, the second-generation catalyst was industrialized in 2010, and was used for the selective hydrogenation of benzene to cyclohexene, as well as its downstream products industrial installations, with the productivity of 100,000 tons per year.

Table 4.12 shows the activity and selectivity of direct hydrogenation over the commercial Ru-Zn-Na_2SiO_3-PEG-10000 catalyst.

Comparing results in Table 4.12 with the direct hydrogenation results of the second generation catalysts shown in Table 4.10, it can be seen that it took 3–5 min more to achieve the same conversion of benzene over the commercial Ru-Zn-Na_2SiO_3-PEG-10000, and the activity slightly decreased; whereas at the same benzene conversion, the cyclohexene selectivity increased by 2–3 percentage points.

Table 4.12 Activity and selectivity of direct hydrogenation over commercial Ru-Zn-Na_2SiO_3-PEG-10000 catalyst

t/min	T/°C	C_{BZ}/%	S_{HE}/%	Y_{HE}/%
5	149	41.47	84.37	34.99
10	146	60.53	80.54	48.75
15	144	73.36	75.85	55.64
20	143	80.16	71.74	57.51
25	143	85.96	67.17	57.74

Note Slurry compositions: 1.96 g Ru-Zn catalyst, 280 mL of 0.007887 mol/L Na_2SiO_3, 0.2 g of PEG-10000, 9.8 g ZrO2, 45.7 g $ZnSO_4 \cdot 7H_2O$; Hydrogenation conditions: 150 °C, 5 MPa, 1400 r/min, 140 mL benzene

Table 4.13 Hydrogenation activity and selectivity of commercial Ru-Zn-Na_2SiO_3-PEG-10000 catalyst after pretreatment for 22 h

t/min	*T*/°C	C_{BZ}/%	S_{HE}/%	Y_{HE}/%
5	151	20.07	90.88	18.24
10	152	38.10	87.82	33.46
15	150	51.96	84.87	44.10
20	150	63.85	82.02	52.37
25	150	72.21	78.74	56.86

Note Pretreatment conditions: 140 °C, 5 MPa, 800 r/min, 22 h; reaction system: 280 mL of 0.007887 mol/L Na_2SiO_3, 0.2 g PEG-10000, 9.8 g ZrO2, 45.7 g $ZnSO_4 \cdot 7H_2O$; Hydrogenation conditions: 150 °C, 5 MPa, 1400 r/min, 140 mL benzene

Table 4.13 lists the hydrogenation activity and selectivity of commercial Ru-Zn-Na_2SiO_3-PEG-10000 catalyst after pretreatment for 22 h.

Comparing Table 4.13 with Table 4.11, the cyclohexene selectivity of the second generation industrial catalyst was 84.87% at the benzene conversion of 51.96% within 15 min, and it was very close to the experimental results of bench-scale, which showed a cyclohexene selectivity of 85% at the benzene conversion of 50% within 15 min.

References

1. Ye, D.Q., Pang, X.X., Huang, Z.T., et al.: A study on the selective hydrogenation of benzene to cyclohexene by catalytic surface modification. Chem. React. Eng. Technol. **8**, 210–213 (1992)
2. Odenbrand, C.U.I., Lundin, S.T.: Hydrogenation of benzene to cyclohexene on a ruthenium catalyst: influence of some reaction parameters. J. Chem. Tech. Biotech. **30**, 677–687 (1980)
3. Odenbrand, C.U.I., Lundin, S.T.: Hydrogenation of benzene to cyclohexene on an unsupported ruthenium catalyst: effect of poisons. J. Chem. Tech. Biotech. **31**, 660–669 (1981)
4. Ronchin, L., Toniolo, L.: Selective hydrogenation of benzene to cyclohexene using a suspended ru catalyst in a mechanically agitated tetraphase reactor. Catal. Today **48**, 255–264 (1999)
5. Ronchin, L., Toniolo, L.: Selective Hydrogenation of benzene to cyclohexene using a ru catalyst suspended in an aqueous solution in a mechanically agitated tetraphase reactor: a study of the influence of the catalyst preparation on the hydrogenation kinetics of benzene and of cyclohexene. Appl. Catal. A: Gen. **208**, 77–89 (2001)
6. Struijk, J., d'Angremond, M., Lucas-de Regt, W.J.M., et al.: Partial liquid phase hydrogenation of benzene to cyclohexene over ruthenium catalysts in the presence of an aqueous salt solution i. preparation, characterization of the catalyst and study of a number of process variables. Appl. Catal. A: Gen.**83**, 263–295 (1992)
7. Struijk, J., Moene, R., van der Kamp, T., et al.: Partial liquid-phase hydrogenation of benzene to cyclohexene over ruthenium catalysts in the presence of an aqueous salt solution ii. influence of various salts on the performance of the catalyst. Appl. Catal. A: Gen. **89**, 77–102 (1992)
8. Hu, S.C., Chen, Y.W.: Partial hydrogenation of benzene to cyclohexene on ruthenium catalysts supported on La_2O_3-ZnO Binary oxides. Ind. Eng. Chem. Res. **36**, 5153–5159 (1997)
9. Spinace, E.V., Vaz, J.M.: Liquid-phase hydrogenation of benzene to cyclohexene catalyzed by Ru/SiO_2 in the presence of water-organic mixtures. Catal. Commun. **4**, 91–96 (2003)
10. Xie, S.H., Qiao, M.H., Li, H.X., et al.: A Novel Ru-B/SiO_2 amorphous catalyst used in benzene-selective hydrogenation. App. Catal. A: Gen. **176**, 129–134 (1999)

11. Liu, Z., Xie, S.H., Liu, B., et al.: Benzene-selective hydrogenation to cyclohexene over supported ruthenium catalysts prepared by novel method. New J. Chem. **23**, 1057–1058 (1999)
12. Li, H.X., Wang, W.J., Li, H., et al.: Crystallization deactivation of Ni-P/SiO_2 amorphous catalyst and the stabilizing effect of silica support on Ni-P amorphous structure. J. Catal. **194**, 211–221 (2000)
13. Wang, J.Q., Wang, Y.Z., Xie, S.H., et al.: Partial hydrogenation of benzene to cyclohexene on A Ru-Zn/m-ZrO_2 nanocomposite catalyst. Appl. Catal., A: Gen. **272**, 29–36 (2004)
14. Bu, J., Liu, J.L., Chen, X.Y., et al.: Ru/SBA-15 catalysts for partial hydrogenation of benzene to cyclohexene: tuning the Ru crystallite size by Ba. Catal. Commun. **9**, 2612–2615 (2008)
15. Liu, J.L., Zhu, Y., Liu, J., et al.: Discrimination of the roles of $CdSO_4$ and $ZnSO_4$ in liquid phase hydrogenation of benzene to cyclohexene. J. Catal. **268**, 100–105 (2009)
16. Fan, G.Y., Jiang, W.D., Wang, J.B., et al.: Selective hydrogenation of benzene to cyclohexene over RuCoB/γ-Al_2O_3 without additive. Catal. Commun. **10**, 98–102 (2008)
17. Steen, P.J.V., Scholten, J.J.F.: Selectivity to cyclohexene in the liquid phase hydrogenation of benzene and toluene over ruthenium catalysts, as influenced by reaction modifiers. Appl. Catal. A **82**, 277–281 (1992)
18. Sun, H.J., Chen, Z.H., Guo, W., et al.: Effect of organic additives on the performance of nano-sized Ru-Zn catalyst. Chin. J. Chem. **29**, 369–373 (2011)
19. Sun, H.J., Zhang, X.D., Chen, Z.H., et al.: Monolayer dispersed Ru-Zn catalyst and its performance in the selective hydrogenation of benzene to cyclohexene. Chin. J. Catal. **32**, 224–239 (2011)
20. Fan, G.Y., Li, R.X., Li, X.J., et al.: Effect of organic additives on partial hydrogenation of benzene. Catal. Commun. **9**, 1394–1397 (2008)

Chapter 5 Third-Generation Catalyst of Benzene Selective Hydrogenation to Cyclohexene—Ru–M (Zn, Mn, Fe, La, Ce) Nano-bimetallic System

The third-generation catalyst of benzene selective hydrogenation, i.e., the Ru–M (Zn, Mn, Fe, La, Ce) nano-bimetallic system belongs to the nanocrystallite, compared with the first-generation amorphous alloy of Ru–M–B/ZrO_2. Additionally, the promoters are extended to other transition elements and rare earth elements, without Na_2SiO_3-PEG-10000 modifier, when compared with the second-generation Ru–Zn–Na_2SiO_3-PEG-10000 catalyst. The main technical indexes of the third-generation catalyst are as follows: benzene conversion of 60%, cyclohexene selectivity of more than 80%, cyclohexene yield of more than 48%, compared with that of the second-generation catalyst, benzene conversion increases by 10 percentage points, cyclohexene yield increases by 8 percentage points.

On the basis of extensive research on transition elements and rare earth elements, transition elements Zn, Mn, Fe, and rare earth elements La, Ce are found to have a promoting effect on the catalytic performance of Ru catalyst for selective hydrogenation of benzene, significantly improving selectivity and yield of cyclohexene. They follow the same general mechanism of action. In the Ru–M (Zn, Mn, Fe, La, and Ce) catalysts before hydrogenation, Ru is present in the metallic state, and the promoter M exists in the form of ZnO, Mn_3O_4, Fe_3O_4, $La(OH)_3$, and CeO, respectively. The catalyst shows a high activity and low selectivity when reacting in the absence of $ZnSO_4$. After pretreatment, the ZnO, Mn_3O_4, Fe_3O_4, $La(OH)_3$ and CeO on the catalyst surface react with $ZnSO_4$ in the slurry to generate a basic salt, which is chemisorbed on the catalyst surface, and thus, the electronic properties and geometry of Ru active center, as well as surface properties of the catalyst, are modified, playing a key role in improving the cyclohexene selectivity and yield.

The main contents of this chapter are as follows: effect of transition elements and rare earth elements on the catalytic performance of Ru-based catalyst, the third-generation catalyst of benzene selective hydrogenation, i.e., the Ru–M (Zn, Mn, Fe, La, Ce) nano-bimetallic system, and main technical indicators of the third-generation catalyst.

Z. Liu et al., *Catalytic Technology for Selective Hydrogenation of Benzene to Cyclohexene*,
https://doi.org/10.1007/978-981-15-6411-6_5

5.1 Effect of Transition Elements and Rare Earth Elements on the Catalytic Performance of Ru-Based Catalyst

The Ru nano-bimetallic catalyst of benzene selective hydrogenation has been a research hotspot due to its good catalytic performance. The typical preparation methods and conditions of Ru-based crystalline bimetallic catalysts were reported in the literature [1–10], so are the existence form of the second metal and its effect on the catalyst performance [11–14]. Based on extensive research on transition elements and rare earth elements [15–18], the third-generation catalyst of benzene selective hydrogenation has been developed.

5.1.1 Preparation and Characterization of Ru–M (Transition Elements) Catalyst

Ru–M (M = Cr, Mn, Fe, Co, Ni, Cu, and Zn) nanocatalysts are prepared by sol-gel method. The $RuCl_3 \cdot 3H_2O$ are used as the precursor of the active component Ru, and $Cr(NO_3)_3 \cdot 9H_2O$, $MnSO_4 \cdot H_2O$, $FeSO_4 \cdot 7H_2O$, $CoCl_2 \cdot 6H_2O$, $NiSO_4 \cdot 6H_2O$, $CuSO_4 \cdot 5H_2O$, and $ZnSO_4 \cdot 7H_2O$ are, respectively, used as the precursor of the transition element M. A mixed solution of Ru and M precursors is prepared according to the theoretical amount, NaOH solution is added with stirring. The resulting sol is then transferred to an autoclave and reduced at 150 °Cunder a hydrogen pressure of 5.0 MPa and at a stirring speed of 800 r/min. The precipitate is washed with distilled water until neutral and dried under vacuum. The amount of M precursor is adjusted to obtain Ru–M(x) catalysts with different M contents, and x is the actual molar percentage of M in the catalyst measured by X-ray fluorescence spectrometry (XRF). The M-free Ru catalyst is denoted as Ru(0). The activity and selectivity of the catalyst are tested in a 1 L Hastelloy autoclave. The reaction conditions are 150 °C, 5 MPa, 1200 r/min, 140 mL benzene, 280 mL H_2O, 9.8 g ZrO_2, and 49.2 g $ZnSO_4 \cdot 7H_2O$. After the start of time, by sampling for every 5 min, the product composition is determined by gas chromatography. The sample after hydrogenation is denoted as Ru–M(x) AH (after hydrogenation). Ru cat. and Ru cat. AH are pure Ru catalysts without the promoter M before and after hydrogenation, respectively, and they are used as references for comparison.

Figure 5.1 displays the XRD patterns of Ru–M(x) samples before and after hydrogenation.

From Fig. 5.1, it can be seen that all the Ru–M catalysts before and after hydrogenation show the diffraction peaks of metallic Ru (JCPDS 01-070-0274) at $2\theta \approx 44°$, indicating that the active component mainly exists as the metallic Ru. The Ru–M crystallite sizes of the catalyst derived by using Scherrer equation are given in the last two columns of Table 5.1, and they tend to decrease compared with that of the catalyst without M.

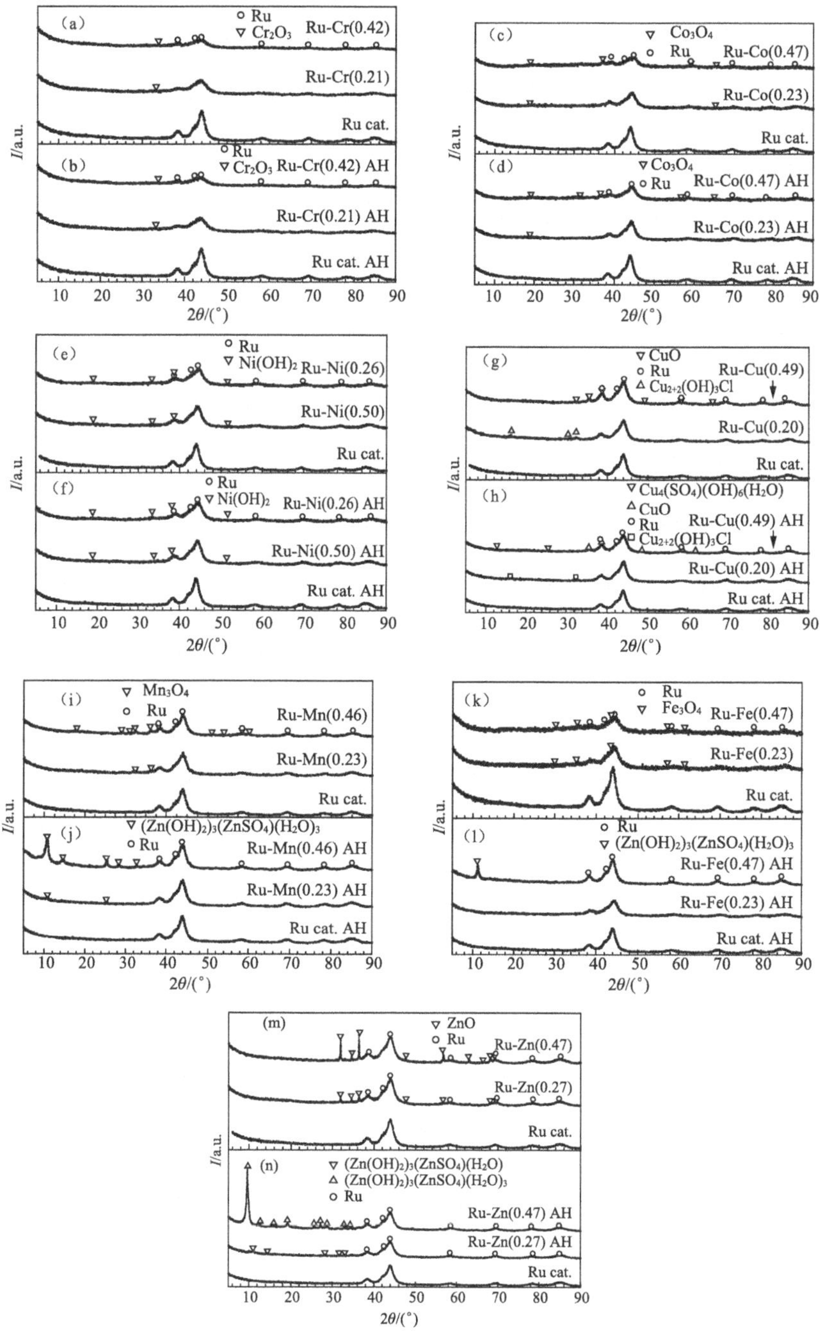

Fig. 5.1 XRD patterns of Ru–M(x) samples before and after hydrogenation. **a**, **b** Ru–Cr(x); **c**, **d** Ru–Co(x); **e**, **f** Ru–Ni(x); **g**, **h** Ru–Cu(x); **i**, **j** Ru–Mn(x); **k**, **l** Ru–Fe(x); **m**, **n** Ru–Zn(x)

Table 5.1 Elemental compositions and Ru crystallite size of Ru–M(*x*) catalysts before and after hydrogenation[a]

Catalyst	M/Ru AH/ (mol/mol)	Zn/Ru AH/ (mol/mol)	S/Ru AH/ (mol/mol)	Ru crystallite size/nm[c]	Ru AH crystallite size/nm[d]
Ru(0)[b]	0	0	0	4.7	4.6
Ru(0)	0	0.0313	0.0026	4.7	4.5
Ru–Cr(0.21)	0.2016	0.0333	0.0032	2.9	2.8
Ru–Cr(0.42)	0.4106	0.0245	0.0019	2.8	3.2
Ru–Co(0.23)	0.2048	0.0335	0.0032	3.8	3.4
Ru–Co(0.47)	0.4539	0.0463	0.0053	3.2	3.6
Ru–Ni(0.26)	0.2587	0.0357	0.0038	3.3	3.6
Ru–Ni(0.50)	0.4890	0.0385	0.0041	3.6	3.1
Ru–Cu(0.20)	0.2016	0.0350	0.0031	4.3	4.8
Ru–Cu(0.49)	0.4906	0.0282	0.0186	4.0	3.9
Ru–Mn(0.23)	0.0207	0.2010	0.0149	3.9	4.2
Ru–Mn(0.46)	0.0206	0.5085	0.0619	4.1	4.5
Ru–Fe(0.23)	0.1662	0.0921	0.0086	4.2	4.5
Ru–Fe(0.47)	0.3322	0.2314	0.0240	3.6	3.6
Ru–Zn(0.27)	0.3775	0.3775	0.0343	5.5	4.1
Ru–Zn(0.47)	0.6153	0.6153	0.0952	4.2	4.3

[a]Hydrogenation conditions: reaction temperature 150 °C, H_2 pressure 5.0 MPa and stirring rate 1400 r/min, 1.96 g Ru–M(*x*) catalysts, 49.2 g $ZnSO_4 \cdot 7H_2O$, 280 mL H_2O, 140 mL benzene
[b]In the absence of $ZnSO_4 \cdot 7H_2O$
[c]Crystallite sizes of Ru–M before hydrogenation
[d]Crystallite sizes of Ru–M after hydrogenation, calculated by Scherrer equation

From Fig. 5.1a, b, it can be seen that Cr is presented as Cr_2O_3 (JCPDS 01-082-0184) before and after hydrogenation, which is consistent with the reported existence form of Cr in the Ru–Cr–B catalyst prepared by chemical reduction [19].

From Fig. 5.1c, d, it can be seen that Co is presented as Co_3O_4 (JCPDS 00-001-1152) before and after hydrogenation, which is consistent with the reported existence form of Co in the RuCoB/γ-Al_2O_3 catalyst prepared by impregnation-reduction method [20].

It can be seen from Fig. 5.1e, f that Ni is presented as $Ni(OH)_2$ in the catalyst before and after hydrogenation.

Figure 5.1g, h shows that before hydrogenation, Cu is presented as low-valency (+1) oxidation state in the Ru–Cu(0.20) catalyst with a low Cu content, and thus the diffraction peak of $Cu_{2+2}(OH)_3Cl$ (JCPDS 00-019-0389) is detected. On the contrary, Cu is presented as two oxidation states (+1 and +2) in the Ru–Cu(0.49) catalyst with a Cu content, and thus the diffraction peaks of CuO (JCPDS 00-041-0254) and $Cu_{2+2}(OH)_3Cl$ (JCPDS 00-019-0389) are detected. After hydrogenation, CuO disappears, only the diffraction peak of $Cu_4(SO_4)(OH)_6(H_2O)$ (JCPDS 00-020-0357) is observed.

Figure 5.1i, j shows that the existence form of Mn changes after hydrogenation. Before hydrogenation, Mn is presented as Mn_3O_4 (JCPDS 00-001-1127), and this is consistent with the result reported in the literature that Mn is presented as Mn_3O_4 in the $Co/Mn/TiO_2$ catalyst when the reduction temperature is below 300 °C [21]. However, Mn_3O_4 disappears after hydrogenation, and instead, a diffraction peak of $(Zn(OH)_2)_3(ZnSO_4)(H_2O)_3$ (JCPDS 01-078-0247) is observed in the Ru–Mn(0.46) catalyst with high Mn content. This is due to the abundance of $ZnSO_4$ in the hydrogenation slurry, and the reactions are taken as below:

$$Mn_3O_4 + 4H_2O = Mn(OH)_2 + 2Mn(OH)_3 \quad (5.1.1)$$

$$Mn(OH)_2 + ZnSO_4 = Zn(OH)_2 + MnSO_4 \quad (5.1.2)$$

$$2Mn(OH)_3 + 3ZnSO_4 = Mn_2(SO_4)_3 + 3Zn(OH)_2 \quad (5.1.3)$$

$$3\,Zn(OH)_2 + ZnSO_4 + 3H_2O = (Zn(OH)_2)_3(ZnSO_4)(H_2O)_3 \downarrow \quad (5.1.4)$$

Figure 5.1k, l shows that the existence form of Fe changes after hydrogenation. Before hydrogenation, Fe is presented as Fe_3O_4 (JCPDS 00-001-1111), which is in consistence with the result that Fe was presented as Fe_3O_4 in the Ru–Fe/C catalyst prepared by chemical reduction [22]. After hydrogenation, Fe_3O_4 disappears, a diffraction peak of $3Zn(OH)_2{\cdot}ZnSO_4{\cdot}3H_2O$ is detected in the Ru–Fe(0.47) catalyst with high Fe content, and the same chemical reactions are taken as those over the Ru–Mn(x) catalyst:

$$Fe_3O_4 + 4H_2O = Fe(OH)_2 + 2Fe(OH)_3 \quad (5.1.5)$$

$$Fe(OH)_2 + ZnSO_4 = Zn(OH)_2 + FeSO_4 \quad (5.1.6)$$

$$2Fe(OH)_3 + 3ZnSO_4 = Fe_2(SO_4)_3 + 3Zn(OH)_2 \quad (5.1.7)$$

$$3\,Zn(OH)_2 + ZnSO_4 + 3H_2O = (Zn(OH)_2)_3(ZnSO_4)(H_2O)_3 \downarrow \quad (5.1.8)$$

From Fig. 5.1m, n, it can be seen that Zn is presented as ZnO (JCPDS 01-074-0534) in the catalyst before hydrogenation, but disappears after hydrogenation. Instead, a diffraction peak of $(Zn(OH)_2)_3(ZnSO_4)(H_2O)_x$ ($x = 1, 3$) (JCPDS 00-039-0690) is observed in the Ru–Zn(0.47) catalyst with high Zn content. This is in full agreement with previous findings [4, 23, 24].

In summary, in the Ru–M(x) catalyst, M exists as an oxide or hydroxide before hydrogenation, but disappears after hydrogenation, and meanwhile a new phase

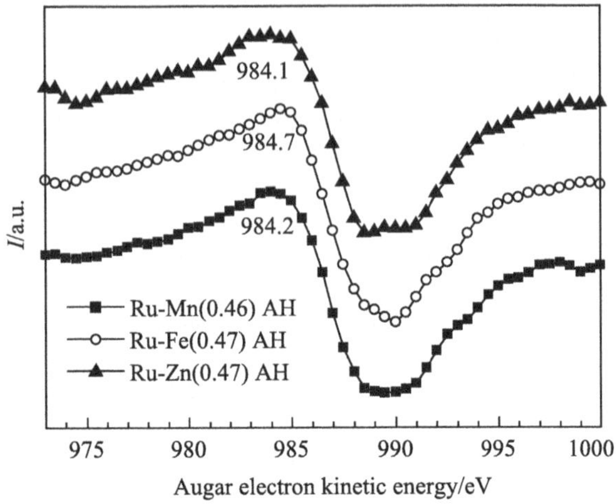

Fig. 5.2 AES Zn LMM spectra of Ru–Zn(0.47), Ru–Fe(0.47) AH, and Ru–Mn(0.46) AH

appears in the catalyst with high M content. In the Ru–Mn(0.46), Ru–Fe(0.47), and Ru–Zn(0.47) catalysts with high M content, the diffraction peaks of poorly soluble basic salts appear after hydrogenation. However, since the content of basic salt is below the threshold of spontaneous monolayer dispersion and is highly dispersed in the catalyst surface as molecules, the peaks are not observed in the Ru–M(x) catalysts with low M content after hydrogenation.

Figure 5.2 shows the Zn LMM Auger electron spectra of Ru–Mn(0.46), Ru–Fe(0.47), and Ru–Zn(0.47) after hydrogenation was recorded with Ar^+ sputtering for 1 min.

From Fig. 5.2, it can be seen that the Zn LMM kinetic energies of Ru–Mn(0.46) AH, Ru–Fe(0.47) AH, and Ru–Zn(0.47) are 984.2 eV, 984.7 eV, and 984.1 eV, respectively, which is consistent with that of Zn in oxidation states [21, 22]. All these indicate that Zn mainly exists in the oxidation states even under the conditions of 150 °C and 5 MPa. These Zn species are mainly from $(Zn(OH)_2)_3{\cdot}(ZnSO_4)(H_2O)_x$ as confirmed by XRD.

Table 5.1 lists the elemental compositions of the Ru–M(x) catalysts before and after hydrogenation measured by X-ray fluorescence (XRF).

As shown in Table 5.1, after hydrogenation, in addition to the Ru and M (Cr, Co, Ni, Cu, Mn, Fe, and Zn), the Ru–M(x) catalyst also contains S from the chemisorbed $ZnSO_4$ in the slurry and $(Zn(OH)_2)_3(ZnSO_4)(H_2O)_x$. The composition of the Ru–M(x) catalysts with the same Ru content after hydrogenation is reflected by the molar ratios of M/Ru and S/Ru. In the absence of $ZnSO_4$ in the slurry, the Ru(0) catalyst contains only Ru after hydrogenation. In the presence of $ZnSO_4$ in the slurry, elements of Zn and S are detected in the Ru(0) catalyst after hydrogenation, and the molar ratio of Zn/Ru is much higher than that of S/Ru, indicating that the Zn species which are chemisorbed on the surface are in the form of both $ZnSO_4$ and $(Zn(OH)_2)_3(ZnSO_4)(H_2O)_x$. However, no diffraction peaks of the basic salt are observed on the XRD pattern of Ru(0) catalyst after hydrogenation because the content of the basic salt is far lower than the threshold of monolayer dispersion on the Ru–M surface.

It can be seen from Table 5.1 that the Cr/Ru ratio of Ru–Cr(x) catalyst is almost unchanged after hydrogenation, which is consistent with the XRD result that Cr_2O_3 acts as the promoter before and after hydrogenation. The Ru/Cr(0.21) AH sample shows similar Zn/Ru and S/Ru molar ratios as the Ru(0) AH catalyst. However, the Zn/Ru and S/Ru molar ratios of Ru–Cr(0.42) catalyst decrease after hydrogenation, indicating that the increase of Cr_2O_3 is not conducive to the formation and chemisorption of $(Zn(OH)_2)_3(ZnSO_4)(H_2O)_x$.

For the Ru–Co(x) and Ru–Ni(x) catalysts, the molar ratios of Co/Ru and Ni/Ru after hydrogenation are similar to that before hydrogenation, which is consistent with the XRD results. This also indicates that both Co_2O_3 and $Ni(OH)_2$ are kept unchanged after hydrogenation. The molar ratios of Zn/Ru and S/Ru increase slightly with the increase of Co_2O_3 and $Ni(OH)_2$ after hydrogenation, indicating that their presence is conducive to the formation of $(Zn(OH)_2)_3(ZnSO_4)(H_2O)_x$.

For the Ru–Cu(0.20) catalyst, the Cu/Ru ratio after hydrogenation is close to that before hydrogenation due to the fact that $Cu_{2+2}(OH)_3Cl$ did not change after hydrogenation. The Zn/Ru and S/Ru molar ratios of Ru(0) catalyst after hydrogenation increase slightly, indicating that $Cu_{2+2}(OH)_3Cl$ has little effect on the formation of basic salts. Although XRD results showed that $Cu_4(SO_4)(OH)_6(H_2O)$ was formed from the interaction between part of CuO and $ZnSO_4$ in the slurry, the Cu/Ru ratio of Ru–Cu(0.49) catalyst was close to that before hydrogenation. This indicates that $Cu_4(SO_4)(OH)_6(H_2O)$ is readily chemisorbed on the catalyst, so the S/Ru molar ratio increases significantly and the Zn/Ru molar ratio decreases, indicating that the presence of $Cu_4(SO_4)(OH)_6(H_2O)$ inhibits the formation of basic salts.

For the Ru–Mn(x) catalyst, only traces of Mn were detected after hydrogenation. The molar ratios of Zn/Ru and S/Ru increase significantly with the increase of Mn_3O_4 content, indicating that almost all of Mn_3O_4 reacted with $ZnSO_4$ in the slurry generating basic zinc sulfate salt, as shown in the reaction equations (5.1.1)–(5.1.4), and this is consistent with the XRD results.

For the Ru–Fe(x) catalyst, the molar ratio of Fe/Ru after hydrogenation decreases obviously compared with that before hydrogenation, and the molar ratios of Zn/Ru and S/Ru increase obviously with the increase of Fe_3O_4 content, indicating that Fe_3O_4 reacts with $ZnSO_4$ to form the basic zinc sulfate salt, as shown in the reaction equations (5.1.5)–(5.1.8), and this is consistent with the XRD results.

For the Ru–Zn(x) catalyst, the molar ratios of Zn/Ru and S/Ru after hydrogenation increase significantly compared with that before hydrogenation, and the molar ratios of Zn/Ru and S/Ru increase with increasing Zn content. This indicates that ZnO on the Ru–Zn(x) catalyst surface reacts with $ZnSO_4$ in the slurry to form a basic salt, which is in accordance with the XRD results.

In summary, the elemental compositions of the Ru–M(x) catalyst measured by XRF are in agreement with the XRD results.

Figure 5.3 shows the TEM images of Ru–Mn(x) and Ru–Fe(x) catalysts before and after hydrogenation.

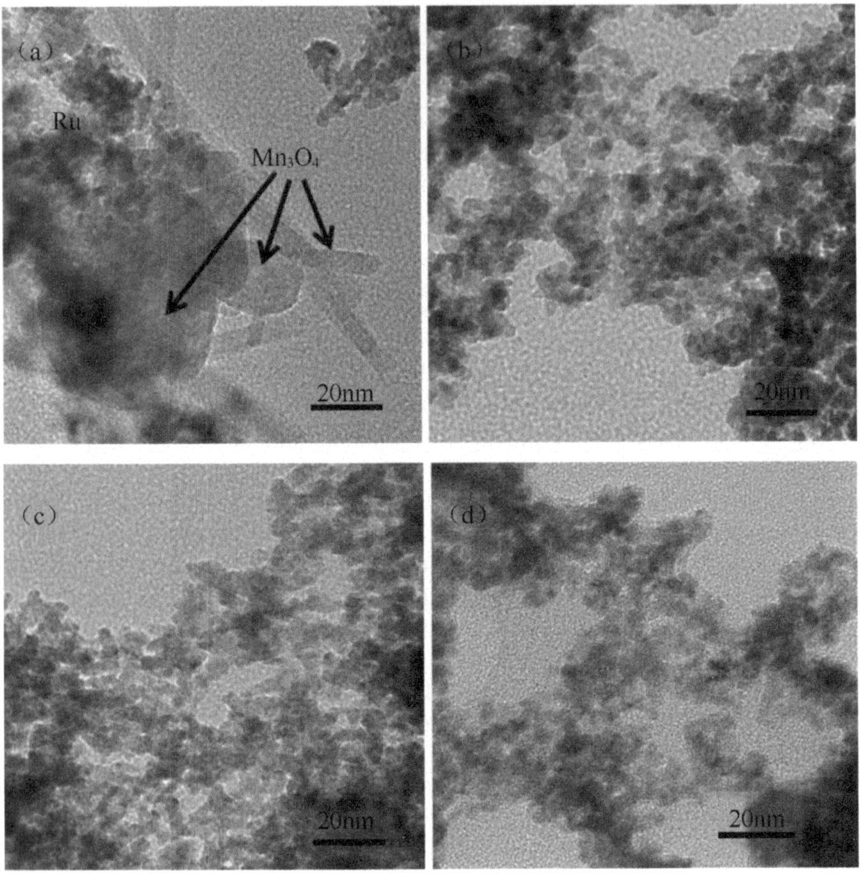

Fig. 5.3 TEM images of Ru–Mn(x) and Ru–Fe(x) catalysts before and after hydrogenation. **a** Ru–Mn(0.46) catalyst; **b** Ru–Mn(0.46) AH; **c** Ru–Fe(0.47) catalyst; **d** Ru–Fe(0.47) AH

TEM images in Fig. 5.3 show that the Ru crystallite sizes of Ru–Mn(x) AH and Ru–Fe(x) AH samples center on about 4 nm, which is consistent with the XRD results. In particular, the spherical and pillar-shaped Mn_3O_4 crystallites were observed in Fig. 5.3a.

No images of basic zinc sulfate salts were observed in Fig. 5.3b, indicating that the salts were highly dispersed on the surface of Ru–Mn(0.46) catalyst.

The images of Fe_3O_4 and basic zinc sulfate salts were not observed in Fig. 5.3c and d, indicating that Fe_3O_4 and basic salts were, respectively, highly dispersed on the Ru crystallite before and after hydrogenation [25].

Table 5.2 lists texture properties of the Ru–M(x) catalysts before and after hydrogenation.

As can be seen from Table 5.2, the BET surface areas S_{BET}, pore volumes V_{p}, and average pore diameters d_{p} of Ru–Cr(x), Ru–Co(x), Ru–Ni(x), and Ru–Cu(x)

Table 5.2 BET surface areas S_{BET}, pore volumes V_p, and average pore diameters d_p of Ru–Fe(***x***) catalysts before and after hydrogenation

Sample	$S_{BET}/(m^2/g)$	$V_p/(cm^3/g)$	d_p/nm
Ru(0)	59	0.18	10.63
Ru(0) AH	56	0.16	10.44
Ru–Cr(0.21)	159	0.42	10.42
Ru–Cr(0.21) AH	147	0.39	10.10
Ru–Cr(0.42)	166	0.25	6.05
Ru–Cr(0.42) AH	163	0.28	6.79
Ru–Co(0.23)	86	0.18	8.22
Ru–Co(0.23) AH	79	0.17	8.04
Ru–Co(0.47)	91	0.23	10.04
Ru–Co(0.47) AH	89	0.21	10.22
Ru–Ni(0.26)	91	0.19	8.50
Ru–Ni(0.26) AH	92	0.20	8.63
Ru–Ni(0.50)	112	0.29	10.37
Ru–Ni(0.50) AH	110	0.28	10.87
Ru–Cu(0.20)	79	0.17	8.73
Ru–Cu(0.20) AH	68	0.16	9.63
Ru–Cu(0.49)	77	0.20	10.47
Ru–Cu(0.49) AH	73	0.16	10.37
Ru–Mn(0.23)	57	0.20	13.74
Ru–Mn(0.23) AH	46	0.15	11.03
Ru–Mn(0.46)	59	0.13	8.82
Ru–Mn(0.46) AH	41	0.11	10.98
Ru–Fe(0.23)	68	0.16	9.44
Ru–Fe(0.23) AH	58	0.14	9.55
Ru–Fe(0.47)	64	0.16	10.37
Ru–Fe(0.47) AH	53	0.13	11.67
Ru–Zn(0.27)	58	0.18	12.30
Ru–Zn(0.27) AH	26	0.13	11.10
Ru–Zn(0.47)	41	0.09	9.2
Ru–Zn(0.47) AH	27	0.05	8.6

catalysts were almost unchanged before and after hydrogenation. On the contrary, the specific surface area and pore volume of Ru–Zn(x) and Ru–Fe(x) catalysts after hydrogenation decreased significantly after hydrogenation, and this was because the basic zinc sulfate salts were chemisorbed in the catalyst surface and pores.

5.1.2 Activity and Selectivity of Ru–M (Transition Elements) Catalyst

Tables 5.3 lists the benzene conversion, cyclohexene selectivity and yield within 15 min, and cyclohexene maximum yield, as well as pH values of liquid phase at room temperature, after hydrogenation over Ru–M(x) catalyst.

Table 5.3 Benzene conversion C_{BZ}, cyclohexene selectivity S_{HE} and yield Y_{HE} within 15 min, cyclohexene maximum yield Y_{max}, and pH values of the aqueous phase after hydrogenation over Ru–M($\boldsymbol{x}$) catalyst[a]

Catalyst	C_{BZ}/%[b]	S_{HE}/%[b]	Y_{HE}/%[b]	Y_{max}/%[c]	pH[d]
Ru(0)	99.8	14.7	14.7	33.0	5.5
Ru(0)[e]	100	0	0	0	6.9
Ru–Cr(0.21)	98.2	17.6	17.3	35.8	4.4
Ru–Cr(0.42)	99.2	5.5	5.4	26.0	5.0
Ru–Cr(0.42)[e]	100	0	0	0	7.2
Ru–Co(0.23)	97.8	24.3	23.8	32.6	4.5
Ru–Co(0.47)	95.5	43.6	41.6	43.2	6.1
Ru–Co(0.47)[e]	100	0	0	0	6.8
Ru–Ni(0.26)	98.0	24.2	23.7	29.7	5.9
Ru–Ni(0.50)	93.1	34.1	32.1	36.1	5.2
Ru–Ni(0.50)[e]	100	0	0	0	7.2
Ru–Cu(0.20)	97.1	32.2	31.3	36.5	4.9
Ru–Cu(0.49)	93.6	47.3	44.3	46.1	5.9
Ru–Cu(0.49)[e]	100	0	0	0	7.1
Ru–Mn(0.23)	88.5	62.5	55.3	55.3	5.9
Ru–Mn(0.46)	34.4	86.9	30.0	43.2	6.0
Ru–Mn(0.46)[e]	100	0	0	0	6.9
Ru–Fe(0.23)	95.2	53.8	43.6	47.3	5.9
Ru–Fe(0.47)	77.4	68.9	53.4	56.7	5.9
Ru–Fe(0.47)[e]	100	0	0	0	7.1
Ru–Zn(0.27)	53.8	81.9	44.2	53.4	5.5
Ru–Zn(0.47)	7.22	97.2	7.02	13.1	5.7
Ru–Zn(0.47)[e]	100	0	0	0	6.9

[a]Hydrogenation conditions: reaction temperature of 150 °C, H_2 pressure of 5.0 MPa and stirring rate of 1400 r/min, 1.96 g Ru–Mn(x) catalysts, 49.2 g $ZnSO_4 \cdot 7H_2O$, 280 mL H_2O, 140 mL benzene
[b]Benzene conversion C_{BZ}, cyclohexene selectivity S_{HE} and yield Y_{HE} within 15 min
[c]Cyclohexene maximum yield Y_{max}
[d]pH values of the aqueous phase measured by pH meter
[e] In the absence of $ZnSO_4$

As can be seen from Table 5.3, for the Ru–Cr catalyst, the benzene conversion remains constant as the Cr content increases, whereas the selectivity of cyclohexene decreases, indicating that Cr is not a favorable promoter of Ru-based catalysts.

For Ru–Co, Ru–Ni, and Ru–Cu catalysts, the benzene conversion slightly decreases and cyclohexene selectivity increases slightly with the increase of Co, Ni, and Cu contents, indicating that Co, Ni, and Cu have little promoting effect on the catalysis of Ru-based catalysts.

However, for Ru–Mn, Ru–Fe, and Ru–Zn catalysts, the conversion of benzene decreases obviously, and the selectivity of cyclohexene increases significantly with the increase of Mn, Fe, and Zn contents. Ru–Zn(0.47) catalyst shows a benzene conversion of 7.2% at 15 min, and the cyclohexene selectivity is 97.2%, i.e., the catalyst has a high cyclohexene selectivity at a low benzene conversion. Ru–Zn(0.46) catalyst shows a benzene conversion of 34.4% at 15 min, and the cyclohexene selectivity is 86.9%. Ru–Zn(0.27) catalyst shows a benzene conversion of 53.8% at 15 min, and the cyclohexene selectivity is 81.9%, indicating that a high benzene conversion and high cyclohexene selectivity are achieved at the same time. The maximum cyclohexene yields of Ru–Mn(0.23), Ru–Fe(0.47), and Ru–Zn(0.27) catalysts are 55.3%, 56.7%, and 53.4%, respectively, all of which are higher than the best results currently reported in the literatures [13, 26–29]. Therefore, Mn, Fe, and Zn are good promoters of Ru catalysts for benzene selective hydrogenation among the transition metals in the fourth period.

The electron configurations of Mn, Fe, and Zn are $[Ar]3d^5 4s^2$, $[Ar]3d^6 4s^2$, and $[Ar]3d^{10} 4s^2$, respectively. The d orbitals and valence electrons are, respectively, half-filled, nearly half-filled, and fully filled. According to the Maxted rule [11], these d orbitals do not form strong chemical bonds with the valence electron orbit $[Kr]4d^7 5s^1$ of Ru, but produce moderate chemisorption, which is the essential reason that Mn, Fe, and Zn have a good promoting effect on the catalysis.

Table 5.3 also shows that when the catalyst contains no promoter M, with nSO_4 containing in the slurry, or the catalyst containing the promoter M, without $ZnSO_4$ containing in the slurry, benzene completely reacts within 15 min over all the Ru–M(x) catalysts, and the yield of cyclohexene is very low, indicating that the cyclohexene selectivity could not effectively increase by the promoter such as Cr, Co, Ni, Cu, Mn, Fe, and Zn alone, or $ZnSO_4$ in the slurry. The cyclohexene selectivity and yield can only be significantly improved by the coexistence of the promoter and $ZnSO_4$. Moreover, the existence of basic zinc sulfate salt on the surface of Ru catalyst has been proved for Ru–Mn(0.23), Ru–Fe(0.47), and Ru–Zn(0.27) catalysts, indicating that the chemisorbed basic salt plays a crucial role in improving the cyclohexene selectivity.

Figure 5.4 shows the Zn/Ru and S/Ru molar ratios to benzene conversion and cyclohexene selectivity at 15 min over Ru/M(x) (M = Mn, Fe, Zn) catalysts.

As can be seen from Fig. 5.4, the benzene conversion of Ru–M(x) (M = Mn, Fe, Zn) catalysts decreases monotonically, while the cyclohexene selectivity monotonically increases with increasing molar ratios of Zn/Ru and S/Ru. The increase of the Zn/Ru and S/Ru molar ratios after hydrogenation indicates that the amount of $(Zn(OH)_2)_3(ZnSO_4)(H_2O)_x$ generated on the catalyst surface is increased. Thus,

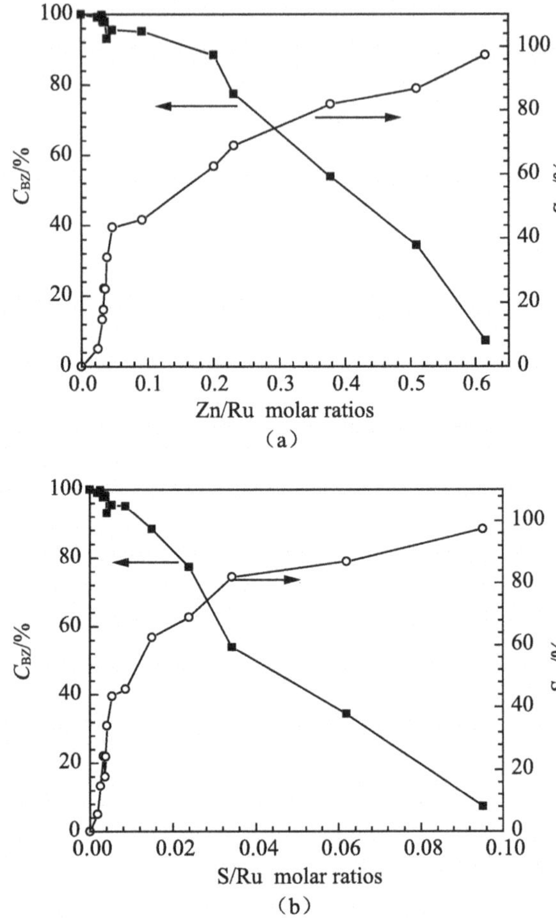

Fig. 5.4 Relationship between the Zn/Ru and S/Ru molar ratios and benzene conversion as well as cyclohexene selectivity over Ru/M(*x*) (M = Mn, Fe, Zn) catalysts. Hydrogenation conditions: reaction temperature of 150 °C, H_2 pressure of 5.0 MPa and stirring rate of 1400 r/min, 1.96 g of Ru–M(*x*) catalysts, 49.2 g of $ZnSO_4 \cdot 7H_2O$, 280 mL of H_2O, 140 mL of benzene, pretreated for 22 h

it can be concluded that promoters M, the presence of $ZnSO_4$ in the slurry, and the pretreatment play an important role in reducing the catalyst activity and increasing the cyclohexene selectivity as well. The essential reason is that the basic salt which is chemisorbed on the catalyst surface modifies the electronic properties, geometric arrangement of the Ru active center, and the surface properties of the catalyst, thereby inhibiting the direct hydrogenation of benzene to cyclohexane in one step and reducing the benzene conversion. When the stepwise hydrogenation of benzene to cyclohexene is dominant, the cyclohexene selectivity increases if cyclohexene, an intermediate product, is rapidly desorbed once it is generated and is difficult to reabsorb.

5.1.3 Preparation and Characterization of Ru–M (Rare Earth Elements) Catalyst

Rare earth elements mainly refer to the common elements La and Ce. The Ru–Ce catalyst discussed here consists of the mono-metal Ru with the addition of CeO_2 prepared under different conditions.

CeO_2 was prepared by the process of precipitation, drying, and roasting. The conditions are as follows: 100 mL of 5% NaOH solution was added to 100 mL of 0.27 mol/L $Ce(NO_3)_3$ solution at 353 K, and stirred for 4 h. The obtained solid was then washed until neutral, and dried under vacuum at 323 K, and calcined for 3 h at 373 K and 473 K, respectively. The calcined samples were denoted as CeO_2 (C373), and CeO_2 (C473), respectively. Similarly, the uncalcined samples were denoted as CeO_2 (NC) (N for No and C for Calcination). M Ru + CeO_2 (NC) − x P (A) was an Ru–Ce catalyst, where M Ru represented a mono-metal Ru, x represented a theoretical molar ratio of CeO_2/Ru, P represented a hydrogenated sample in the presence of $ZnSO_4$, and A represented a hydrogenated sample in the absence of $ZnSO_4$.

Figure 5.5 shows the XRD patterns of CeO_2, mono-metal Ru, and Ru–Ce catalysts M Ru + CeO_2 (NC) − x P.

As can be seen from Fig. 5.5a, the intensity of diffraction peaks of CeO_2 (JCPDS 00-004-0593) increases with increasing calcination temperature, indicating an increase in crystallinity. As can be seen from Fig. 5.5b, only the diffraction peaks of metal Ru (JCPDS 01-078-0246) were observed for the mono-metal M Ru catalyst. As can be seen from Fig. 5.5c, at a CeO_2/Ru molar ratio of 0.20, the intensity of diffraction peaks of $(Zn(OH)_2)_3(ZnSO_4)(H_2O)_3$ (JCPDS 00-078-0247) increases with decreasing calcination temperature. When the theoretical molar ratio of CeO (NC)/Ru increases from 0.15 to 0.20, the intensity of diffraction peaks of $(Zn(OH)_2)_3(ZnSO_4)(H_2O)_3$ increases. From Fig. 5.5d, it can be seen that the molar ratio of CeO (NC)/Ru is 0.15, in the absence of $ZnSO_4$, not only the diffraction peaks of Ru, but also the diffraction peaks of CeO_2 are detected in the Ru catalyst after hydrogenation.

Table 5.4 presents the compositions of the catalyst and pH values of liquid phase after hydrogenation.

In Table 5.4, only the Ru element was detected in the absence of $ZnSO_4$ for the mono-metal Ru catalyst M Ru A. The Zn and S elements were detected in the presence of $ZnSO_4$ for M Ru P catalyst, and the Zn/Ru molar ratio was larger than S/Ru, indicating Zn species contains $(Zn(OH)_2)_3(ZnSO_4)(H_2O)_3$.

When the molar ratio of CeO_2/Ru is 0.20 and 0.15, the molar ratio of Ce/Ru decreases, whereas the molar ratios of Zn/Ru and S/Ru increase, and the pH value of the slurry increases with the decreasing calcination temperature of CeO_2 in the presence of $ZnSO_4$. This indicates that the uncalcined CeO_2 readily reacts with $ZnSO_4$ to form $(Zn(OH)_2)_3(ZnSO_4)(H_2O)_3$, and Ce enters the slurry as ions. The higher the calcination temperature of CeO_2 is, the better the crystallization is, and thus CeO_2 is more difficult to react with $ZnSO_4$ to form $(Zn(OH)_2)_3(ZnSO_4)(H_2O)_3$.

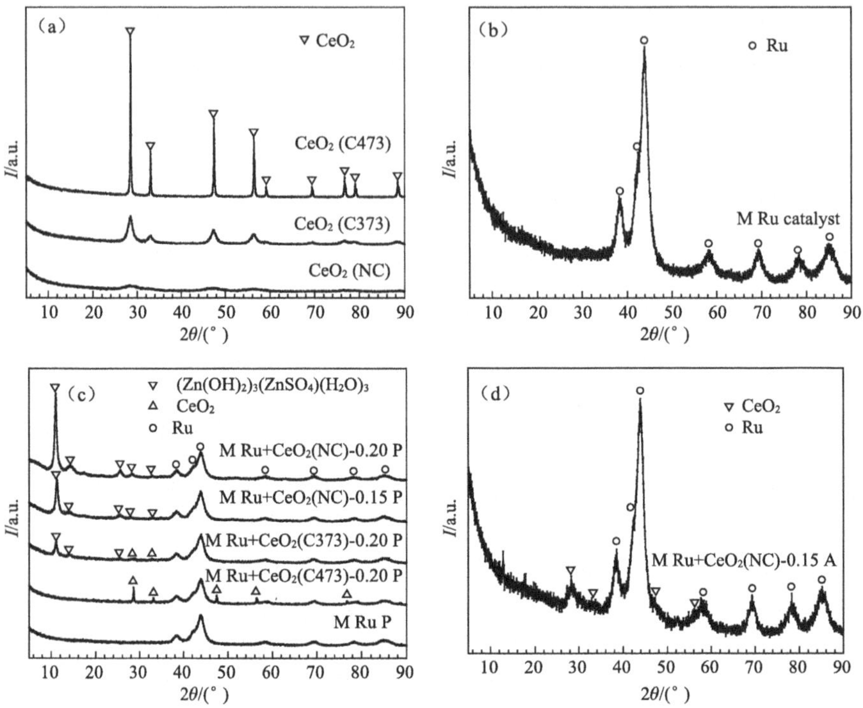

Fig. 5.5 XRD patterns of CeO_2, mono-metal Ru, and Ru–Ce catalysts M Ru + CeO_2 (NC) − *x* P. **a** CeO_2 calcined at different temperatures; **b** mono-metal Ru; **c** M Ru + CeO_2 (NC) − *x* P represents the Ru–Ce catalysts after hydrogenation in the presence of $ZnSO_4$; **d** sample after hydrogenation in the absence of $ZnSO_4$ at a theoretical CeO (NC)/Ru molar ratio of 0.15

Table 5.4 Compositions of the catalysts and pH values of liquid phase after hydrogenation

Condition	Ce/Ru AH (molar ratio)[a]	Zn/Ru AH (molar ratio)[a]	S/Ru AH (molar ratio)[a]	pH
M Ru A	0	0	0	7.10
M Ru P	0	0.0213	0.0026	5.53
M Ru + CeO_2 (C473) − 0.20 P	0.1749	0.1115	0.0071	5.64
M Ru + CeO_2 (C373) − 0.20 P	0.1129	0.2515	0.0305	5.87
M Ru + CeO_2 (NC) − 0.20 P	0.0093	0.5014	0.0603	6.47
M Ru + CeO_2 (NC) − 0.15 P	0.0048	0.3670	0.0440	6.27
M Ru + CeO_2 (NC) − 0.15 A	0.1424	0	0	7.12

[a]Measured by XRF; AH represents after hydrogenation

From the pH of the liquid phase after hydrogenation, it can be seen that the slurry is acidic in the presence of $ZnSO_4$, and the formation mechanism of $(Zn(OH)_2)_3(ZnSO_4)(H_2O)_3$ is inferred as follows: $Zn(OH)_2$ generated from the hydrolysis of $ZnSO_4$ reacts with a large number of $ZnSO_4$ to form insoluble $(Zn(OH)_2)_3(ZnSO_4)(H_2O)_3$, which is then chemisorbed on the surface of Ru catalyst. CeO_2 is an alkaline oxide that reacts with H^+ generated by the hydrolysis of $ZnSO_4$, resulting in an increase in the degree of hydrolysis of $ZnSO_4$, and an increase in the amount of $Zn(OH)_2$ generated, thus, the amount of $(Zn(OH)_2)_3(ZnSO_4)(H_2O)_3$ increased ultimately. The reaction equations are as follows:

$$Zn^{2+} + 2H_2O \rightarrow Zn(OH)_2 + 2H^+ \quad (5.1.9)$$

$$CeO_2 + 4H^+ \rightarrow Ce^{4+} + 2H_2O \quad (5.1.10)$$

$$3Zn(OH)_2 + ZnSO_4 + 3H_2O \rightarrow (Zn(OH)_2)_3(ZnSO_4)(H_2O)_3 \downarrow \quad (5.1.11)$$

Figure 5.6 shows the AES Zn LMM spectrum of Ru catalyst after hydrogenation in the presence of $ZnSO_4$ at CeO_2 (NC)/Ru theoretical molar ratio of 0.15.

As can be seen from Fig. 5.6, the Auger electron kinetic energy of Zn LMM is 987.0 eV, which is in agreement with the reported value of Zn in the oxidation state [12, 13], and this is a mutual confirmation that the Zn species exists as basic salt in the hydrogenated Ru catalyst as determined by XRD.

Figure 5.7 shows TEM images of Ru catalysts and Ru + CeO_2.

As can be seen from Fig. 5.7, the mono-metal M Ru and Ru + CeO_2 (NC) – 0.15P catalysts after hydrogenation in the presence of $ZnSO_4$ are composed of ellipsoid and spherical microcrystals with a size distribution centering around 4.5 nm, and the basic salt is highly dispersed on the surface of the catalyst.

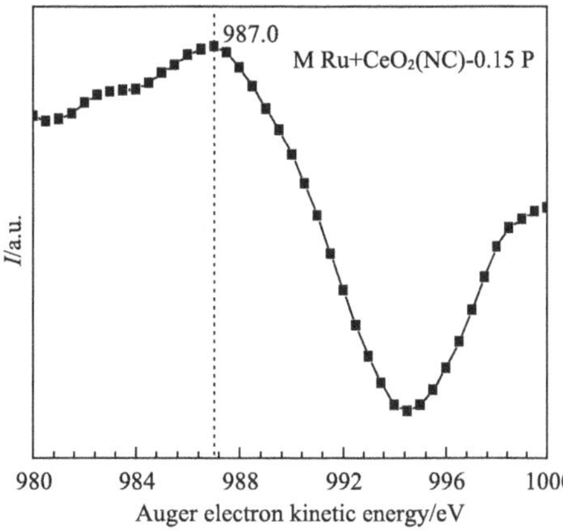

Fig. 5.6 AES Zn LMM spectrum of Ru catalyst after hydrogenation in the presence of $ZnSO_4$ at CeO_2 (NC)/Ru theoretical molar ratio of 0.15

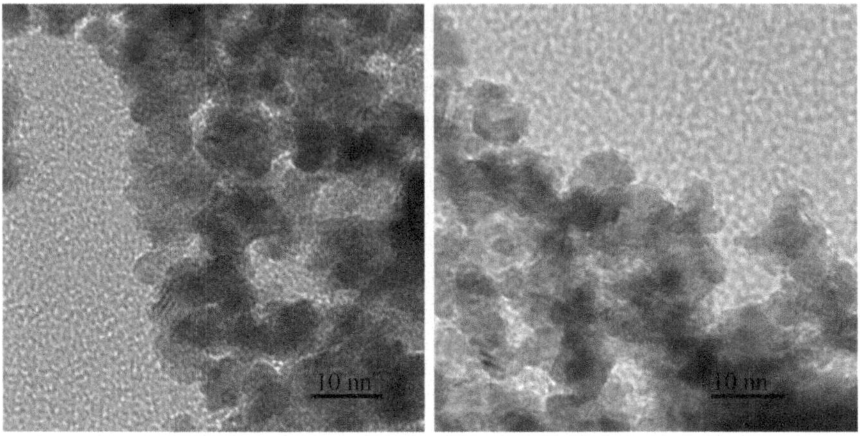

(a) (b)

Fig. 5.7 TEM images of mono-metal Ru and Ru + CeO_2 catalysts. **a** M Ru (mono-metal Ru); **b** Ru + CeO_2 (NC) – 0.15P (the Ru–Ce catalyst after hydrogenation in the presence of $ZnSO_4$ at CeO_2 (NC)/Ru theoretical molar ratio of 0.15)

5.1.4 Activity and Selectivity of Ru–M (Rare Earth Elements) Catalyst

The activity and selectivity of benzene selective hydrogenation to cyclohexene over Ru and Ru–Ce catalysts are displayed in Fig. 5.8.

From Fig. 5.8, benzene was completely converted at 5 min over the Ru catalyst in the absence of $ZnSO_4$ at a CeO_2 (NC)/Ru molar ratio of 0.15, and no cyclohexene was detected, indicating that CeO_2 alone and the $Ce(SO_4)_2$ produced by the reaction of CeO_2 with H^+ cannot improve the cyclohexene selectivity of Ru catalyst. In the presence of $ZnSO_4$ at a CeO_2 (NC)/Ru molar ratio of 0.15, the cyclohexene yield was 58.5% at a benzene conversion of 73.0% within 20 min, obtaining a high yield of cyclohexene, indicating that the formation of cyclohexene mainly occurred on the catalyst surface with chemisorbed basic salts.

Figure 5.9 shows the mechanism of $(Zn(OH)_2)_3(ZnSO_4)(H_2O)_3$ chemisorbed on the Ru surface to enhance cyclohexene selectivity.

As shown in Fig. 5.9a, Zn^{2+} in the $(Zn(OH)_2)_3(ZnSO_4)(H_2O)_3$ chemisorbed on the Ru surface can cover a part of Ru active sites, and leads to the geometrical rearrangement of Ru active sites. The density of the Ru active site is thus reduced, reducing the ability to adsorb and dissociate hydrogen, and the coverage of H atoms around the adsorbed activated benzene molecules. Moreover, the ability to chemisorb cyclohexene is reduced by the deactivation of the surface, and this is conducive to the desorption of cyclohexene, inhibiting the further hydrogenation of cyclohexene to cyclohexane [30, 31].

As shown in Fig. 5.9b, the $(Zn(OH)_2)_3(ZnSO_4)(H_2O)_3$ salt chemisorbed on the Ru surface is rich in crystallization water, and forms a stable water-stagnant layer

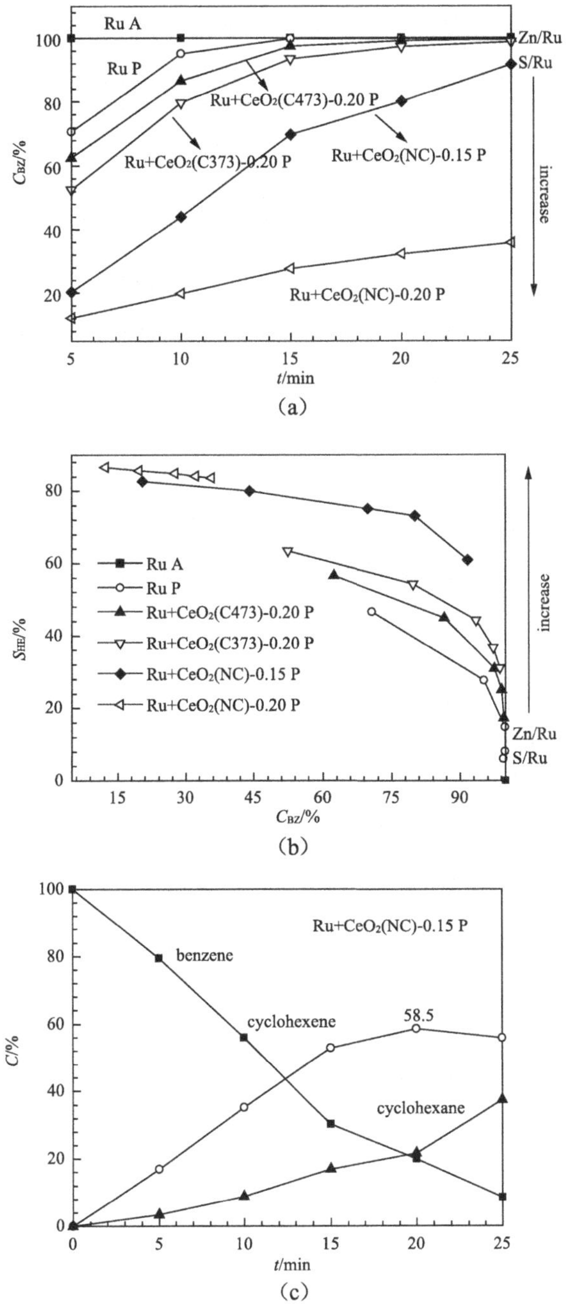

Fig. 5.8 Activity and selectivity of benzene selective hydrogenation to cyclohexene over Ru and Ru–Ce catalysts. **a** C_{BZ}-t curve; **b** S_{HE}-C_{BZ} curve; **c** C-t curve

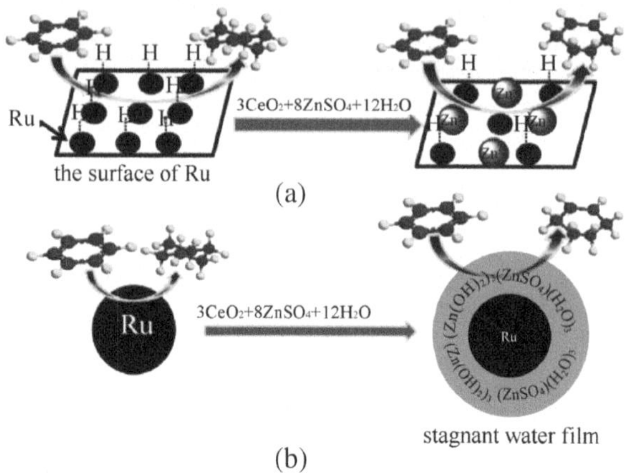

Fig. 5.9 Mechanism of $(Zn(OH)_2)_3(ZnSO_4)(H_2O)_3$ chemisorbed on the Ru surface

on the Ru surface. The lower solubility of cyclohexene in water than benzene is conducive to accelerate the desorption of cyclohexene and inhibits the reabsorption of the desorbed cyclohexene, hindering the further hydrogenation of cyclohexene, and thereby improving the selectivity of cyclohexene.

The Zn^{2+} in the $(Zn(OH)_2)_3(ZnSO_4)(H_2O)_3$ that is chemisorbed on the Ru surface can form loose bonds with cyclohexene, and stabilize the cyclohexene formed on the Ru surface [32]. More Ru active sites are covered by larger amount of $(Zn(OH)_2)_3(ZnSO_4)(H_2O)_3$, and thus results in a lower activity but higher cyclohexene selectivity of the catalyst. The amount of $(Zn(OH)_2)_3(ZnSO_4)(H_2O)_3$ approaches the monolayer dispersion threshold when the molar ratio of CeO_2 (NC)/Ru is 0.15, and the highest cyclohexene yield is obtained over the Ru catalyst. When the molar ratio of CeO_2 (NC)/Ru continues to increase, the amount of $(Zn(OH)_2)_3(ZnSO_4)(H_2O)_3$ continues to increase, which results in not only the decrease of total Ru active sites, but also the decrease of active sites responsible for the formation of cyclohexene. Ultimately, the activity of the catalyst decreases and the cyclohexene selectivity only slightly increases.

CeO_2 is difficult to form $(Zn(OH)_2)_3(ZnSO_4)(H_2O)_3$ with $ZnSO_4$ on the catalysts after the process of calcination and high temperature reduction, and this is an important reason for the low selectivity of cyclohexene [33].

Figure 5.10 is a schematic diagram showing that Zn^{2+} in the slurry can form a stable complex [8] with cyclohexene.

In Fig. 5.10, since Zn^{2+} can form a stable complex with cyclohexene, cyclohexene desorption can be accelerated and cyclohexene reabsorption can be suppressed, leading to a rise in cyclohexene selectivity.

Figure 5.11 shows the reusability of Ru–Ce catalyst in the presence of $ZnSO_4$ at CeO_2 (NC)/Ru molar ratio of 0.15.

From Fig. 5.11, it can be seen that without any addition during the repeated use, the benzene conversion and cyclohexene selectivity are kept above 75.7% and 75.0%, respectively, in the four recycles. The cyclohexene yield is stable above

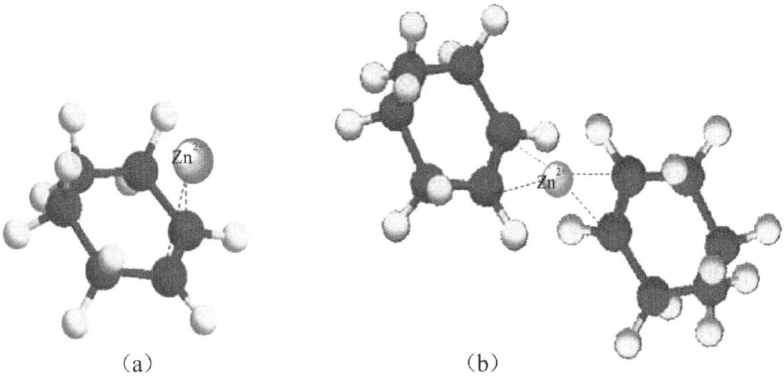

Fig. 5.10 Complex of cyclohexene with Zn^{2+}. **a** Complex formed by a Zn^{2+} and a cyclohexene molecule; **b** complex formed by a Zn^{2+} and two cyclohexene molecules

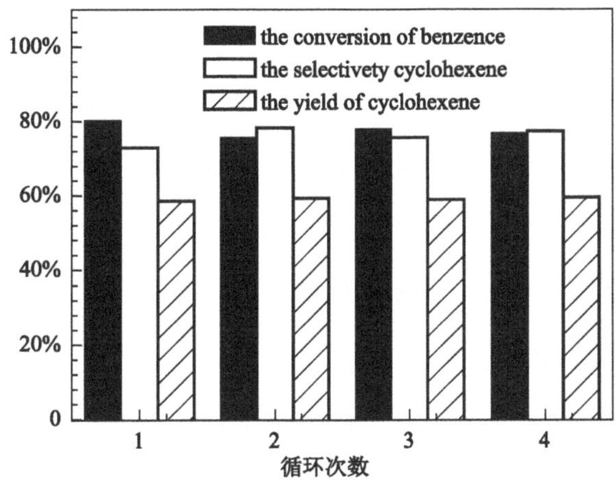

Fig. 5.11 Reusability of Ru–Ce catalyst in the presence of $ZnSO_4$ at CeO_2 (NC)/Ru molar ratio of 0.15

58.0%, indicating a good stability and good prospects for industrial application of the catalyst.

On the basis of extensive research on transition elements and rare earth elements, transition elements Zn, Mn, Fe and rare earth elements La, Ce are found to have a promoting effect on the catalysis of Ru catalyst for selective hydrogenation of benzene, and the third-generation catalysts for selective hydrogenation of benzene Ru–M (Zn, Mn, Fe, La, Ce) nano-bimetallic system are developed.

5.2 Third-Generation Catalysts for Selective Hydrogenation of Benzene—Ru–M (Zn, Mn, Fe, La, Ce) Nano-bimetallic System

5.2.1 *Nano Ru–Zn Catalyst*

(1) Characterization of Ru–Zn catalyst

The texture properties of Ru–Zn(x) catalysts before and after hydrogenation are listed in Table 5.5 and Table 5.6, respectively.

Table 5.5 BET surface area S_{BET}, pore volume V_p, average pore diameter d_p, and crystallite size d of Ru–Zn(x) catalysts before hydrogenation

Catalyst	$S_{BET}/(m^2/g)$	$V_p/(cm^3/g)$	d_p/nm	d/nm[a]
Ru–Zn(0)	88	0.18	4.1	4.1
Ru–Zn(2.6%)	77	0.18	4.8	4.3
Ru–Zn(5.2%)	67	0.15	4.6	3.8
Ru–Zn(7.7%)	74	0.17	4.5	3.7
Ru–Zn(8.6%)	77	0.16	4.0	3.9
Ru–Zn(9.6%)	55	0.15	5.3	4.3
Ru–Zn(12.4%)	66	0.27	8.2	4.0
Ru–Zn(14.9%)	64	0.27	8.4	3.8
Ru–Zn(29.1%)	57	0.22	7.8	3.8

[a]Crystallite sizes of Ru–Zn(x) catalysts calculated by using full width at half maximum (FWHM) of the strongest diffraction peak in XRD patterns and Scherrer equation

Table 5.6 BET surface area S_{BET}, pore volume V_p, and average pore diameter d_p of Ru–Zn(x) catalysts after hydrogenation

Catalyst	$S_{BET}/(m^2/g)$	$V_p/(cm^3/g)$	d_p/nm
ZrO_2	34	0.13	7.80
Ru–Zn(0%) AH[a]	44	0.19	17.0
Ru–Zn(2.6%) AH	42	0.17	19.8
Ru–Zn(5.2%) AH	43	0.18	20.8
Ru–Zn(7.7%) AH	40	0.16	16.4
Ru–Zn(8.6%) AH	41	0.28	26.5
Ru–Zn(9.6%) AH	43	0.35	33.0
Ru–Zn(12.4%) AH	43	0.27	29.7
Ru–Zn(14.9%) AH	38	0.25	25.9
Ru–Zn(29.1%) AH	42	0.23	21.8

[a]AH represents the samples after hydrogenation

As can be seen from Table 5.5, with increasing Zn content, the specific surface area of the fresh Ru–Zn(x) catalyst decreases, while the pore volume and average pore diameter increase. Low specific surface area means fewer micropores with the pore diameter below 2 nm in the catalyst. Desorption of the intermediate cyclohexene is favorable to occur on the outer surface and in the macropores of the catalyst. Zn is presented in the form of ZnO on the surface of Ru–Zn(x) catalyst as confirmed by TEM-EDS, and this is the main reason for the influence of Zn content on the texture properties of Ru–Zn(x) catalyst.

As can be seen from Table 5.6, the Ru–Zn(x) AH samples show a decrease in the specific surface area, pore volume, and average pore diameter than that before hydrogenation. TEM results confirm that the hydrogenated Ru–Zn(x) AH sample is a physical mixture of Ru–Zn(x) and dispersant ZrO_2, in which the mass ratio of ZrO_2 to Ru–Zn(x) is 5:1. The average specific surface area of the catalyst after hydrogenation is thus reduced due to the relatively low specific surface area of ZrO_2.

Figure 5.12 shows the XRD patterns of Ru–Zn catalysts before and after hydrogenation.

Figure 5.12a shows that in the Ru–Zn(x) catalyst, the diffraction peaks at $2\theta =$ 38.4°, 44.0°, 58.3°, 69.4°, 78.4°, and 84.7° are assigned to the metal Ru (JCPDS 01-070-0274). The half-width of the strongest diffraction peak of Ru at $2\theta = 44.0°$ and Scherrer equation were used to obtain the Ru crystallite size, which is in the range of 3.7–4.3 nm (listed in the last column of Table 5.5), indicating that the Ru–Zn bimetallic catalyst belongs to the nanocrystalline, and the Zn content shows little effect on the Ru crystallite size. It has been reported in the literature that the Ru crystallite size of a series of supported Ru–Zn/SiO_2 catalysts with different Zn contents prepared by impregnation method increased with the increase of Zn content [4], indicating that Zn content has different effect on the Ru crystallite size for the catalysts prepared by different preparation methods. The diffraction peaks of Ru–Zn(29.1%) catalyst at $2\theta =$ 31.8°, 34.4°, 36.3°, 47.5°, 56.6°, 62.9°, and 68.0° are attributed to ZnO (JCPDS 01-070-2551), indicating that Zn exists as ZnO. When the Zn content is low, ZnO randomly distributes between the Ru lattices to form a solid solution. When the Zn content is high, ZnO is enriched in the catalyst surface as a separate phase.

In Fig. 5.12b, all the diffraction peaks except the peaks at $2\theta =$ 8.1°, 16.2°, and 44.0° are assigned to the dispersant ZrO_2 in the Ru–Zn(x) AH samples. The faint diffraction peak appearing near 44.0° is attributed to the metallic Ru (JCPDS 01-070-0274); moreover, the shortening and broadening of the peak indicate that the crystallite size of Ru becomes smaller. The diffraction peaks at 8.09° and 16.2° are attributed to $3Zn(OH)_2{\cdot}ZnSO_4{\cdot}5H_2O$ (JCPDS 01-078-0246), a slightly complex $ZnSO_4$ basic double salt. With the increase of Zn content, the diffraction peak of the basic salt increases from scratch and gradually increases, indicating that the content gradually increases and the crystallite gradually grows. When Zn content is less than 9.6% (mass fraction), no characteristic peak of basic salt is observed. According to the principle of spontaneous monolayer dispersion, the content of the basic salt in the Ru–Zn(8.6%) catalyst is determined to be the monolayer dispersion threshold of the Ru–Zn catalyst. Below this threshold, the salt is monolayer-dispersed, whereas

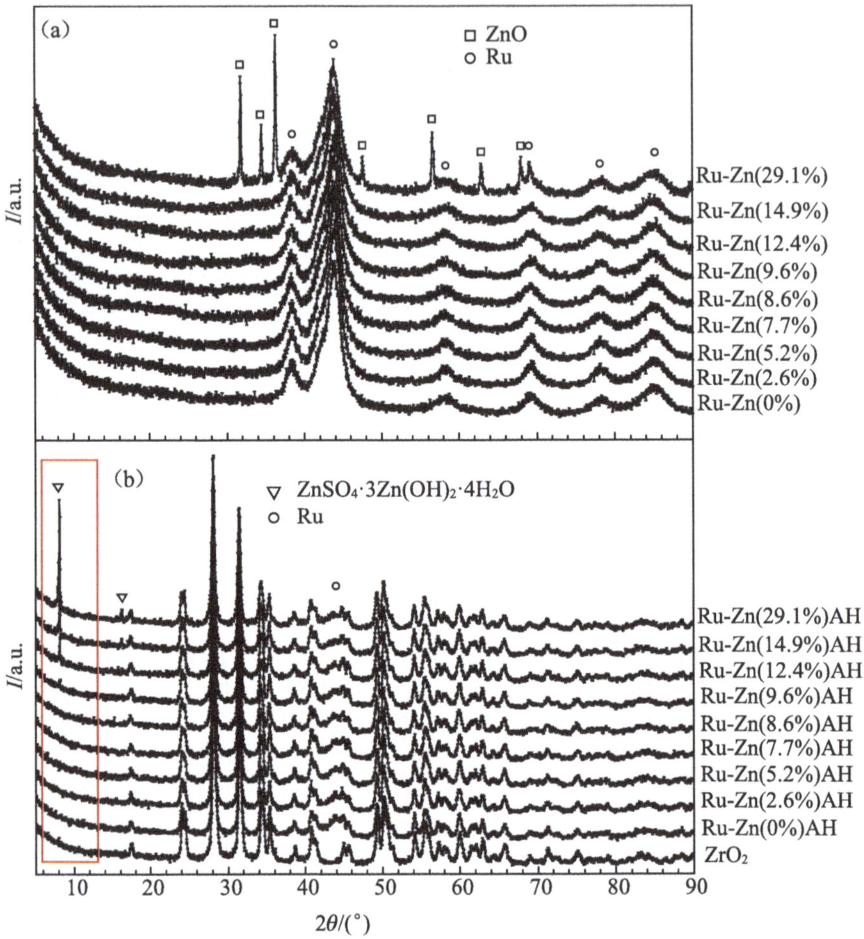

Fig. 5.12 XRD patterns of Ru–Zn catalysts before and after hydrogenation. **a** Ru–Zn(x) catalyst; **b** Ru–Zn(x) AH

the salt alone turns into a phase when the content is higher than the threshold. XRD directly proves the presence of the basic salt on the surface of the Ru–Zn catalyst after hydrogenation.

Figure 5.13 shows the XPS and AES spectra of Ru–Zn catalysts before and after hydrogenation.

From Fig. 5.13a, it can be seen that the electron binding energy (BE) of Zn $2p_{3/2}$ over the Ru–Zn(8.6%) catalyst is 1021.7 eV, which is very close to the BE (1021.6 eV) of Zn $2p_{3/2}$ of ZnO [17]. As can be seen from Fig. 5.13c, the Zn LMM Auger kinetic energy (KE) of Ru–Zn(8.6%) catalyst is 986.0 eV when Ar^+ sputtering time is 0 and 30 s, which increased slightly to 986.5 eV at the Ar^+ sputtering time of 1 min, but lower than that of Zn(II) species in the PtZn/C catalyst (KE 988.1 eV

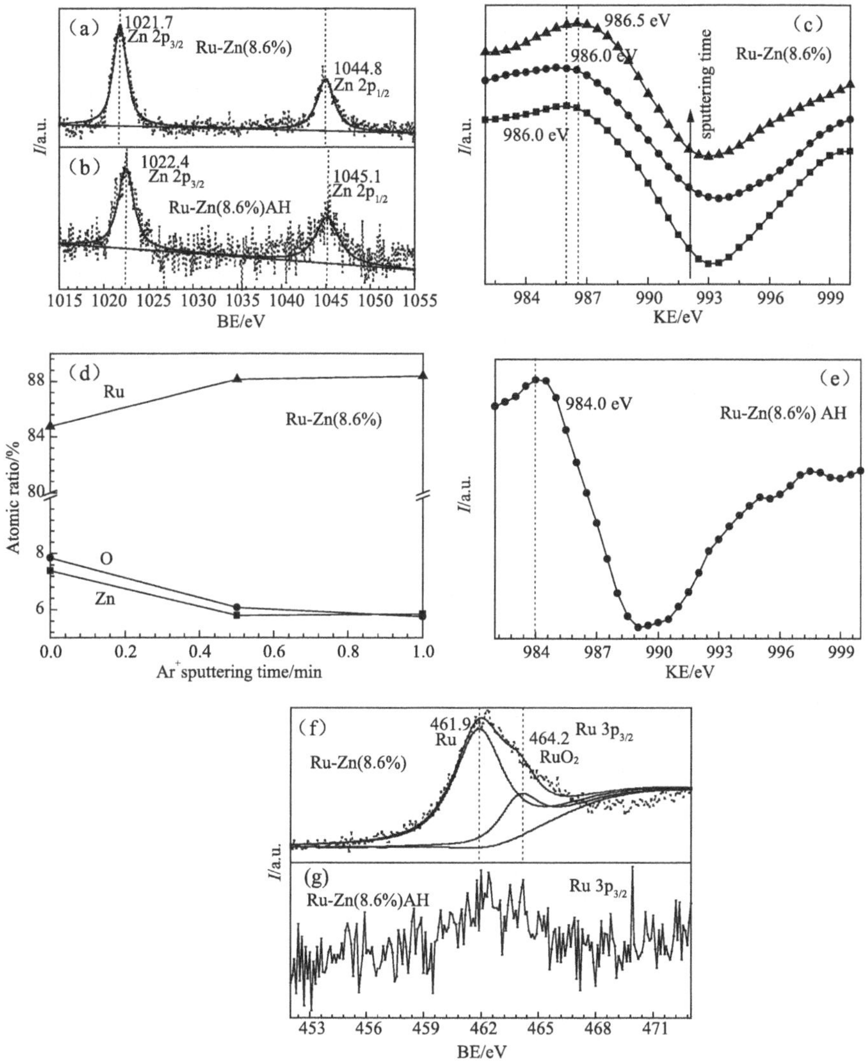

Fig. 5.13 XPS and AES spectra of Ru–Zn catalyst before and after hydrogenation. **a** Zn 2p and **f** XPS of the Ru–Zn(8.6%) catalyst before hydrogenation; **b** Zn 2p, **c** Zn LMM, **e** Zn LMM, **g** Ru 3p3/2, and **d** AES depth profiles of the Ru–Zn(8.6%) AH sample after hydrogenation

[14]) and Pt/Cr–ZnO catalyst without reduction (KE 987.0 eV [28]). Therefore, Zn in the Ru–Zn(8.6%) catalyst is mainly present as ZnO. Ru $3d_{3/2}$ is used to discuss the electronic state of Ru, since the energy spectrum of Ru 3d coincides with that of C 1 s, which is easily disturbed by reactants and products [11]. From the XPS spectra of Ru–Zn(8.6%) catalyst shown in Fig. 5.13f, it can be seen that Ru $3p_{3/2}$ consists of two peaks at 461.9 eV and 464.2 eV, corresponding to the metal Ru and RuO_2

[34], respectively, indicating that a small portion of the catalyst surface is oxidized to RuO_2.

The AES depth profiles of Ru–Zn(8.6%) catalyst (Fig. 5.13d) show that the concentration of Ru increases, and the Zn concentration as well as the O concentration decrease with the Ar^+ sputtering time, indicating the enrichment of ZnO on catalyst surface. Moreover, the atomic ratio of Zn to O is 0.93 on the top surface of the catalyst, indicating the oxidation of a small part of metallic Ru to RuO_2, which is in accordance with the XPS results. The content of RuO_2 decreases gradually and the atomic ratio of Zn/O approaches 1 with the sputtering time, indicating that the bulk phase of the catalyst is composed of the metallic Ru and ZnO, which is in conformity with the XRD results.

From Fig. 5.13b, it can be seen that the binding energy of Zn $2p_{3/2}$ over the Ru–Zn(8.6%) AH is 1022.4 eV, which is 0.7 eV higher than that for the catalyst before hydrogenation. The valence state of Zn is difficult to be determined by XPS, because the chemical shift of metallic Zn relative to the Zn in ZnO is only 0.1 eV. From Fig. 5.13e, it could be seen that the Zn LMM KE (984.0 eV) of Ru–Zn(8.6%) AH is 2.0 eV lower than that (986.0 eV) of Ru–Zn(8.6%) catalyst. All these indicate that a new Zn(II) species was formed on the catalyst surface during the hydrogenation process, which is in accordance with valence state of Zn in $3Zn(OH)_2 \cdot ZnSO_4 \cdot 5H_2O$ as confirmed by XRD.

Unfortunately, the peak of Ru $3p_{3/2}$ for Ru–Zn(8.6%) AH (Fig. 5.13g) becomes too dispersive to discriminate between the metallic Ru and RuO_2 due to the dispersion of ZrO_2. However, the peak of Ru $3p_{3/2}$ becomes complex, indicating the variation of the coordination environment of Ru species due to the formation of the $3Zn(OH)_2 \cdot ZnSO_4 \cdot 5H_2O$ salt on the surface of the catalyst. For this reason, it is difficult to directly judge whether the Zn(II) species donated some electrons to metallic Ru or sulfate since the sulfate is more effective than Ru in drawing electrons. However, the composition of Ru–Zn(8.6%) AH was measured by the XRF instrument and it is found that the molar ratios of Zn/Ru and S/Ru are 0.44 and 0.06, respectively. This suggests that the amount of sulfate is much lower than that of Zn(II) species on the surface of Ru–Zn(8.6%) AH. Moreover, Struijk et al. [11] found that the binding energy of S 2p for Ru catalyst after hydrogenation in the presence of $ZnSO_4$ was 169.9 eV, which was 0.7 eV higher than the standard value of $ZnSO_4$ [35]. All of these indicate that the sulfate gets a few electrons from the Zn(II) species. These prompt us to suggest that partial electrons might transfer from the Zn(II) species to the metallic Ru, forming the electron-rich $Ru^{(\delta-)}$ active center. In addition, it is found that the pH values of the liquid phase after hydrogenation at room temperature are around 6.0 due to the hydrolysis of $ZnSO_4$. It is well known that increasing temperature favors the hydrolysis. This means that the acidity of liquid phase is much higher at the reaction temperature of 150 °C due to the increase of hydrolysis degree of $ZnSO_4$. As we know, it is difficult for the metallic Zn to exist in the acid solution, which is consistent with no clear evidence of the presence of metallic Zn observed in the XPS and AES results.

Figure 5.14 shows the TEM images of Ru–Zn catalysts before and after hydrogenation.

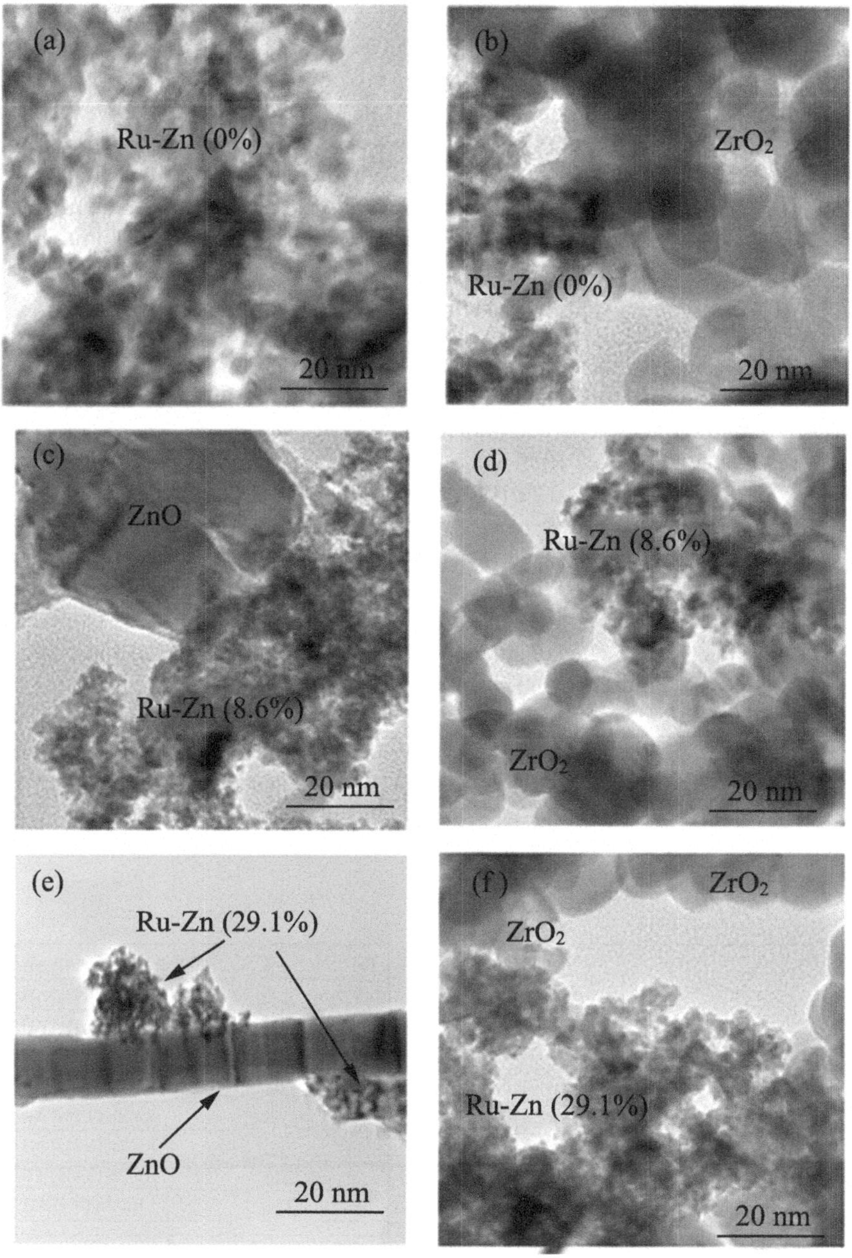

Fig. 5.14 TEM images of Ru–Zn catalyst before and after hydrogenation. **a** Ru–Zn(0%) catalyst; **b** Ru–Zn(0%) AH; **c** Ru–Zn(8.6%) catalyst; **d** Ru–Zn(8.6%) AH; **e** Ru–Zn(29.1%) catalyst; **f** Ru–Zn(29.1%) AH

TEM images in Fig. 5.14 show that the Ru crystallite sizes of Ru–Zn(0%) catalyst, Ru–Zn(8.6%) catalyst, and Ru–Zn(29.1%) catalyst center on about 4 nm, which is consistent with the XRD results.

A piece of thing is found on the surface of Ru–Zn(8.6%) catalyst in Fig. 5.14c, which is confirmed to be ZnO by EDS. This also indicates that ZnO is rich on the surface and cannot be highly dispersed. Although the results of XPS, AES-Ar^+, and TEM all show the existence of ZnO, no ZnO phase is detected in XRD pattern of the Ru–Zn(8.6%) catalyst. This indicates that ZnO exists in the amorphous form on the surface of Ru. Eo et al. [36] found that ZnO with heat treatment at 100–200 °C for 60 min still showed an amorphous phase. Thus, it is usual that no crystalline ZnO phases are detected when Zn content is lower than 29.1%.

Pillars are found on the surface of Ru–Zn(29.1%) catalyst in Fig. 5.14e, which is confirmed to be hexagonal phases of ZnO by EDS, indicating its oriented growth under the catalyst preparation conditions.

However, neither pieces ZnO nor pillars ZnO are detected on Ru–Zn(8.6%) AH in Fig. 5.14d and Ru–Zn(29.1%) AH in Fig. 5.14f. This is probably because ZnO on the catalyst surface reacted with $ZnSO_4$ in the slurry, forming the $3Zn(OH)_2{\cdot}ZnSO_4{\cdot}5H_2O$ salt, which is a monolayer dispersed on the catalyst surface [37]. The testing results of activity and selectivity over catalysts show that the properties of Ru–Zn(8.6%) catalyst had significant difference under the presence or absence of $ZnSO_4$ in the slurry, suggesting that $3Zn(OH)_2{\cdot}ZnSO_4{\cdot}5H_2O$ played a key role in improving the selectivity to cyclohexene.

The catalyst particles of Ru–Zn(0%) AH, Ru–Zn(8.6%) AH, and Ru–Zn(29.1%) AH samples are separated and isolated by ZrO_2, indicating that ZrO_2 can reduce the chance of the collision of different catalyst particles and suppress the agglomeration of the catalyst, which are beneficial for the extension of the catalyst life [26].

Figure 5.15 displays the H_2-TPR profiles of Ru–Zn catalysts before and after hydrogenation.

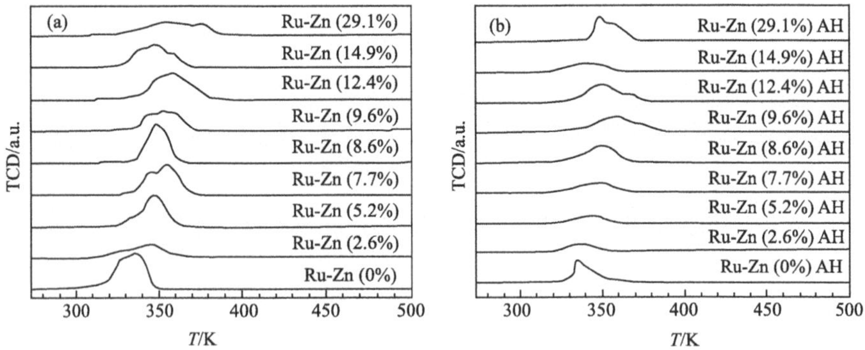

Fig. 5.15 H_2-TPR profiles of Ru–Zn catalysts before and after hydrogenation. **a** Ru–Zn(x) catalyst; **b** Ru–Zn(x) AH

From Fig. 5.15a, it can be seen that a H_2 consumption peak with a shoulder is observed between 300 and 400 K for the Ru–Zn(x) catalyst. The H_2 consumption peak is attributed to the reduction of RuO_2 to metallic Ru ($Ru^{4+} \rightarrow Ru^{3+} \rightarrow Ru^{2+} \rightarrow Ru^0$), so the shoulders consist of two peaks or three peaks. Besides, the absence of any additional peaks indicates that the promoter ZnO cannot be reduced to metallic Zn within 500 K, which is consistent with the results of XPS and XRD. Although bulk ZnO reduction is thermodynamically feasible, temperatures as high as 923 K are required.

As can be seen from Fig. 5.15b, a reduction shoulder peak also appears in 300–400 K for Ru–Zn(x) AH, and the peak is assigned to the step-by-step reduction of RuO_2 to metallic Ru, which is similar to the fresh Ru–Zn(x) catalyst, suggesting that the addition of ZrO_2 and the formation of $3Zn(OH)_2 \cdot ZnSO_4 \cdot 5H_2O$ have little effect on the reduction of RuO_2. The temperatures of complete reduction for all Ru–Zn(x) AH samples are lower than 150 °C, indicating only the existence of metallic Ru under the hydrogenation reaction conditions.

The reduction temperature of Ru–Zn(x) AH is lower than that of the Ru–Ce/SBA-15 (343–573 K) and Ru–Zn/SiO_2 (393–543 K) catalysts, indicating the weak interaction between the Ru–Zn(x) catalyst and the dispersant ZrO_2, which helps to improve the hydrogenation activity of Ru–Zn(x) catalyst.

(2) Performance of Ru–Zn catalyst for selective hydrogenation of benzene

Table 5.7 presents the benzene conversion, cyclohexene selectivity and yield, maximum yield of cyclohexene, and pH values of the aqueous phase after hydrogenation.

Figure 5.16 presents the C_{BZ}-t and S_{HE}-C_{BZ} curves of selective hydrogenation of benzene over the Ru–Zn catalysts with different Zn contents.

From Fig. 5.16a, b, it can be seen that in the presence of $ZnSO_4$, the benzene conversion over the Ru–Zn(x) catalysts decreased with the increase of Zn content, and the selectivity of cyclohexene increased monotonically, but the yield and maximum yield of cyclohexene both increased first and deceased afterwards, reaching the maximum yield value of 50.9% over the Ru–Zn(8.6%) catalyst.

The selectivity and the yield of cyclohexane reached 38.2% and 22.1%, respectively, at benzene conversion of 57.9% on Ru–Zn(0%) catalyst at 5 min in the presence of $ZnSO_4$. However, in the absence of $ZnSO_4$ with only 5 min, benzene was totally consumed on this catalyst. This indicates that the reaction of the cyclohexene formation only happens on the surface of the Ru catalysts which can make contact with $ZnSO_4$. For Ru–Zn(8.6%) catalyst, the conversion of benzene reached 86.7% within 5 min in the absence of $ZnSO_4$. The yield of cyclohexene was only 1.6% with the corresponding selectivity to cyclohexene of 1.9%. This implies that the promoter ZnO alone cannot enhance the selectivity to cyclohexene of Ru–Zn(8.6%) catalyst. However, this catalyst afforded a selectivity to cyclohexene of 76.2% and a cyclohexene yield of 35.1% with the corresponding benzene conversion of 46.0% at 5 min in the presence of $ZnSO_4$, indicating that the synergistic effect of ZnO and $ZnSO_4$ enhances the selectivity to cyclohexene of the catalyst. Namely, the

Table 5.7 Benzene conversion C_{BZ}, cyclohexene selectivity S_{HE} and yield Y_{HE}, cyclohexene maximum yield Y_{max}, and pH of the aqueous phase after hydrogenation over Ru–Zn catalyst[a]

Catalyst	$C_{BZ}/\%$[b]	$S_{HE}/\%$[b]	$Y_{HE}/\%$[b]	$Y_{max}/\%$[b]	pH
Ru–Zn(0%)	57.9	38.2	22.1	23.1	5.7
Ru–Zn(2.6%)	68.1	47.2	32.1	32.1	5.8
Ru–Zn(5.2%)	54.4	69.5	37.8	48.0	5.9
Ru–Zn(7.7%)	50.5	71.1	35.9	47.4	5.9
Ru–Zn(8.6%)	46.0	76.2	35.1	50.9	6.1
Ru–Zn(9.6%)	26.4	84.1	22.2	50.4	5.9
Ru–Zn(12.4%)	20.7	86.0	17.8	49.8	6.1
Ru–Zn(14.9%)	13.1	87.5	11.5	34.2	5.8
Ru–Zn(29.1%)	6.8	90.4	6.2	19.0	6.0
Ru–Zn(0%)[d]	100	0	0	0	7.2
Ru–Zn(8.6%)[d]	86.7	1.9	1.6	1.6	7.3

[a]Reaction conditions: reaction temperature of 150 °C, H_2 pressure of 5.0 MPa, stirring rate of 1400 r/min, 1.96 g of catalyst, 49.2 g of $ZnSO_4 \cdot 7H_2O$, 9.8 g of ZrO_2, 140 mL of benzene, and 280 mL of water
[b]Benzene conversion, cyclohexane selectivity, and yield at 5 min
[c]The maximum cyclohexane yield at 25 min
[d]In the absence of $ZnSO_4$

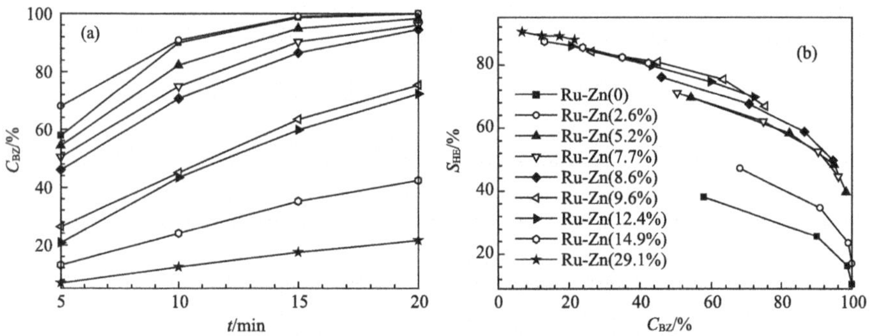

Fig. 5.16 C_{BZ}-t and S_{HE}-C_{BZ} curves of selective hydrogenation of benzene over the Ru–Zn catalysts with different Zn contents. Reaction conditions: reaction temperature of 150 °C, H_2 pressure of 5.0 MPa, stirring rate of 1400 r/min, 1.96 g of catalyst, 49.2 g of $ZnSO_4 \cdot 7H_2O$, 9.8 g of ZrO_2, 140 mL of benzene, and 280 mL of water

$3Zn(OH)_2 \cdot ZnSO_4 \cdot 5H_2O$ salt formed by the ZnO on the surface of the catalyst reacting with $ZnSO_4$ plays a key role in improving the selectivity to cyclohexene of the catalyst.

Pretreatment of the catalyst is carried out to make the catalytic system operate for a certain time under catalyst operation conditions before the addition of benzene,

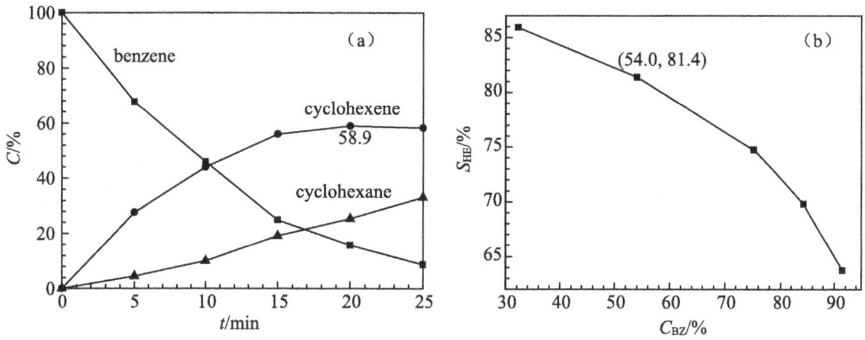

Fig. 5.17 Curves of concentrations versus reaction time, and cyclohexene selectivity versus benzene conversion during benzene selective hydrogenation over the Ru–Zn(8.6%) catalyst after pretreating for 22 h. **a** *C*-*t* curve; **b** S_{HE}-C_{BZ} curve. Pretreatment conditions: reaction temperature of 140 °C, H_2 pressure of 5.0 MPa, stirring rate of 800 r/min, 1.96 g of catalysts, 49.2 g of $ZnSO_4 \cdot 7H_2O$, 9.8 g of ZrO_2, 280 mL of water, and time of 22 h

and the purpose was to decrease the catalyst activity, improve the selectivity of cyclohexane, and ensure the stability of the catalyst.

Figure 5.17 shows the curves of concentrations versus time, and cyclohexene selectivity versus benzene conversion during benzene selective hydrogenation over the Ru–Zn(8.6%) catalyst after pretreating for 22 h.

From Fig. 5.17a, it can be seen that the pretreated Ru–Zn(8.6%) catalyst afforded a maximum cyclohexene yield of 58.9% after reacting for 20 min, which was among the best results reported so far. Moreover, a cyclohexene selectivity of 81.4% at a benzene conversion of 54.0% was achieved after reacting for 10 min over the pretreated catalyst as shown in Fig. 5.17b. The cyclohexene selectivity of more than 80% was requested at a benzene conversion of 40%. Obviously, the performance of Ru–Zn(8.6%) catalyst meets this need.

XRF results revealed that the Zn/Ru atomic ratio of Ru–Zn(8.6%) catalyst after pretreatment was 0.48, which was higher than the value (0.44) before the pretreatment, indicating that Zn^{2+} in the slurry was partially transferred to the catalyst surface. The transfer involved the chemical adsorption of Zn^{2+} and the interaction of ZnO on the catalyst surface with $ZnSO_4$ in the slurry to form a basic salt. In the absence of $ZnSO_4$ in the slurry, the selectivity to cyclohexene was 5% at 5 min over Ru–Zn catalyst after pretreatment, indicating the correctness of the above conclusions. In the absence of $ZnSO_4$, the catalyst was pretreated in a nitrogen atmosphere, no improvement in the selectivity to cyclohexene was observed, indicating that during the pretreatment process, hydrogen was absorbed on the catalyst surface, prior to occupying those strongest active centers, and this was also one of the reasons for the improvement of cyclohexene selectivity.

The benzene selective hydrogenation to cyclohexene was carried out over Ru–Zn(8.6%) catalyst, and the catalyst was pretreated for 22 h in 0.6 mol/L $ZnSO_4$

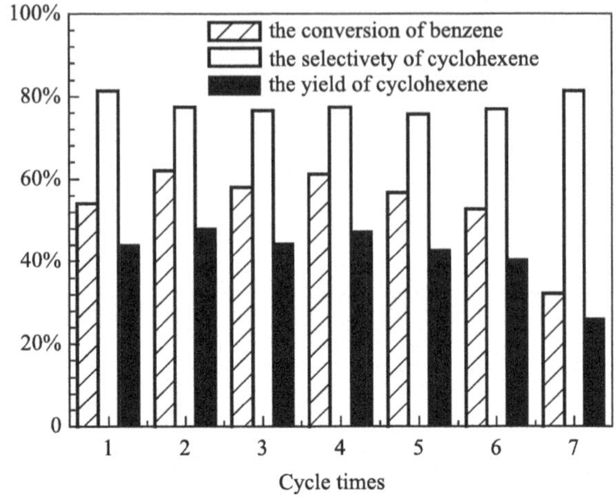

Fig. 5.18 Reusability of the Ru–Zn(8.6%) catalyst. Reaction conditions: reaction temperature of 150 °C, H_2 pressure of 5.0 MPa, stirring rate of 1400 r/min, 1.96 g of catalysts, 49.2 g of $ZnSO_4 \cdot 7H_2O$, 280 mL of water, 140 mL of benzene, and initial hydrotreating pretreatment of 22 h

solution. The pretreated catalyst was separated and recycled without any more pretreatment and additions.

Figure 5.18 shows the activity, cyclohexane selectivity, and yield of the catalyst after being recycled for 7 times.

As can be seen in Fig. 5.18, the benzene conversions are stable above 50%, and the cyclohexene selectivity and yields are kept above 76% and 40% in the first six recycles, respectively, indicating a good stability and good prospects for industrial application of the catalyst.

5.2.2 *Nano Ru–Mn Catalyst*

(1) Structure, morphology, and texture of catalyst

Figure 5.19 displays the XRD patterns of Ru–Mn catalyst before and after hydrogenation.

Figure 5.19 shows that all the Ru–Mn(x) catalysts displayed the diffraction peaks of the hexagonal phases of metallic Ru (JCPDS 01-070-0274) at $2\theta = 38.5°$, 42.3°, 44.0°, 58.3°, 69.2°, 78.4°, and 85.2°. The Ru–Mn(8.0%) and Ru–Mn(10.8%) catalysts showed the diffraction peaks of Mn_3O_4 (JCPDS 00-001-1127) at $2\theta = 18.0°$, 32.7°, 36.1°, 38.1°, 58.9°, and 69.3°. The reduction temperature of the catalyst was only 150 °C, and this was in accordance with the result reported in the literature that Mn mainly existed as Mn_3O_4 in the temperature below 300 °C. When the Mn contents were in the range of 3.4–5.4%, the diffraction peaks of Mn_3O_4 were not observed in the XRD patterns of the Ru–Mn(x) catalysts, indicating that Mn_3O_4 was highly dispersed.

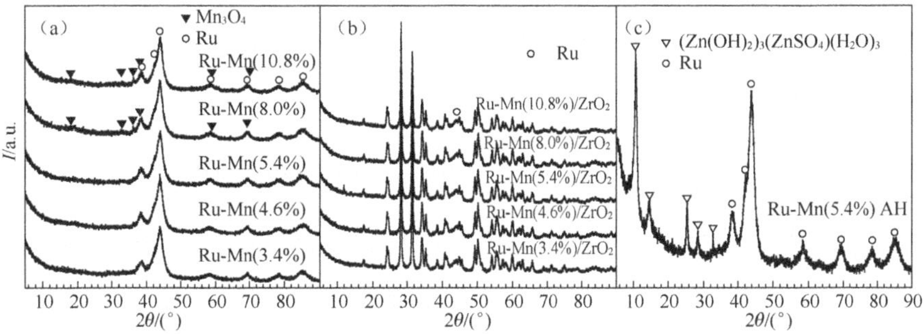

Fig. 5.19 XRD patterns of Ru–Mn catalyst before and after hydrogenation. **a** Ru–Mn(x) catalyst; **b** Ru–Mn(x)/ZrO_2 AH; **c** Ru–Mn(5.4%) AH

Table 5.8 BET surface area S_{BET}, average pore diameter d_p, pore volume V_p, and crystallite size d of Ru–Mn(x) catalyst

Catalyst	$S_{BET}/(m^2/g)$	d_p/nm	$V_p/(cm^3/g)$	d/nm
Ru–Mn(3.4%)	72	12.2	0.22	3.7
Ru–Mn(4.6%)	71	11.8	0.21	4.3
Ru–Mn(5.4%)	62	11.6	0.18	3.8
Ru–Mn(8.0%)	68	9.4	0.16	4.0
Ru–Mn(10.8%)	56	11.2	0.16	4.5

The last column of Table 5.8 presents the Ru crystallite sizes of the catalyst calculated from the strongest peak broadening at $2\theta = 44.0°$ using the Scherrer equation. As can be seen, the Ru crystallite sizes of the Ru–Mn(x) catalysts were distributed in the narrow range of 3.7–4.5 nm indicating that the introduction of Mn_3O_4 had little influence on the Ru crystallite size.

From Fig. 5.19c, it can be seen that not only the diffraction peaks of metallic Ru, but also the diffraction peaks of $(Zn(OH)_2)_3(ZnSO_4)(H_2O)_3$ (JCPDS 01-078-0247) at $2\theta = 10.6°, 14.7°, 25.4°, 28.4°$, and $32.8°$ were detected in the Ru–Mn(5.4%) catalyst without the dispersant ZrO_2 after hydrogenation. This implied that in the presence of $ZnSO_4$, a new insoluble $(Zn(OH)_2)_3(ZnSO_4)(H_2O)_3$ salt was formed in the Ru–Mn(5.4%) catalyst after hydrogenation, whereas Fig. 5.19b showed that the diffraction peaks of the salt were not observed in all the Ru–Mn(x) catalysts with the addition of ZrO_2, indicating $(Zn(OH)_2)_3(ZnSO_4)(H_2O)_3$ was below the monolayer dispersion capacity [38].

Figure 5.20 is the TEM images and crystallite size distribution of Ru–Mn catalyst.

From Fig. 5.20a, the Ru–Mn(5.4%) catalyst was composed of nanoscale spherical and ellipsoidal crystallites, and the crystallite size was concentrated at around 4.8 nm, which was consistent with the XRD results. Figure 5.20b shows that the ZrO_2 dispersant mainly consisted of ZrO_2 crystallites of size about 20 nm. The catalyst particles were separated and isolated by ZrO_2 after hydrogenation, with the crystallite

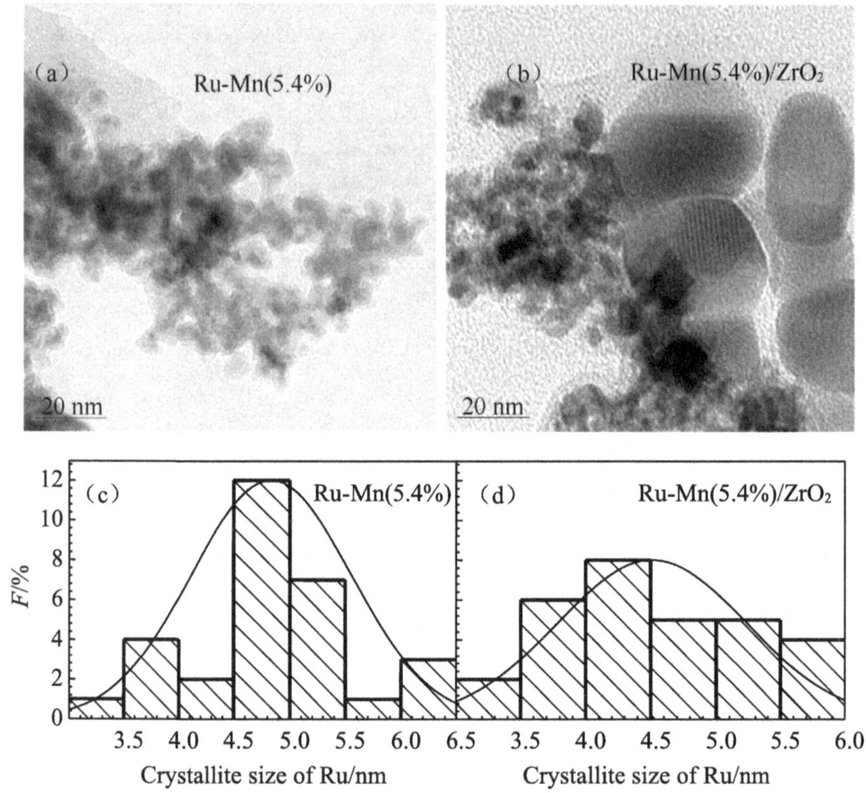

Fig. 5.20 TEM images and crystallite size distributions of Ru–Mn catalyst. **a**, **b** TEM images; **c**, **d** Crystallite size distributions

size being concentrated at around 4.5 nm, which was lower than that of the catalyst before hydrogenation, indicating that a suitable amount of ZrO_2 could significantly suppress agglomeration and growth of Ru–Mn crystallites.

Table 5.8 lists the textural properties of Ru–Mn(x) catalyst.

As can be seen from Table 5.8, the BET surface areas, average pore diameters, and pore volumes generally decreased with increasing Mn content of the catalysts. It was reported in the literature that the specific surface area of TiO_2 support of MnO_x/TiO_2 catalyst was low, and this was because amorphous MnO_x was dispersed on the TiO_2 surface and blocked some of the catalyst pores [39]. Similarly, it is proposed that Mn_3O_4 was dispersed on the surface of the catalysts and could block some of the catalyst pores, resulting in the decrease of surface areas, pore volumes, and pore diameters. In the Ru–Mn(x) catalyst after hydrogenation, the mass ratio of ZrO_2 to catalyst was 5:1, and the texture properties of Ru–Mn(x)/ZrO_2 catalyst after hydrogenation were similar to that of ZrO_2 [40].

Table 5.9 Compositions of Ru–Mn(x) catalyst with the dispersant ZrO_2 after hydrogenation, and concentrations of the metallic ions in the aqueous phase at room temperature

Sample	Mn precursor	Atomic ratio (mol/mol)[a]			Ion concentration (mol/L)[b]		pH[c]
		Mn/Ru	Zn/Ru	Zr/Ru	Zn^{2+}	Mn^{2+}	
Ru–Mn(3.4%)/ZrO_2	$MnSO_4$	0.02	0.34	5.16	0.44	3.11×10^{-3}	5.6
Ru–Mn(4.6%)/ZrO_2	$MnSO_4$	0.02	0.42	5.12	0.43	4.11×10^{-3}	5.4
Ru–Mn(5.4%)/ZrO_2	$MnSO_4$	0.02	0.46	4.98	0.42	6.36×10^{-3}	6.0
Ru–Mn(8.0%)/ZrO_2	$MnSO_4$	0.02	0.52	5.07	0.41	9.82×10^{-3}	5.9
Ru–Mn(10.8%)/ZrO_2	$MnSO_4$	0.02	0.55	5.26	0.37	1.0×10^{-2}	6.2
Ru–Mn(5.2%)/ZrO_2	$Mn(NO_3)_2$	0.02	0.45	5.04	–	–	–
Ru–Mn(5.6%)/ZrO_2	$MnCl_2$	0.02	0.47	5.17	–	–	–
Ru–Mn(5.4%) AH	$MnSO_4$	0.02	0.49	0	–	–	6.0

[a]XRF
[b]AAS
[c]pH at room temperature

Table 5.9 shows the compositions of the Ru–Mn(x) catalyst with the dispersant ZrO_2 after hydrogenation and concentrations of the metallic ions in the aqueous phase at room temperature.

As shown in Table 5.9, the Ru content in the Ru–Mn(x) catalyst with various Mn contents was different, thus, the Mn/Ru, Zn/Ru and Zr/Ru atomic ratios clearly reflect the variations in the catalyst compositions. The Mn/Ru molar ratios of the Ru–Mn(x)/ZrO_2 catalysts after hydrogenation were all 0.02, indicating there were trace amounts of Mn in Ru–Mn(x)/ZrO_2 catalyst after hydrogenation. The Zn/Ru atomic ratios of Ru–Mn(x)/ZrO_2 catalyst after hydrogenation increased, and the concentrations of Mn^{2+} (slurry is colorless, where Mn existed as Mn^{2+}) in the aqueous phase increased and the concentrations of Zn^{2+} decreased with Mn content. All of these results indicated that the Mn_3O_4 on the surfaces of Ru–Mn(x) catalysts had reacted with $ZnSO_4$ in the slurry to form a new Zn species and Mn(II) species. Therein, the Zn species were chemisorbed on the surface of Ru catalyst, and the Mn(II) species were dissolved into the slurries. Only a trace amount of Mn was also detectable in the Ru–Mn(5.4%) catalyst without the addition of ZrO_2 after hydrogenation, and the Zn/Ru atomic ratio was close to the catalyst with ZrO_2 after hydrogenation. In addition, the S/Ru atomic ratio of the solid catalyst after hydrogenation was 0.06, and S was derived from the basic salt chemisorbed on the catalyst surface, which was consistent with the XRD results. The addition of ZrO_2 lowered the S contents of the Ru–Mn(x) catalysts after hydrogenation to below the detection limit of the XRF instrument. All these results implied that the Zn species existed as the $[Zn(OH)_2]_3(ZnSO_4)(H_2O)_3$ salt, and the produced amount of the salt increased with Mn_3O_4 content, which resulted in an increase in the Zn/Ru atomic ratio. Therefore, it can be concluded that Mn_3O_4 was mainly presented on the Ru surface, because the $ZnSO_4$ in the slurry could only react with Mn_3O_4 on the surface of catalyst, i.e., the $[Zn(OH)_2]_3(ZnSO_4)(H_2O)_3$ salt form could only be chemisorbed on the Ru surface. The pH values of the aqueous solutions at room temperature after hydrogenation increased generally with Mn content of the catalyst, due to the decrease of Zn^{2+} concentration and hydrolyzing degree in the slurry. Since the amount of ZrO_2 was fixed, the Zr/Ru atomic ratio was stable at around 5.

A series of Ru–Zn(x)/ZrO_2 catalysts were prepared from different precursors, namely $MnSO_4$, $Mn(NO_3)_2$ and $MnCl_2$, in equimolar quantities. The mass percentages of Mn in the catalysts, measured by atomic absorption spectrometry, were 5.4%, 5.2%, and 5.6%, respectively. In Table 5.9, though the precursors of Mn were different, the Mn/Ru, Zn/Ru and Zr/Ru atomic ratios of Ru–Mn(x)/ZrO_2 catalysts after hydrogenation were similar, indicating similar chemical compositions of the Ru–Mn(x)/ZrO_2 catalysts after hydrogenation.

The Zn species chemisorbed on the surface of the Ru crystallite played an important role in improving the cyclohexene selectivity of the Ru catalysts. However, there were different opinions about the Zn valence. In the co-precipitation of $RuCl_2$ and $ZrOC1_2$ by ammonia, and the preparation of Ru–Zn/m-ZrO_2 catalyst reduced by H_2 in $ZnSO_4$ solution, the valence of Zn determined by XPS was close to the binding energy of Zn $2p_{3/2}$ of metallic Zn. This was because Zn^{2+} in the $ZnSO_4$ was partially absorbed on the surface of the catalyst, and further reduced by the hydrogen atoms that spilled from the surface of Ru catalyst [3]. According to the XPS results, certain authors proposed that Zn^{2+} could be reduced to metallic Zn [12, 13], whereas some thought that the chemisorbed Zn^{2+} could not be reduced to metallic Zn; it would be deposited on the surface of catalyst once reduced, covering the Ru active sites [41]. Since the difference in binding energies of Zn $2p_{3/2}$ in ZnO and metallic Zn was 0.1 eV, it was very difficult to assess the valences of Zn based on the Zn $2p_{3/2}$ binding energy. This drawback can be overcome by using the Zn LMM Auger transition, as the Auger shift between Zn(II) and metallic Zn was higher than 4.6 eV [42].

Figure 5.21 displays the Zn LMM Auger electron spectroscopy (AES) of Ru–Mn(5.4%) catalyst with the dispersant ZrO_2 after hydrogenation.

The spectra shown in Fig. 5.21 are recorded after Ar^+ sputtering for 1 min to avoid interruptions of the surface oxidation of the catalyst. As can be seen, the kinetic energy of Zn LMM for Ru–Mn(5.4%)/ZrO_2 after hydrogenation is 984.2 eV, while the Zn LMM Auger electron kinetic energy of metallic Zn commonly appears in the range of 991–995 eV, indicating that Zn is presented in oxidation state in the catalyst, and this is the same with the valence of Zn in the insoluble basic salt formed by the reaction of ZnO on the catalyst surface and $ZnSO_4$ in the slurry which is confirmed by XRD [14]. However, this can only show that most of Zn exists in the form of ZnO in the catalyst, and enriched on the surface, but could not completely exclude that a trace amount of Zn^{2+} adsorbed on the catalyst surface is reduced to metallic Zn by spillover hydrogen. Quantum chemical calculations show that a small amount

Fig. 5.21 Zn LMM AES spectrum of Ru–Mn(5.4%)/ZrO_2 catalyst

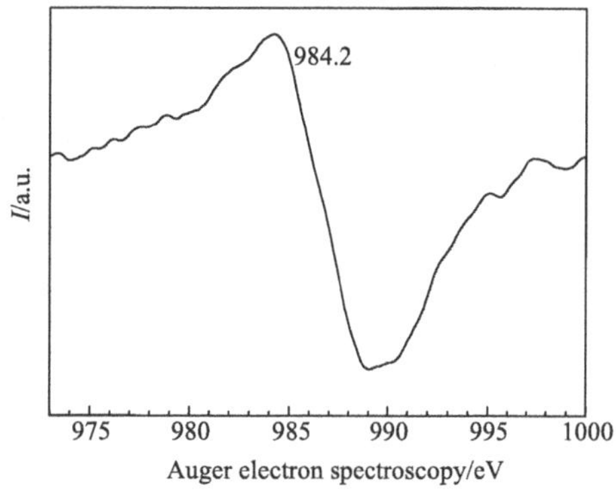

of metal Zn played an important role in improving the selectivity to cyclohexene [12, 13].

The chemisorbed $(Zn(OH)_2)_3(ZnSO_4)(H_2O)_3$ salt is rich in crystallization water. Chemisorption of the salt on the surface of the catalyst, therefore, results in the Ru catalyst being surrounded by a stagnant water layer, constraining the diffusion of hydrogen, reducing the surface coverage of hydrogen on the catalyst. The existence of a stagnant water layer on the catalyst surface could accelerate desorption and hinder re-adsorption of cyclohexene for further hydrogenation to cyclohexane. The $(Zn(OH)_2)_3(ZnSO_4)(H_2O)_3$ salt could selectively cover the most reactive sites of the catalyst, which could reduce the adsorption enthalpies of benzene and cyclohexene, and thereby results in a decrease in the reaction rate of benzene hydrogenation and an increase in the cyclohexene desorption rate. This could intervene with the benzene hydrogenation process, reduce the possibility of the direct hydrogenation of benzene to cyclohexane, and make the step-by-step hydrogenation of benzene to cyclohexene be dominated. Moreover, cyclohexene would be desorbed quickly once generated, and thus could significantly improve the cyclohexene selectivity and yield [43].

(2) Performance of Ru–Mn(x)/ZrO_2 catalyst

Figure 5.22 shows the change of benzene conversion C_{BZ} versus time t (C_{BZ}-t curve), the change of selectivity to cyclohexane S_{HE} versus benzene conversion C_{BZ} (S_{HE}-C_{BZ} curve), and the change of cyclohexane yield Y_{HE} versus time t (Y_{HE}-t curve) over Ru–Mn(x)/ZrO_2 catalysts with different Mn contents.

It can be seen from Fig. 5.22 that with the increase of Mn content, the activity of the catalyst decreased gradually, and the selectivity to cyclohexene increased at the same benzene conversion. When Mn content increased from 3.4 to 5.4%, the cyclohexene yield gradually increased to 61.3%, which was one of the best reported results. When Mn content continued to increase, the cyclohexene yield decreased. The selectivity of cyclohexene was 82.7% at the benzene conversion of 59.4% over the Ru–Mn(5.4%) catalyst without the dispersant ZrO_2 within 15 min. After the addition of the dispersant ZrO_2, the cyclohexene selectivity was 80.7% at the benzene

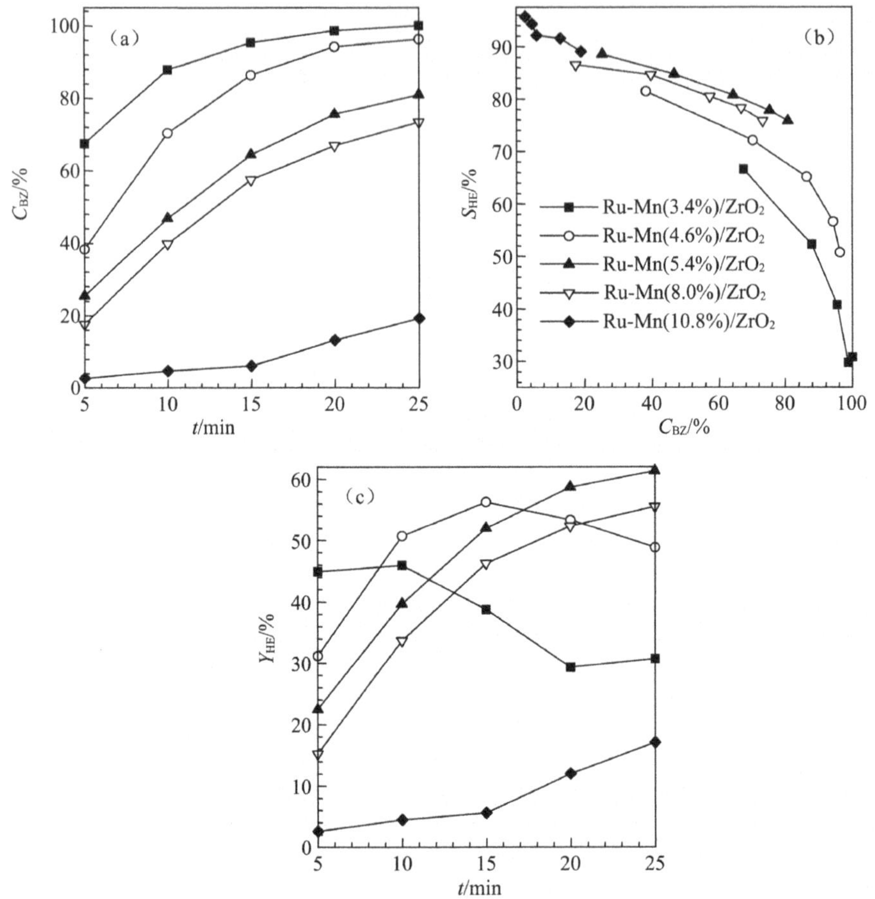

Fig. 5.22 **a** C_{BZ}-t curve; **b** S_{HE}-C_{BZ} curve; **c** Y_{HE}-t curve of Ru–Mn(5.4%)/ZrO_2 catalysts with different Mn contents. Reaction conditions: reaction temperature of 150 °C, H_2 pressure of 5.0 MPa, stirring rate of 1400 r/min, 1.96 g of Ru–Mn(x) catalysts, 49.2 g of $ZnSO_4 \cdot 7H_2O$, 9.8 g of ZrO_2, 140 mL of benzene, and 280 mL of water

conversion of 64.4%, indicating that the dispersion effect of ZrO_2 on the Ru catalyst improved the activity of catalyst, and this was consistent with the TEM results.

Figure 5.23 shows the change of benzene conversion C_{BZ} versus time t (C_{BZ}-t curve), the change of selectivity to cyclohexane S_{HE} versus benzene conversion C_{BZ} (S_{HE}-C_{BZ} curve), and the change of cyclohexane yield Y_{HE} versus time t (Y_{HE}-t *curve*) over Ru–Mn(5.4%) catalysts with the dispersant ZrO_2 prepared by different precursors.

As can be seen from Fig. 5.23, the activities, cyclohexane selectivity, and yields of Ru–Mn(5.4%) catalysts prepared from different precursors were very similar. The characterization results confirmed that these Ru–Mn(5.4%)/ZrO_2 catalysts had

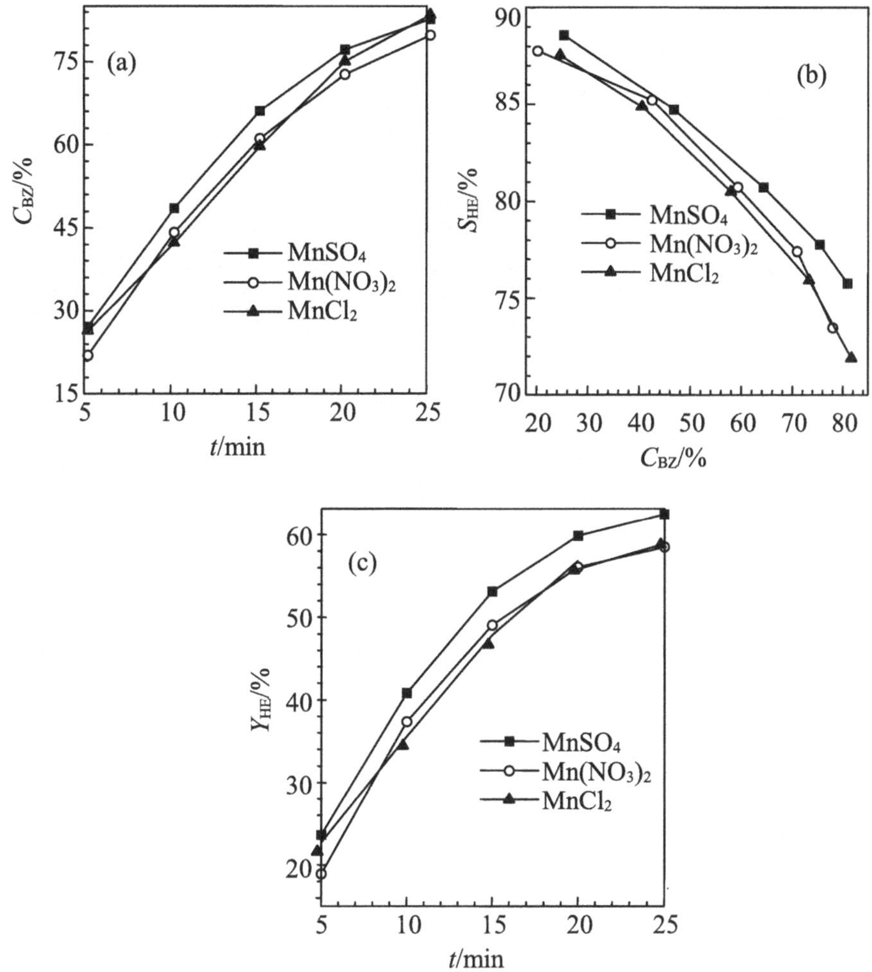

Fig. 5.23 **a** C_{BZ}-t curve; **b** S_{HE}-C_{BZ} curve; **c** Y_{HE}-t curve of Ru–Mn(5.4%)/ZrO_2 catalysts prepared by different Mn precursors. Reaction conditions: reaction temperature of 150 °C, H_2 pressure of 5.0 MPa, stirring rate of 1400 r/min, 1.96 g of Ru–Mn(x) catalysts, 49.2 g of $ZnSO_4{\cdot}7H_2O$, 9.8 g of ZrO_2, 140 mL of benzene, and 280 mL of water

similar chemical compositions. This could be responsible for the similar performances of these catalysts. The results also suggested that SO_4^{2-}, NO_3^-, and Cl^- ions of the Mn precursors had little effect on the performances of these catalysts.

Figure 5.24 shows the benzene conversion, the cyclohexane selectivity, and yield of Ru–Mn(5.4%)/ZrO_2 catalyst in 8 recycles.

As can be seen, benzene conversion and cyclohexene selectivity were stable at above 60.6% and 77.5%, respectively, and the cyclohexene yields remained above 48.5% in the first four cycles. From the fifth cycle to the seventh cycle, the benzene

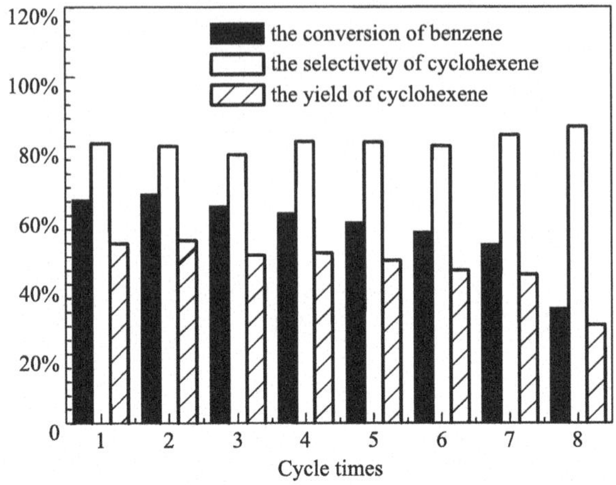

Fig. 5.24 Reusability of the Ru–Mn(5.4%)/ZrO_2 catalyst. Reaction conditions: reaction temperature of 150 °C, H_2 pressure of 5.0 MPa, stirring rate of 1400 r/min, 1.96 g of Ru–Mn(x) catalysts, 49.2 g of $ZnSO_4 \cdot 7H_2O$, 9.8 g of ZrO_2, 140 mL of benzene, and 280 mL of water

conversion gradually decreased. However, benzene conversion was still above 50%, and the cyclohexene selectivity and yields remained above 80% and 40%, respectively. This indicated that Ru–Mn(5.4%)/ZrO_2 catalysts had good stability and good prospects for industrial application. The benzene conversion declined drastically to 33.1% in the eighth cycle. However, the cyclohexene selectivity remained as high as 85.4%. The main reasons for the deactivation of the catalyst were the loss of catalyst and the absence of regeneration for long recycling times.

5.2.3 *Nano Ru–Fe Catalyst*

(1) Characterization of Ru–Fe catalyst

The XRD patterns of Ru–Fe catalyst before and after hydrogenation are displayed in Fig. 5.25.

From Fig. 5.25a, it can be seen that the diffraction peaks of hexagonal metallic Ru (JCPDS 01-070-0247) are detected at 2θ of 38.1°, 42.6°, 43.7°, 58.3°, 69.7°, 78.5°, and 85.6° in the Ru–Fe(x) catalyst before hydrogenation. It can be seen from Fig. 5.25b that the diffraction peaks of metallic Ru are also detected at 2θ of 38.7°, 42.6°, 44.5°, 59.1°, 70.1°, 78.5°, and 85.8° in the Ru–Fe(x) AH, indicating that metallic Ru is the active phase of the catalyst. In addition, the diffraction peaks of Fe_3O_4 (JCPDS 00-001-1111) are detected at 2θ of 18.3°, 35.7°, 42.6°, 53.4°, 57.4°, 71.4°, 79.1°, and 86.7° in the Ru–Fe(0.55) and Ru–Fe(0.66) catalysts, indicating that Fe mainly exists as Fe_3O_4 in the catalyst after reduction. As reported in the literature, Fe existed as Fe_3O_4 in the Ru–Fe/C catalyst prepared by chemical reduction method [22]. Moreover, the $2p_{3/2}$ binding energy of Fe in the Ru–Fe/C catalyst prepared by impregnation method and Fe_3O_4 were the same [44], and was also the same with the conclusion that Fe existed as Fe_3O_4 in the Ru–Fe/γ-Al_2O_3 catalyst

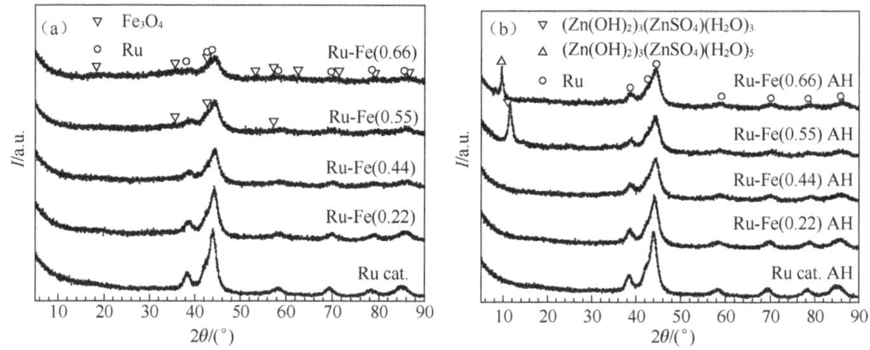

Fig. 5.25 XRD patterns of Ru–Fe catalyst before and after hydrogenation. **a** Ru–Fe(x) catalyst; **b** Ru–Fe(x) AH

prepared by the stepwise impregnation method [45]. When the Fe/Ru molar ratio was 0.22 and 0.44, no diffraction peaks of Fe_3O_4 were observed due to low content of Fe. However, the diffraction peaks of $(Zn(OH)_2)_3(ZnSO_4)(H_2O)_3$ (JCPDS 01-078-0247) and $(Zn(OH)_2)_3(ZnSO_4)(H_2O)_5$ (JCPDS 01-078-0246) were detected at 2θ of 11.0°, and 9.5°, respectively, in the Ru–Fe(0.55) AH and Ru–Fe(0.66) AH, but the characteristic peak of Fe_3O_4 disappeared, indicating that Fe_3O_4 in the catalyst reacted with $ZnSO_4$ in the slurry to produce an insoluble $(Zn(OH)_2)_3(ZnSO_4)(H_2O)_x$ ($x = 3$ or 5) salt, which was the same with the species on the Ru–Zn and Ru–Mn catalysts after hydrogenation [40, 43]. The diffraction peaks of the basic salt were not detected in the Ru–Fe(0.22) AH and Ru–Fe(0.44) AH, indicating that its content was lower than the monolayer dispersion capacity, which showed the same rule as that of Ru–Zn and Ru–Mn catalysts.

Table 5.10 shows the textural parameters of the Ru–Fe catalyst before and after hydrogenation.

Table 5.10 BET surface area S_{BET}, pore volume V_p, and average pore diameter d_p of Ru–Fe catalyst before and after hydrogenation

Sample	S_{BET}/(m^2/g)	V_p/(cm^3/g)	d_p/nm
Ru(0)	59	0.18	10.63
Ru–Fe(0.22)	68	0.16	9.44
Ru–Fe(0.44)	64	0.13	8.37
Ru–Fe(0.55)	69	0.16	9.50
Ru–Fe(0.66)	55	0.19	13.94
Ru(0) AH	56	0.16	10.4
Ru–Fe(0.22) AH	58	0.14	9.55
Ru–Fe(0.44) AH	53	0.16	11.67
Ru–Fe(0.55) AH	59	0.11	7.51
Ru–Fe(0.66) AH	43	0.17	15.50

Table 5.11 Fe/Ru, Zn/Ru and S/Ru atomic ratios of Ru–Fe(*x*) catalyst after hydrogenation and crystallite size before hydrogenation

Catalyst	Fe/Ru AH (mol/mol)	Zn/Ru AH (mol/mol)	S/Ru AH (mol/mol)	Ru crystalline size before hydrogenation/nm	Ru crystalline size after hydrogenation/nm
Ru(0)	0	0.0313	0.0026	4.7	4.5
Ru–Fe(0.22)	0.1662	0.0921	0.0056	4.2	4.5
Ru–Fe(0.44)	0.3322	0.2314	0.0140	3.6	3.6
Ru–Fe(0.55)	0.3677	0.3071	0.0356	3.4	3.0
Ru–Fe(0.66)	0.4467	0.3832	0.0419	4.1	3.0
Ru–Fe(0.55)[a]	0.5890	0	0	3.4	3.3

[a]Without $ZnSO_4·7H_2O$

From Table 5.10, it can be seen that the specific surface area, pore volume, and average pore diameter maintained almost unchanged when the Fe/Ru atomic ratio increased from 0 to 0.66. The specific surface area decreased from 55–68 m^2/g to 43–59 m^2/g, pore volume decreased from 0.13–0.19 cm^3/g to 0.11–0.17 cm^3/g, but the pore volume kept unchanged, indicating that the addition of Fe had little influence in the texture properties of Ru catalyst.

Table 5.11 lists the Fe/Ru, Zn/Ru and S/Ru atomic ratios of Ru–Fe(*x*) catalyst after hydrogenation, and the Ru crystallite sizes calculated by using full width at half maximum (FWHM) of the strongest diffraction peak in XRD patterns and Scherrer equation.

In Table 5.11, the Ru content of Ru–Fe(*x*) catalyst with various Fe contents was the same, so the change of Fe/Ru, Zn/Ru and S/Ru atomic ratios over Fe content reflected the variation in the composition of the catalyst. For the hydrogenation in slurry with $ZnSO_4$, the Ru(0) catalyst after hydrogenation contained the elements Zn and S. The Ru–Fe(0.55) catalyst was hydrogenated in the slurry without $ZnSO_4$, and after hydrogenation it did not contain Zn and S elements. The atomic ratio of Zn/Ru was far higher than that of S/Ru in the Ru–Fe(*x*) catalyst after hydrogenation, indicating that the Zn species chemisorbed on the surface was not only $ZnSO_4$, but also the basic salt of $ZnSO_4$. The crystallite size of Ru was centered on 3.4–4.7 nm, whereas it was 3.0–4.5 nm after hydrogenation, indicating that the Ru crystallite size decreased slightly after hydrogenation in the slurry with $ZnSO_4$, which was consistent with the XRD results.

In the Ru–Zn catalyst, Zn mainly existed as ZnO, under the conditions of pretreatment and hydrogenation; the ZnO rich on the surface of the catalyst reacted with $ZnSO_4$ in the slurry to form an insoluble salt and was chemisorbed in the catalyst. The Ru/Fe atomic ratio of Ru–Fe(*x*) catalyst was significantly decreased after hydrogenation, and the atomic ratios of Zn/Ru and S/Ru increased significantly with the increase of Fe content, suggesting that Fe_3O_4 in the Ru–Fe(*x*) catalyst partially reacted with $ZnSO_4$ in the slurry to form $(Zn(OH)_2)_3(ZnSO_4)(H_2O)_x$ ($x = 3$ or 5) salt, and iron ions entered the slurry.

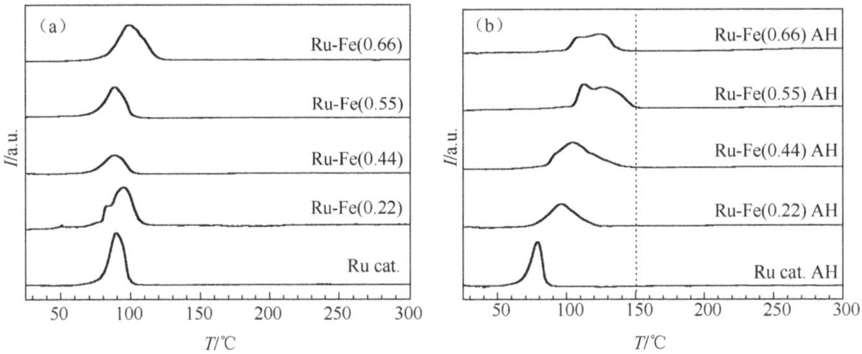

Fig. 5.26 TPR profiles of Ru–Fe(*x*) catalysts before and after hydrogenation. **a** Ru–Fe(*x*) catalyst; **b** Ru–Fe(*x*) AH

Figure 5.26 displays the TPR profiles of Ru–Fe(*x*) catalysts before and after hydrogenation.

As can be seen from Fig. 5.26a, only one H_2 consumption peak was detected between 50 and 100 °C for the Rn–Fe(*x*) catalyst, which was attributed to the reduction of RuO_x, indicating that Ru of the catalyst was partially oxidized. From Fig. 5.26b, it can be seen that with an increasing amount of Fe_3O_4, a shoulder also appeared in the temperature range of 50–100 °C for the Rn–Fe(*x*) AH, and the peak shifted to the high temperature. This was due to the formation of the basic salt in the sample after hydrogenation, which was then chemisorbed on the catalyst surface. The reduction process of RuO_x became complicated, and difficult to be reduced. However, the end temperature of reduction for all samples was less than the hydrogenation reaction temperature of 150 °C, indicating that Ru was in metallic state under reaction conditions, playing a role in providing active sites.

In summary, Fe was presented in the form of Fe_3O_4 in Ru–Fe(*x*) catalyst and was enriched on the surface. In the process of pretreatment and hydrogenation, Fe_3O_4 reacted with $ZnSO_4$ in the slurry to produce an insoluble basic salt, which was then chemisorbed on the surface of Ru, playing a key role in improving selectivity to cyclohexene.

(2) Performance of Ru–Fe catalyst

Figure 5.27 displays the change of benzene conversion C_{BZ} versus time t (C_{BZ}-t curve), and the change of selectivity to cyclohexane S_{HE} versus benzene conversion C_{BZ} (S_{HE}-C_{BZ} curve) over Ru–Fe(*x*) catalyst with different Fe contents.

From Fig. 5.27, it can be seen that with the increase of Fe content, the benzene conversion decreased at the same time, and the cyclohexene selectivity increased at the same benzene conversion. The Fe/Ru atomic ratio increased from 0.22 to 0.44, and the cyclohexene yield gradually increased to 56.7% and then decreased when Fe content continued to increase.

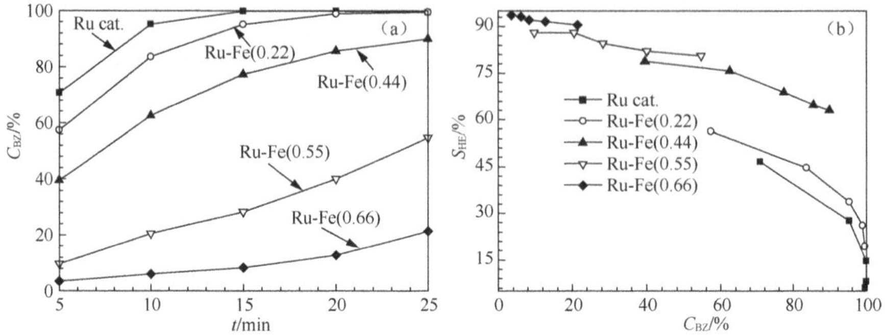

Fig. 5.27 **a** C_{BZ}-t curve; **b** S_{HE}-C_{BZ} curve of Ru–Fe(x) catalysts with different Fe contents. Reaction conditions: reaction temperature of 150 °C, H_2 pressure of 5.0 MPa, stirring rate of 1400 r/min, 1.96 g of Ru–Fe(x) catalysts, 49.2 g of $ZnSO_4 \cdot 7H_2O$, 9.8 g of ZrO_2, 140 mL of benzene, and 280 mL of water

Table 5.12 Benzene conversion C_{BZ}, cyclohexene selectivity S_{HE}, and pH values of aqueous solution at room temperature under the maximum yield Y_{HE} of Ru–Fe(x) catalyst

Catalyst	C_{BZ}/%	S_{HE}/%	Y_{HE}/%	t/min	pH
Ru cat.	70.7	46.7	33.0	5	5.53
Ru–Fe(0.22)	83.7	44.6	37.3	10	5.89
Ru–Fe(0.44)	89.9	63.1	56.7	25	5.90
Ru–Fe(0.55)	54.8	80.6	44.1	25	5.73
Ru–Fe(0.66)	21.5	90.6	19.5	25	5.55
Ru–Fe(0.55)[a]	100	0	0	5	6.74

[a]Without $ZnSO_4$ in the slurry of hydrogenation

Table 5.12 lists the benzene conversion, cyclohexene selectivity, and the pH values of aqueous solution at room temperature under the maximum yield of Ru–Fe(x) catalyst.

From Table 5.12, the cyclohexene selectivity and the maximum yield was 63.1% and 56.7% at the benzene conversion of 89.9% in 25 min for the Ru–Fe(0.44) catalyst, which was the best result as reported for the Ru–Fe based catalyst [46–51].

Compared with Ru–Zn and Ru–Mn catalysts, the Ru–Fe bimetallic catalyst showed high activity and cyclohexane selectivity, and Zn, Mn, and Fe were all good promoters. With increasing content of the promoter, the catalyst activity had the same change trend, and they had an optimal value. Although they existed in different forms in the catalyst, one thing in common was that in the reaction system with $ZnSO_4$, under hydrogenation conditions, the ZnO, Mn_3O_4, and Fe_3O_4 rich on the surface of the pretreated catalyst could react with $ZnSO_4$ in the slurry to form an insoluble basic complex salt, which was then chemisorbed on the catalyst surface, playing a vital role in improving the selectivity to cyclohexene. The difference is that ZnO reacted directly with $ZnSO_4$, whereas Mn_3O_4 and Fe_3O_4 were first hydrolyzed to

OH^-, and then reacted with $ZnSO_4$, which provided sufficient evidence to reveal the heterogeneous catalytic mechanism and nature of science of the high selectivity and yield of cyclohexene.

5.2.4 Nano Ru–La Catalyst

(1) Characterization of Ru–La catalyst

Figure 5.28 shows N_2 adsorption-desorption isotherms and pore size distribution curves of Ru–La(x) catalysts before and after hydrogenation.

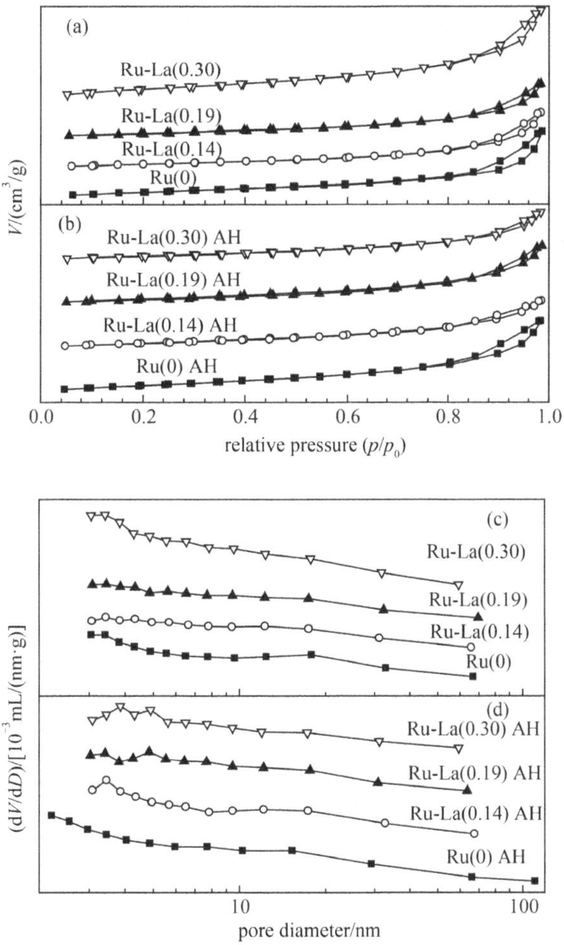

Fig. 5.28 Absorption-desorption isotherms (**a**, **b**) and pore size distribution curves (**c**, **d**) of Ru–La(x) catalysts before and after hydrogenation. **a** Ru–La(x); **b** Ru–La(x) AH; **c** Ru–La(x); **d** Ru–La(x) AH

Figure 5.28a, b shows that the adsorption isotherm of Ru–La catalyst before and after hydrogen is close to type-V with H1 hysteresis loops, according to IUPAC classification. The adsorption isotherm reflects the surface properties and pore distribution of the solid, and the interaction between the adsorbate and the solid surface, while the hysteresis loop reflects the properties of the pore structure of the catalyst. The adsorption isotherms of both type-IV and type-V belong to the mesoporous structure, but type-IV reflects the strong interaction between the adsorbate and adsorbent, whereas the type-V reflects the weak interaction between the adsorbate and adsorbent. The H1 hysteresis loop suggests that the pore structure of the catalyst is mainly composed of the pores accumulated by the spherical particles with uniform size. An obvious feature for the adsorption isotherm of Ru–La catalyst is that the adsorption curves and desorption curves are completely the same at the relative pressure p/p^0 below 0.8, indicating that the adsorption and desorption is completely reversible, corresponding to the monolayer adsorption, for which adsorption and desorption occurred on the outer surface of catalyst. The relative pressure at the separate region in the adsorption curves and desorption curves is more than 0.8, indicating a catalyst with a bigger pore diameter. Figure 5.28c, d shows that most of the pores in the Ru–La catalyst before and after hydrogenation are mainly distributed in the range of 2–50 nm, which is favorable for desorption of cyclohexene.

Table 5.13 shows the textural parameters of the Ru–La(x) catalyst before and after hydrogenation.

From Table 5.13, it can be seen that the Ru–La(x) catalyst before hydrogenation had a BET surface area of 52–59 m^2/g, a pore volume of 0.15–0.18 cm^3/g, an average pore diameter of 10.6–11.0 nm, and the Ru crystallite size was 3.6–4.7 nm. The Ru–La(x) catalyst after hydrogenation had a specific surface area of 43–56 m^2/g, pore volume of 0.09–0.16 cm^3/g, average pore diameter of 8.7–10.4 nm, and the Ru crystallite size was 3.8–4.5 nm. Considering the measurement error, the promoter La has no significant effect on the textural properties of Ru–La catalyst, indicating that Ru and La show good compatibility due to similar physical and chemical properties. But the general trend is that the specific surface area, pore volume, average pore size,

Table 5.13 BET surface area S_{BET}, pore volume V_p, average pore diameter d_p, and Ru crystallite size d of Ru–La catalyst before and after hydrogenation

Sample	$S_{BET}/(m^2/g)$	$V_p/(cm^3/g)$	d_p/nm	d/nm
Ru(0)	59	0.18	10.63	4.7
Ru–La(0.14)	52	0.15	11.67	4.5
Ru–La(0.19)	53	0.15	10.96	4.7
Ru–La(0.30)	57	0.17	10.89	3.6
Ru(0) AH	56	0.16	10.44	4.5
Ru–La(0.14) AH	55	0.14	9.13	4.4
Ru–La(0.19) AH	52	0.16	9.76	4.7
Ru–La(0.30) AH	43	0.09	8.72	3.8

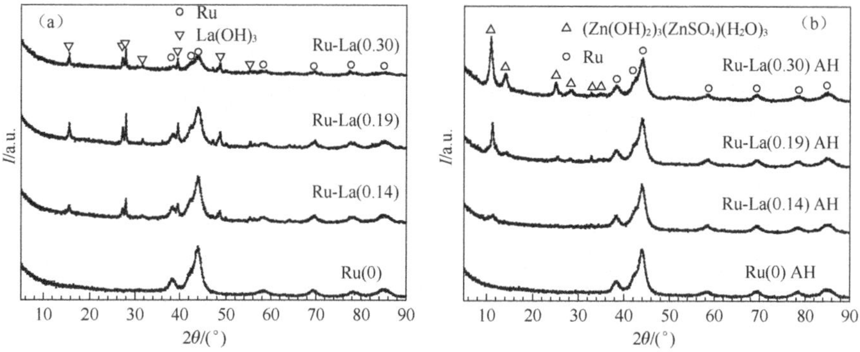

Fig. 5.29 XRD patterns of the Ru–La catalysts before and after hydrogenation. **a** Ru–La(*x*) catalyst; **b** Ru–La(*x*) AH

and crystallite size are reduced after hydrogenation, and this is closely related to the phenomenon that an insoluble basic complex salt is formed and then chemisorbed on the catalyst surface during the pretreatment and hydrogenation process.

Figure 5.29 shows XRD patterns of the Ru–La catalysts before and after hydrogenation.

Figure 5.29a shows that all the Ru–La(*x*) catalysts before hydrogenation showed the diffraction peaks of metallic Ru (JCPDS 01-070-0274), indicating that after hydrogenation Ru in the catalyst mainly existed as metallic Ru, forming active phase of the catalyst, providing the active sites for benzene hydrogenation. Besides, the diffraction peaks of $La(OH)_3$ (JCPDS 00-006-0585) were presented in the XRD patterns of Ru–La(*x*) catalyst, indicating that La mainly existed as $La(OH)_3$. Moreover, the intensity of the diffraction peaks of $La(OH)_3$ increased with the increase of La/Ru ratio, indicating the increment of $La(OH)_3$ contents.

Figure 5.29b shows that the diffraction peaks of metallic Ru were detected in the XRD patterns of Ru–La(*x*) AH, indicating that Ru^0 was an active center of the catalyst under hydrogenation conditions. At the same time the diffraction peak of $La(OH)_3$ disappeared, while the diffraction peaks of $(Zn(OH)_2)_3(ZnSO_4)(H_2O)_3$ were clearly observed, and the diffraction peak at 11.2° of this salt became stronger with the increase of La/Ru ratio, indicating that $La(OH)_3$ on the catalyst surface reacted with $ZnSO_4$ in the slurry to form an insoluble complex salt according to the following reaction equation:

$$2La(OH)_3 + 4\,ZnSO_4 + 3H_2O \rightarrow (Zn(OH)_2)_3(ZnSO_4)(H_2O)_3 \downarrow + La_2(SO_4)_3 \quad (5.2.1)$$

The Ru crystallite size of the Ru–La(*x*) and Ru–La(*x*) AH calculated by using Scherrer equation are given in the last column of Table 5.13. When the La/Ru atomic ratio was less than 0.30, the Ru crystallite sizes of the Ru–La(*x*) catalyst before and after hydrogenation were distributed in the narrow range of 4.4–4.7 nm, and showed small change. When the La/Ru atomic ratio was 0.30, the Ru crystallite size of the

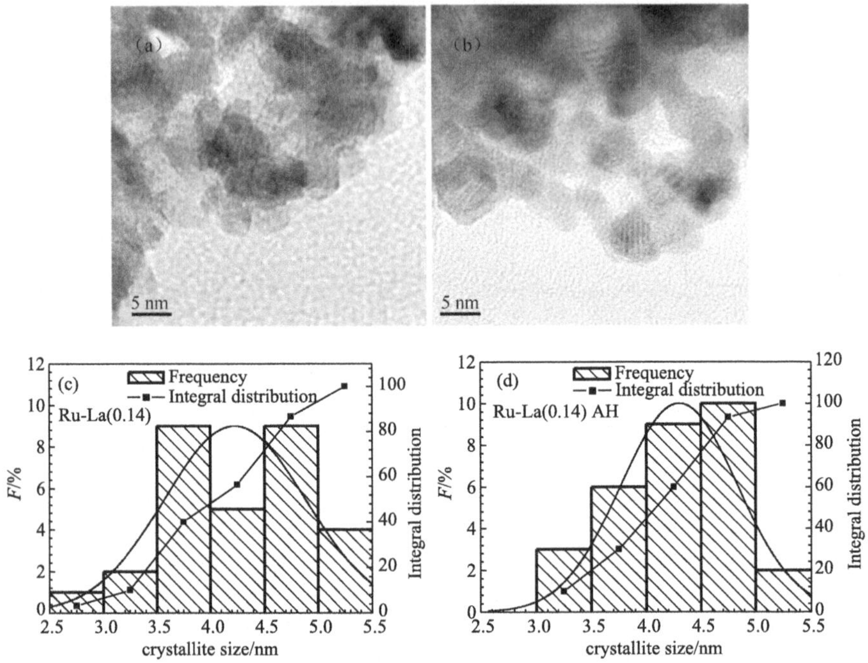

Fig. 5.30 TEM images (**a, b**) and Ru crystallite size distributions (**c, d**) of the Ru–La(0.14) catalysts before and after hydrogenation. **a, c** Ru–La(0.14) catalyst; **b, d** Ru–La(0.14) AH

Ru–La(0.30) catalyst before and after hydrogenation was significantly reduced to 3.6–3.8 nm, indicating that the promoter La played a role in the dispersion of Ru crystallite.

Figure 5.30 displays the TEM images and Ru crystallite size distributions of the Ru–La(0.14) catalysts before and after hydrogenation.

Figure 5.30a, b shows that both Ru–La(0.14) catalyst and Ru–La(0.14) AH consisted of spherical and ellipsoidal crystallites. Figure 5.30c, d shows that the Ru crystallite sizes of Ru–La(0.14) catalyst and Ru–La(0.14) AH were mainly distributed around 4.2 nm, which was consistent with the XRD results, whereas the $(Zn(OH)_2)_3(ZnSO_4)(H_2O)_3$ salt was not observed in the TEM images of Ru–La(x) AH, suggesting that the coverage of the salt was lower than the monolayer dispersion capacity, but the lattice fringes of the (101) plane of the metallic Ru with an average spacing of 0.21 nm was detectable on the images of Ru–La(x) catalyst and Ru–La(x) AH. This might be due to the uniform dispersion of $La(OH)_3$ on Ru–La(x) catalyst and uniform dispersion of $(Zn(OH)_2)_3(ZnSO_4)(H_2O)_3$ salt on Ru–La(x) AH.

Figure 5.31 shows the AES Zn LMM spectra of Ru–La(0.14) AH at different Ar^+ sputtering time.

As can be seen from Fig. 5.31, the Auger kinetic energies of Zn LMM of Ru–La(0.14) AH at Ar^+ sputtering time of 0 min, 4.5 min, 9 min were 983.7 eV, 983.7 eV, and 983.6 eV, respectively, which was close to the reported value of Zn-LMM in ZnO,

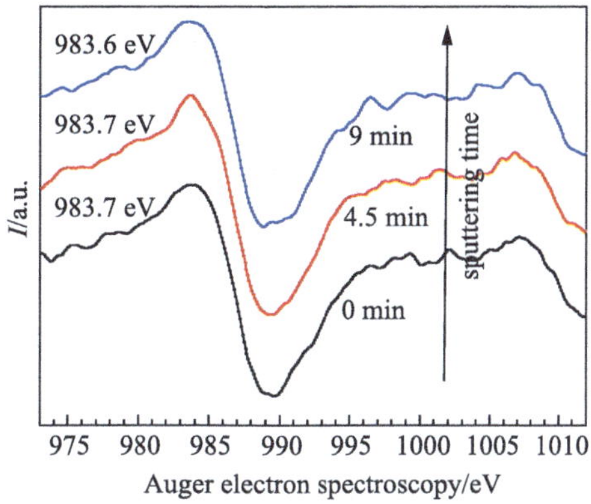

Fig. 5.31 AES Zn LMM spectra of Ru–La(0.14) AH

indicating that Zn was present on the catalyst surface mainly as oxidized Zn even under the reaction conditions of 150 °C and 5 MPa H_2, which was in accordance with the XRD results that Zn on the surface of Ru–La(0.14) AH was present in the form of $(Zn(OH)_2)_3(ZnSO_4)(H_2O)_3$ salt. Unfortunately, AES measurements did not allow discerning any additional contribution from metallic Zn (commonly appearing in the 991–995 eV range).

The compositions of Ru–La(x) AH and pH values of liquid phase at room temperature after hydrogenation are listed in Table 5.14.

As can be seen from Table 5.14, the pH value of the slurry containing $ZnSO_4$ at room temperature after hydrogenation was around 6; at 150 °C due to the increased degree of hydrolysis of $ZnSO_4$, the pH of the slurry would be lower; it was difficult for metallic Zn to exist in acid solutions, and this was consistent with the lack of evidence from the AES results indicating the presence of metallic Zn on the Ru catalyst. After hydrogenation, only trace amounts of La were detected on Ru–La(*x*) AH, indicating that $La(OH)_3$ enriched on the catalyst surface almost completely reacted with $ZnSO_4$ in the slurry to produce a basic salt and La enters the slurry in the form of ions. With

Table 5.14 Compositions of Ru–La(*x*) AH and pH values of liquid phase at room temperature after hydrogenation

Catalyst	La/Ru AH[c]	Zn/Ru AH[c]	S/Ru AH[c]	pH[d]
Ru(0)[a]	0	0.0313	0.0026	5.53
Ru–La(0.14)[a]	0.0017	0.2542	0.0163	6.18
Ru–La(0.19)[a]	0.0018	0.3885	0.0381	6.35
Ru–La(0.30)[a]	0.0032	0.5789	0.0673	6.37
Ru–La(0.14)[b]	0.1427	0	0	6.99

[a]In the presence of 0.6 mol/L $ZnSO_4$
[b]In the absence of 0.6 mol/L $ZnSO_4$
[c]Determined by XRF
[d]Measured by pH meter

the increase in $La(OH)_3$, the atomic ratios of Zn/Ru and S/Ru increased, indicating that the amount of basic complex salt increased, which was consistent with the XRD results.

(2) Activity and selectivity of Ru–La catalyst

Figure 5.32 shows the change of benzene conversion C_{BZ} versus time t (C_{BZ}-t curve), the change of cyclohexane selectivity S_{HE} versus benzene conversion C_{BZ} (S_{HE}-C_{BZ} curve), and the change of cyclohexane yield Y_{HE} versus time t (Y_{HE}-t curve) over Ru–La(x) catalyst.

Figure 5.32a, b shows that the activity decreased and the cyclohexane selectivity of the catalysts increased with the molar ratio of La/Ru in the presence of $ZnSO_4$. Figure 5.32c shows that the cyclohexene yield was increased with the increment of the molar ratio of La/Ru, and then declined at higher doping amounts of La, with the optimum La/Ru molar ratio being 0.14. Ru–La(0.14) catalyst gave the highest cyclohexene yield of 59.5% in the presence of $ZnSO_4$. However, Ru–La(0.14) catalyst gave the poorest selectivity of 2.3% and the weakest yield of 1.9% in the absence of $ZnSO_4$, suggesting that the promoter $La(OH)_3$ alone could not enhance the selectivity to cyclohexene and the cyclohexene yield of Ru catalyst. This indicated that

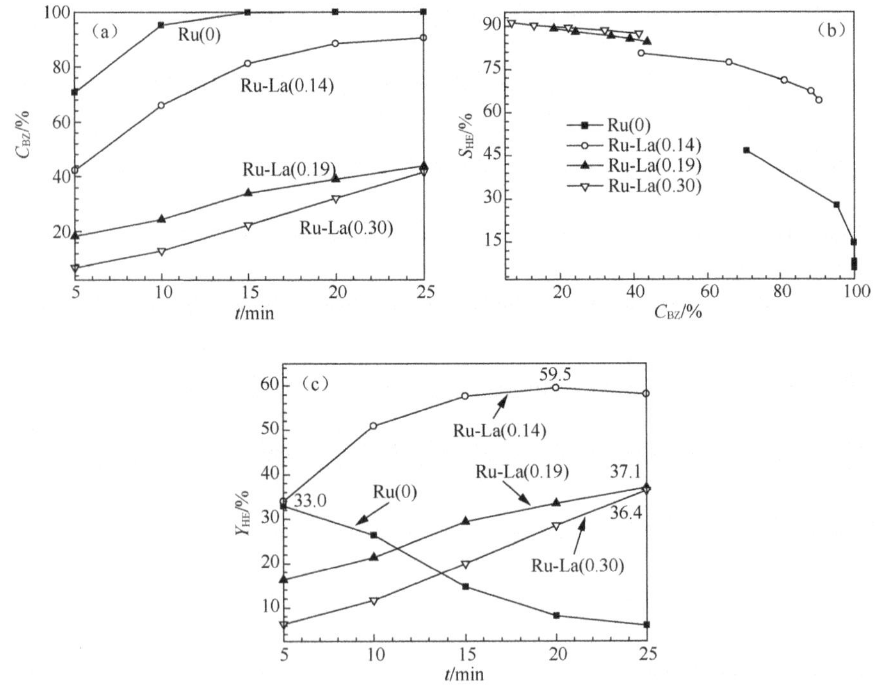

Fig. 5.32 **a** C_{BZ}-t curve; **b** S_{HE}-C_{BZ} curve; **c** Y_{HE}-t curve of Ru–Fe(x) catalyst. Reaction conditions: reaction temperature of 150 °C, H_2 pressure of 5 MPa, stirring rate of 1400 r/min, 1.96 g of Ru–La(x) catalysts, 140 mL of benzene, 280 mL of water, and 49.2 g of $ZnSO_4 \cdot 7H_2O$

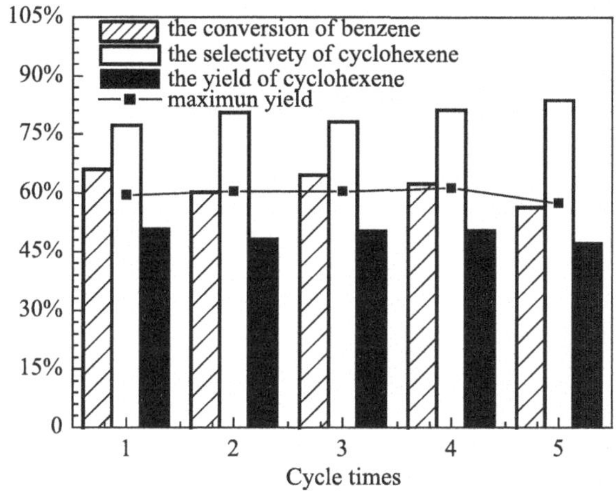

Fig. 5.33 Benzene conversion, cyclohexene selectivity and cyclohexene yield at 10 min as well as the maximum cyclohexene yield in 25 min over Ru–La(0.14) catalyst in five reuse times. Reaction conditions: a share of Ru–La(x) catalysts, 49.2 g of $ZnSO_4{\cdot}7H_2O$, 280 mL of H_2O, 140 mL of benzene, reaction temperature of 150 °C, H_2 pressure of 5.0 MPa, and stirring rate of 1400 r/min

the chemisorbed $(Zn(OH)_2)_3(ZnSO_4)(H_2O)_3$ salt formed by $La(OH)_3$ reacting with $ZnSO_4$ played a key role in improving the selectivity to cyclohexene of the Ru catalyst.

Figure 5.33 displays the benzene conversion, cyclohexene selectivity, and cyclohexene yield at 10 min as well as the maximum cyclohexene yield in 25 min over Ru–La(0.14) catalyst in five reuse times.

From Fig. 5.33, it can be seen that the benzene conversion was stable above 60%, and the selectivity of cyclohexene and cyclohexene yield were kept above 77% and 48%, respectively, in the first 4 reuse times. Moreover, the maximum cyclohexene yields stabilized above 59%. Although the benzene conversion slightly decreased to 56.5% in the 5th reuse time, the selectivity of cyclohexene and maximum cyclohexene yield were as high as 83.7 and 57.5%. All of these suggested that Ru–La catalyst had a good reusability.

5.2.5 *Nano Ru–Ce Catalyst*

(1) Characterization of Ru–Ce catalyst

Figure 5.34 shows the TEM images and crystallite size distribution of Ru–Ce(III, 0.19) catalyst. It is found that the catalyst particles consist of polycrystalline conglomerates of spherical and ellipsoidal crystallites with the crystallite size being in the range of 3.1–4.5 nm and are mainly distributed around 3.8 nm.

Figure 5.35 shows the XRD patterns of Ru–Ce catalysts before and after hydrogenation.

As can be seen in Fig. 5.35a, the XRD patterns of the catalyst before hydrogenation display the diffraction peaks of metallic Ru (JCPDS 01-070-0274), indicating that the Ru in the catalysts are mainly present in metallic Ru. From Table 5.16, it is concluded

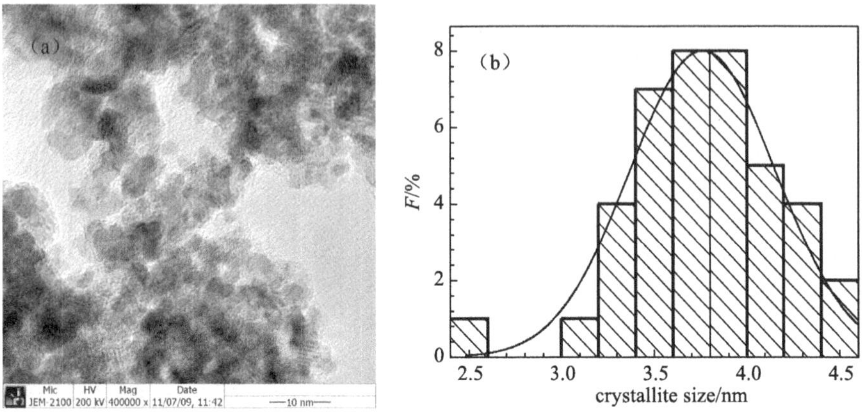

Fig. 5.34 TEM images (**a**) and crystallite size distribution (**b**) of Ru–Ce(III, 0.19) catalyst

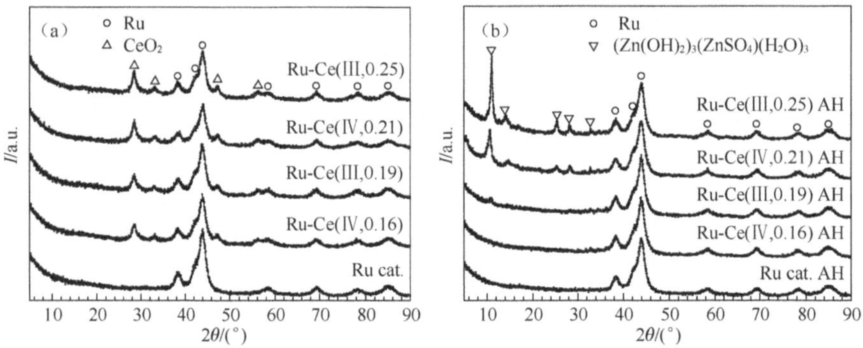

Fig. 5.35 XRD patterns of Ru–Ce catalysts before and after hydrogenation. **a** Ru–Ce(*x*) catalyst; **b** Ru–Ce(*x*) AH

that the Ru crystallite sizes calculated by Scherrer equation are in the range of 4.4–5.3 nm, which are similar to those derived from the TEM result. Besides, the XRD patterns of the catalysts prepared using the Ce precursors with different valences all show the diffraction peaks of CeO_2 (JCPDS: 00-004-0593) without other Ce species, indicating that the Ce species in all catalysts prepared using the Ce precursors with different valences mainly exist as CeO_2. This implies that the valences of the Ce precursors have little effect on the valences of the Ce species in the catalysts, since it is easy for the oxidization of Ce(III) to Ce(IV) in the precipitation process.

Figure 5.35b shows the XRD patterns of different catalysts after hydrogenation. It is interesting to find that the diffraction peaks of CeO_2 disappear and the diffraction peaks of the $(Zn(OH)_2)_3(ZnSO_4)(H_2O)_3$ salt (JCPDS: 01-078-0247) are observed on the XRD patterns of Ru–Ce(III, 0.19), Ru–Ce(IV, 0.21), and Ru–Ce(III, 0.25) catalysts. All these indicate that the promoter CeO_2 on the surface of the catalyst has reacted with the reaction modifier $ZnSO_4$ and H_2O to form an insoluble

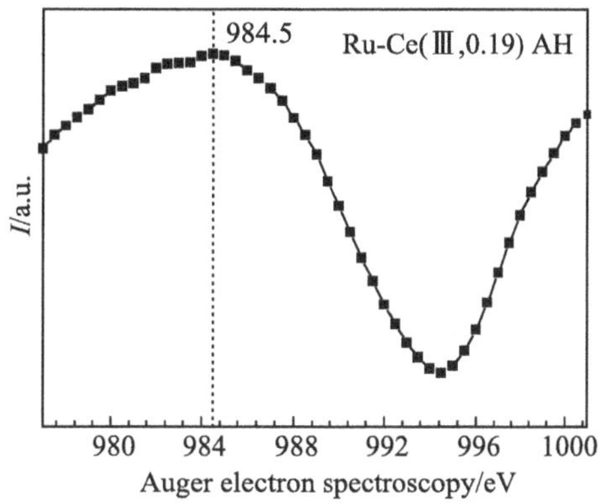

Fig. 5.36 AES Zn LMM spectrum of Ru–Ce(III, 0.19) AH

$(Zn(OH)_2)_3(ZnSO_4)(H_2O)_3$ salt, which is shown in Reaction (5.2.2). The intensity of the diffraction peak at $2\theta = 11.2°$ of this salt increases with the CeO_2 loading, indicating the increase of the amount of the $(Zn(OH)_2)_3(ZnSO_4)(H_2O)_3$ salt formed. However, the diffraction peaks of the $(Zn(OH)_2)_3(ZnSO_4)(H_2O)_3$ salt are not found on the XRD patterns of Ru–Ce(IV, 0.16) catalyst due to its low amount confirmed by the XRF results below. The crystallite sizes of different catalysts after hydrogenation are in the range of 4.6–5.3 nm, similar to those of the corresponding catalysts. This suggests that serious agglomeration of the catalysts does not happen in the hydrogenation processes.

$$2CeO_2 + 8ZnSO_4 + 12H_2O \rightarrow 2(Zn(OH)_2)_3(ZnSO_4)(H_2O)_3 \downarrow + 3Ce(SO_4)_2 \tag{5.2.2}$$

Figure 5.36 shows the AES Zn LMM spectrum of Ru–Ce(III, 0.19) AH. It is worth noting here that the spectrum is recorded after Ar^+ sputtering for 1 min to avoid the interruptions of the surface oxidation of catalysts. As can be seen, the kinetic energy (KE) of Zn LMM of Ru–Ce(III, 0.19) AH is 984.5 eV, which is close to the Zn(II) in ZnO as reported in the literature [14, 28], and this is consistent with the XRD results that the Zn species are mainly present as $(Zn(OH)_2)_3(ZnSO_4)(H_2O)_3$ salt in the Ru–Ce(III, 0.19) AH.

Table 5.15 shows the compositions of different catalysts and pH values of aqueous phase after hydrogenation.

It should be noted that different catalysts have the same Ru contents. Thus the molar ratios of Ce/Ru, Zn/Ru and S/Ru reflect the compositions of the catalysts after hydrogenation. The Ru catalyst after hydrogenation in the absence of $ZnSO_4$ only contains Ru element. However, the Zn and S elements are found on Ru catalyst after hydrogenation in the presence of $ZnSO_4$. Moreover, the molar ratio of Zn/Ru and S/Ru in the Ru–Ce(IV, 0.16) catalyst after hydrogenation is very low, indicating the trace amount of the $(Zn(OH)_2)_3(ZnSO_4)(H_2O)_3$ salt. Therefore, the XRD pattern of Ru catalyst after hydrogenation does not show the diffraction peaks of the $(Zn(OH)_2)_3(ZnSO_4)(H_2O)_3$ salt due to its low amount. With the increase of

Table 5.15 Compositions of different catalysts and pH values of aqueous phase after hydrogenation[a]

Catalyst	Ce/Ru AH	Zn/Ru AH	S/Ru AH	pH
Ru cat.[b]	0	0	0	7.02
Ru cat.	0	0.0313	0.0026	5.53
Ru–Ce(IV, 0.16)	0.0017	0.2561	0.0187	6.27
Ru–Ce(III, 0.19)	0.0025	0.2908	0.0246	6.27
Ru–Ce(III, 0.21)	0.0030	0.3346	0.0275	6.32
Ru–Ce(III, 0.25)	0.0027	0.4206	0.0396	6.20

[a]Reaction conditions: reaction temperature of 150 °C, H_2 pressure of 5.0 MPa and stirring rate of 1400 r/min, 1.96 g of catalysts, 49.2 g of $ZnSO_4 \cdot 7H_2O$, 280 mL of H_2O, 140 mL of benzene
[b]In the absence of $ZnSO_4 \cdot 7H_2O$

the CeO_2 loadings, the diffraction peaks of the $(Zn(OH)_2)_3(ZnSO_4)(H_2O)_3$ salt are detected in the XRD patterns of Ru–Ce(IV, 0.21), Ru–Ce(III, 0.16) catalysts, and Ru catalyst after hydrogenation. However, only trace amount of the Ce is detected in all the catalysts after hydrogenation, indicating that almost all CeO_2 enter the slurry in the form of Ce $(SO_4)_2$ in the hydrogenation process, as shown in Eq. (5.2.2). The molar ratios of Zn/Ru and S/Ru after hydrogenation increase with the CeO_2 loadings of the catalysts, indicating the increased amount of $(Zn(OH)_2)_3(ZnSO_4)(H_2O)_3$ salt.

Table 5.16 shows the texture characters of the catalyst before and after hydrogenation.

It is found that the BET surface areas, pore volumes, and pore diameters of the catalysts change little with the increase of the CeO_2 loadings. This suggests that the addition of the CeO_2 has little effect on the texture properties of the catalysts. However, the BET surface areas, pore volumes, and pore diameters after

Table 5.16 BET surface area S_{BET}, pore volume V_p, average pore diameter d_p, and Ru crystallite size d of Ru–Ce catalyst before and after hydrogenation

Sample	$S_{BET}/(m^2/g)$	$V_p/(cm^3/g)$	d_p/nm	d/nm[a]
Ru cat.	69	0.18	9.63	4.7
Ru–Ce(IV, 0.16)	75	0.18	9.12	4.4
Ru–Ce(III, 0.19)	78	0.18	8.53	4.8
Ru–Ce(IV, 0.21)	74	0.19	9.16	4.9
Ru–Ce(III, 0.25)	72	0.17	8.45	5.3
Ru cat. AH	73	0.18	8.44	4.6
Ru–Ce(IV, 0.16) AH	65	0.18	7.85	4.8
Ru–Ce(III, 0.19) AH	58	0.14	7.83	4.9
Ru–Ce(IV, 0.21) AH	56	0.13	7.73	4.8
Ru–Ce(III, 0.25) AH	51	0.10	7.45	5.3

[a]Measured by XRD

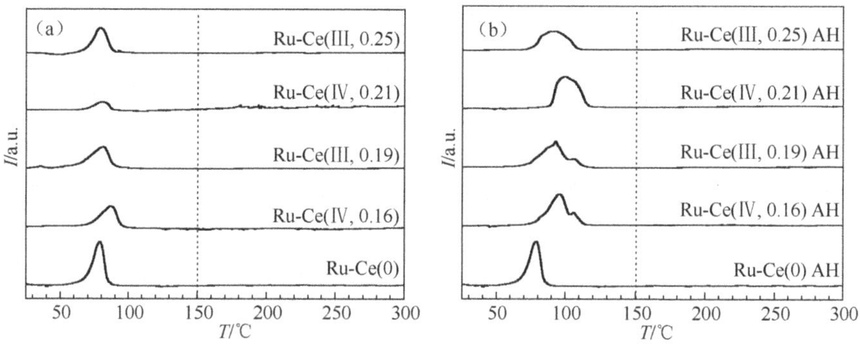

Fig. 5.37 H_2-TPR spectra of catalysts before and after hydrogenation. **a** Ru–Ce(x) catalyst; **b** Ru–Ce(x) AH

hydrogenation decrease with the CeO_2 loadings of the catalysts. This suggests that the $(Zn(OH)_2)_3(ZnSO_4)(H_2O)_3$ salt covers some of the Ru surface and enters the channels of the pores.

Each TPR profile in Fig. 5.37 for the catalysts shows only one peak shoulder between 50 and 100 °C, ascribable to the reduction of RuO_2. The absence of any additional peaks indicates that the promoter CeO_2 could not be reduced within 300 °C. The required temperature of CeO_2 reduction was above 630 °C [33]. TPR profiles for the catalysts after hydrogenation also show peak shoulders between 50 and 100 °C, also attributable to the reduction of RuO_2. Specially, the temperatures of complete reduction for all the catalysts after hydrogenation are much lower than the hydrogenation temperature of 150 °C, indicating only the existence of metallic Ru under the hydrogenation conditions of 150 °C and 5 MPa H_2.

In a word, the TPR results have confirmed that the Ru species mainly exist as metallic Ru in the hydrogenation process. The XRD results, the AES results, and the XRF results have confirmed that the promoter CeO_2 could react with the reaction modifier $ZnSO_4$ to form an insoluble $(Zn(OH)_2)_3(ZnSO_4)(H_2O)_3$ salt.

(2) Activity and selectivity of Ru–Ce catalyst

Figure 5.38 shows the change of benzene conversion C_{BZ} versus time t (C_{BZ}-t curve), the change of cyclohexene selectivity S_{HE} versus benzene conversion C_{BZ} (S_{HE}-C_{BZ} curve), and the concentrations of benzene, cyclohexene, and cyclohexane C versus time t (C-t curve) over Ru–Ce catalyst.

Figure 5.38a, b shows that benzene conversion decreases and cyclohexene selectivity increases with the CeO_2 loading of the catalysts in the presence of $ZnSO_4$. From Fig. 5.38a–c, it can be seen that Ru–Ce(III, 0.19) catalyst gave a cyclohexene selectivity of 63.8% and a maximum cyclohexane yield of 57.4% at the benzene conversion of 90.0% at 20 min, which was among the best results reported so far. However, the benzene conversion reached 100% at 5 min and cyclohexene was barely detectable in the absence of $ZnSO_4$, suggesting that the promoter CeO_2 alone could

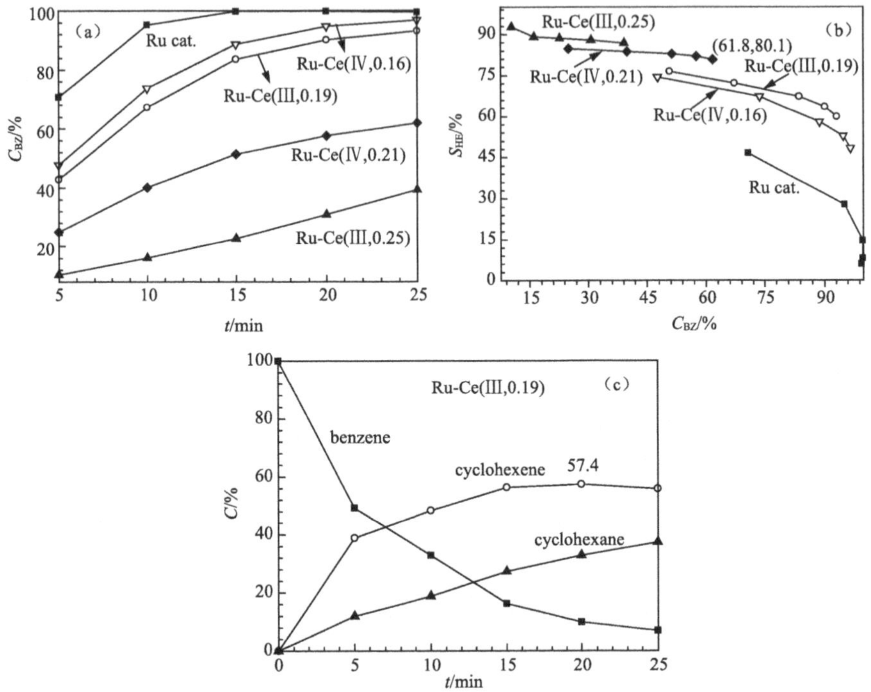

Fig. 5.38 Benzene conversion (**a**) and cyclohexene selectivity over different catalysts (**b**), as well as reaction course of benzene hydrogenation (**c**) over Ru–Ce(III, 0.19) catalyst. Reaction conditions: reaction temperature of 150 °C, H_2 pressure of 5.0 MPa and stirring rate of 1400 r/min, 1.96 g of catalyst, 49.2 g of $ZnSO_4{\cdot}7H_2O$, 280 mL of H_2O, 140 mL of benzene

not improve the selectivity to cyclohexene. This indicated that the selectivity of cyclohexene of the Ru catalyst might be closely related to the $(Zn(OH)_2)_3(ZnSO_4)(H_2O)_3$ salt chemisorbed on Ru surface.

Figure 5.39 shows the relationships between the Zn/Ru or S/Ru molar ratios and benzene conversion as well as cyclohexene selectivity at 5 min. It was found that the benzene conversion monotonically decreased and the selectivity to cyclohexene monotonically increased with the molar ratios of Zn/Ru and S/Ru. This indicated that the $(Zn(OH)_2)_3(ZnSO_4)(H_2O)_3$ salt chemisorbed on the Ru surface directly improved the selectivity to cyclohexene of Ru catalyst.

Figure 5.40 shows the reusability of the Ru–Ce(III, 0.19) catalyst. As can be seen, although the benzene conversion slightly decreased with cycle times, the selectivity of cyclohexene increased with cycle times and the cyclohexene yields were kept above 57% in the five cycles. This indicated that the Ru–Ce(III, 0.19) catalyst had good stability and good prospects for industrial application.

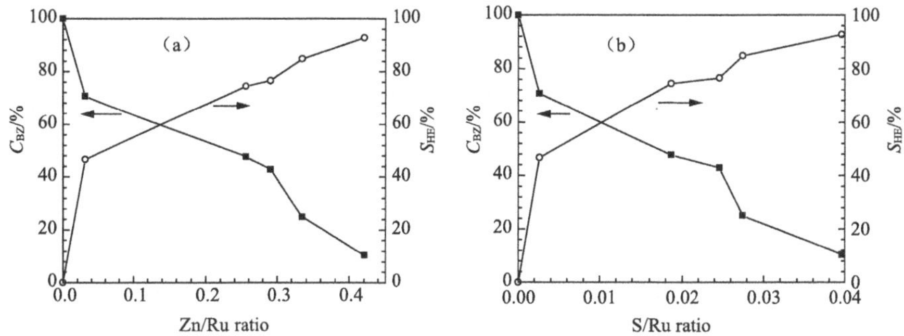

Fig. 5.39 Relationships between the Zn/Ru or S/Ru molar ratios and benzene conversion as well as cyclohexene selectivity at 5 min. Reaction conditions: reaction temperature of 150 °C, H_2 pressure of 5.0 MPa and stirring rate of 1400 r/min, 1.96 g of catalyst, 49.2 g of $ZnSO_4 \cdot 7H_2O$, 280 mL of H_2O, and 140 mL of benzene

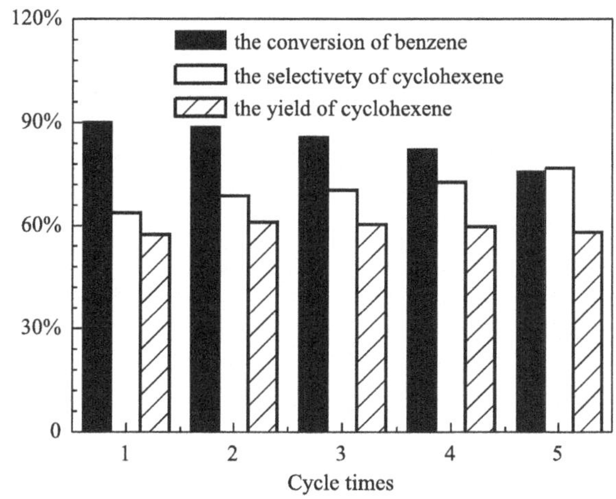

Fig. 5.40 Reusability of the Ru–Ce(III, 0.19) catalyst. Reaction conditions: reaction temperature of 150 °C, H_2 pressure of 5.0 MPa and stirring rate of 1400 r/min, 1.96 g of catalyst, 49.2 g of $ZnSO_4 \cdot 7H_2O$, 280 mL of H_2O, and 140 mL of benzene

5.3 The Main Technical Indicators of the Third-Generation Catalysts

The main technical indicators of the third generation of foreign catalysts for selective hydrogenation of benzene: ① activity index: the minimum γ_{50} is not less than 55; ② cyclohexene selectivity S_{HE}: the minimum value is not less than 75% at the benzene conversion of 50%; ③ Ru crystallite size d_p: the maximum value is not more than 70 Å; ④ Zn content (mass fraction): 5%-15%; ⑤ Fe content: the maximum value does not exceed 1000 ppm.

The determination method of γ_{50} and cyclohexene selective S_{HE} are as follows:

The selective hydrogenation of benzene was performed in a 1 L Hastelloy autoclave. The reaction conditions were as follows: reaction temperature of 150 °C, H_2

pressure of 5.0 MPa and stirring rate of 1400 r/min, 1.96 g of catalyst, 49.2 g of $ZnSO_4{\cdot}7H_2O$, 9.8 g of ZrO_2, 140 mL of benzene, 280 mL of H_2O, and pretreatment time of 22 h. The reaction time was started to be recorded after the addition of benzene, and a small amount of reaction mixture was sampled every 5 min and sent for gas chromatographic analysis with an FID detector, and the benzene conversion and cyclohexene selectivity were calculated. Then, the curves of benzene conversion C_{BZ} (mole fraction, following are the same) versus reaction time t and cyclohexene selectivity S_{HE} (mole fraction, following are the same) versus benzene conversion C_{BZ} were plotted, deriving the C_{BZ}-t curve and S_{HE}-C_{BZ} curve. The reaction time t_{50} can be calculated from C_{BZ}-t curve by using interpolation method, and then the corresponding cyclohexene selectivity S_{HE} under the benzene conversion can be calculated from S_{HE}-C_{BZ} curve. The activity index γ_{50} is then calculated using the following equation:

$$\gamma_{50} = \frac{V_{BZ}(\text{mL}) \times \rho_{BZ} \times C_{BZ}}{t_{50}(\text{h}) \times W_{Cat}(\text{g})} \tag{5.3.1}$$

where V_{BZ} is the number of milliliters of benzene (140 mL), ρ_{BZ} is the density of benzene (0.88 g/cm^3), C_{BZ} is the benzene conversion (0.5), t_{50} (min) is the time when benzene conversion reaching 50%, W_{Cat} (g) is the weight of the catalyst. Substituting the benzene conversion C_{BZ} (0.5) and the time t_{50} (min) reaching the corresponding conversion, γ_{50} can be obtained in units of h^{-1}, representing the grams of benzene converted with per gram of the catalyst per hour at benzene conversion of 50%.

With reference to the above method, the reaction times t_{40}, t_{50}, and t_{60} corresponding to the benzene conversion of 40, 50, and 60% were derived from C_{BZ}-t curve by using the interpolation method. The cyclohexene selectivity S_{40}, S_{50}, and S_{60} corresponding to the benzene conversion of 40, 50, and 60% were derived from S_{HE}-C_{BZ} curve, and the activity index γ_{40}, γ_{50}, and γ_{60} were calculated according to the Eq. (5.3.1).

Tables 5.17, 5.18, 5.19, 5.20, 5.21, 5.22, 5.23 and 5.24 list the test data of activity and selectivity of industrial Ru–Zn catalyst given by the enterprise during February

Table 5.17 Benzene conversion and cyclohexene selectivity of the first batch of Ru–Zn catalyst (2013-02-05, 5.6 kg)

t/min	C_{BZ}/%	S_{HE}/%	Y_{HE}/%
5	11.50	88.98	10.23
15	37.34	84.92	32.05
30	65.76	78.41	51.56
45	81.71	72.41	59.17
60	89.93	65.74	59.12

Reaction conditions: reaction temperature of 150 °C, H_2 pressure of 5.0 MPa and stirring rate of 1400 r/min, 1.96 g of catalyst, 140 mL of benzene, 280 mL of H_2O, 9.8 g of ZrO_2, 45.7 g of $ZnSO_4{\cdot}7H_2O$, and pretreatment time of 22 h; These are the same for Tables 5.18, 5.19, 5.20, 5.21, 5.22, 5.23 and 5.24

Table 5.18 Benzene conversion and cyclohexene selectivity of the second batch of Ru–Zn catalyst (2013-02-05, 5.6 kg)

t/min	C_{BZ}/%	S_{HE}/%	Y_{HE}/%
5	3.40	93.34	3.18
15	14.36	89.97	12.92
30	31.91	86.79	27.70
45	48.82	83.49	40.76
60	62.10	80.05	49.71

Table 5.19 Benzene conversion and cyclohexene selectivity of the third batch of Ru–Zn catalyst (2013-08-12, 5.6 kg)

t/min	C_{BZ}/%	S_{HE}/%	Y_{HE}/%
5	11.12	90.47	10.06
15	37.95	85.77	32.55
30	68.05	79.34	53.99
45	83.80	72.50	60.76
60	91.81	65.34	59.99

Table 5.20 Benzene conversion and cyclohexene selectivity of the fourth batch of Ru–Zn catalyst (2013-08-15, 5.6 kg)

t/min	C_{BZ}/%	S_{HE}/%	Y_{HE}/%
5	13.34	89.94	11.20
15	41.94	85.43	35.83
30	71.81	79.45	57.05
45	86.21	71.00	61.21
60	93.43	63.28	59.12

Table 5.21 Benzene conversion and cyclohexene selectivity of the fifth batch of Ru–Zn catalyst (2013-08-19, 5.6 kg)

t/min	C_{BZ}/%	S_{HE}/%	Y_{HE}/%
5	11.71	90.34	10.58
15	36.88	86.15	31.77
30	65.16	80.38	52.38
45	81.58	74.00	60.36
60	90.04	67.62	60.89

Table 5.22 Benzene conversion and cyclohexene selectivity of the sixth batch of Ru–Zn catalyst (2013-09-04, 5.6 kg)

t/min	C_{BZ}/%	S_{HE}/%	Y_{HE}/%
5	15.21	88.66	13.49
15	48.16	82.96	39.95
30	77.22	74.78	57.75
45	90.66	65.16	59.07
60	96.26	55.64	53.56

Table 5.23 Benzene conversion and cyclohexene selectivity of the seventh batch of Ru–Zn catalyst (2013-09-09, 5.6 kg)

t/min	C_{BZ}/%	S_{HE}/%	Y_{HE}/%
5	16.58	89.10	14.77
15	50.59	83.22	42.10
30	79.52	75.23	59.83
45	91.58	66.65	61.04
60	96.05	59.98	56.65

Table 5.24 Benzene conversion and cyclohexene selectivity of the eighth batch of Ru–Zn catalyst (2013-11-01, 5.6 kg)

t/min	C_{BZ}/%	S_{HE}/%	Y_{HE}/%
5	12.48	89.77	11.21
15	39.43	85.81	33.84
30	66.17	80.41	53.20
45	80.00	74.58	59.66
60	89.95	68.22	61.37

to November 2011. Figures 5.41, 5.42, 5.43, 5.44, 5.45, 5.46, 5.47 and 5.48 give the curves of C_{BZ}-t and S_{HE}-C_{BZ} plotted according to the data given in Tables 5.17, 5.18, 5.19, 5.20, 5.21, 5.22, 5.23 and 5.24, and t_{40}, t_{50}, and t_{60} (min), and S_{40}, S_{50}, and S_{60} were marked in the corresponding curves. Moreover, the values of γ_{40}, γ_{50}, and γ_{60} calculated by using Eq. (5.3.1), and the main technical indicators of each batch of the catalyst, are also listed.

The main technical indicators of the first batch of Ru–Zn catalyst:

$$t_{40} = 17\text{ min},\ \gamma_{40} = 89,\ S_{40} = 84\%$$
$$t_{50} = 22\text{ min},\ \gamma_{50} = 86,\ S_{50} = 82\%$$
$$t_{60} = 27\text{ min},\ \gamma_{60} = 84,\ S_{60} = 80\%$$

The main technical indicators of the second batch of Ru–Zn catalyst:

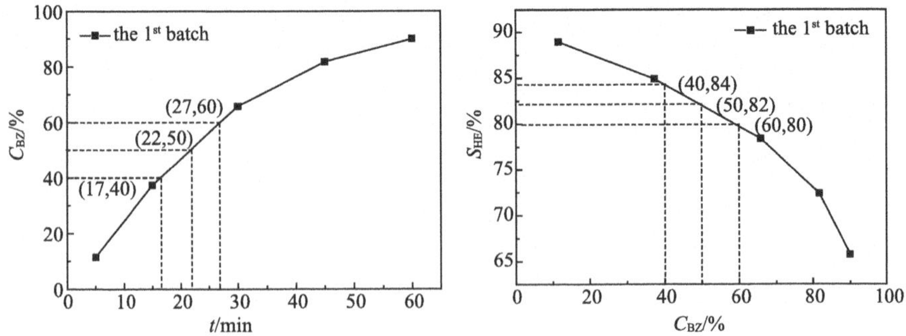

Fig. 5.41 C_{BZ}-t curve and S_{HE}-C_{BZ} curve of the first batch of Ru–Zn catalyst

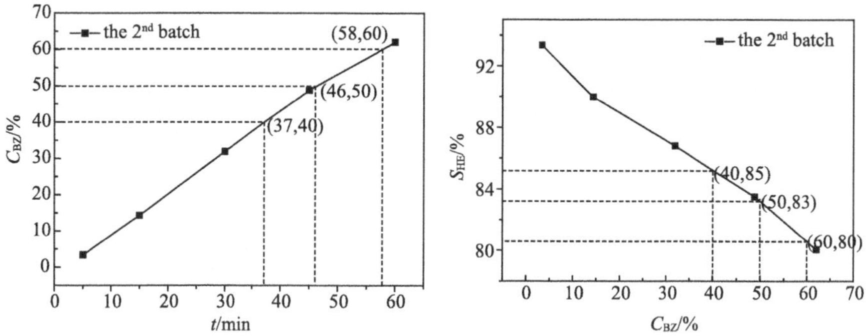

Fig. 5.42 C_{BZ}-t curve and S_{HE}-C_{BZ} curve of the second batch of Ru–Zn catalyst

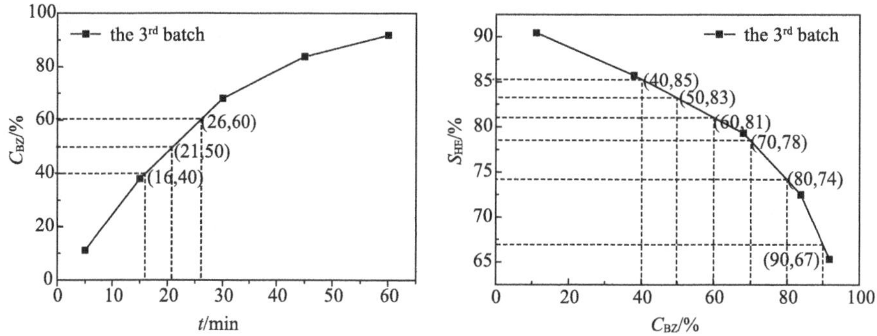

Fig. 5.43 C_{BZ}-t curve and S_{HE}-C_{BZ} curve of the third batch of Ru–Zn catalyst

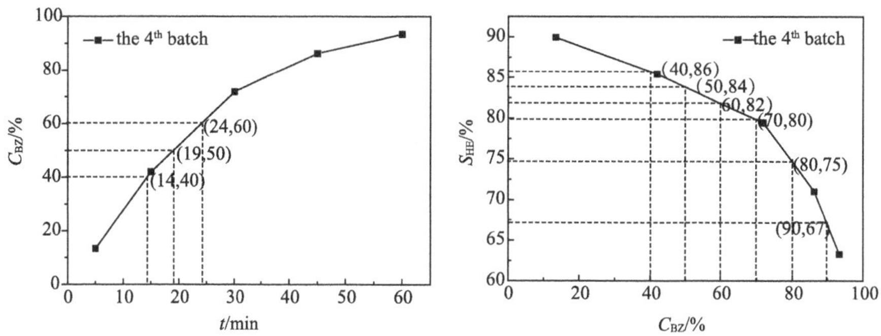

Fig. 5.44 C_{BZ}-t curve and S_{HE}-C_{BZ} curve of the fourth batch of Ru–Zn catalyst

$$t_{40} = 37\ \text{min},\ \gamma_{40} = 41,\ S_{40} = 85\%$$
$$t_{50} = 46\ \text{min},\ \gamma_{50} = 41,\ S_{50} = 83\%$$
$$t_{60} = 58\ \text{min},\ \gamma_{60} = 39,\ S_{60} = 81\%$$

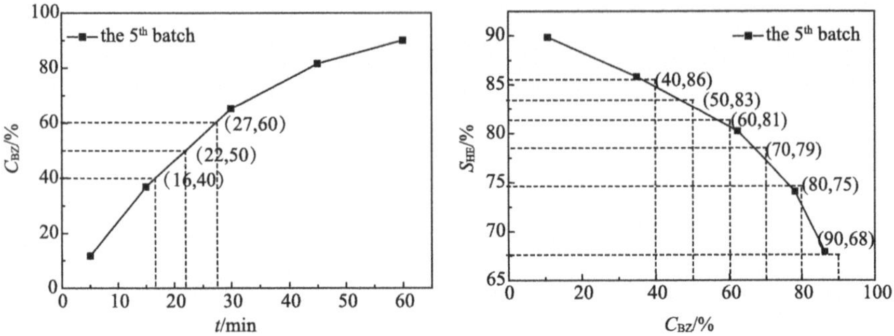

Fig. 5.45 C_{BZ}-t curve and S_{HE}-C_{BZ} curve of the fifth batch of Ru–Zn catalyst

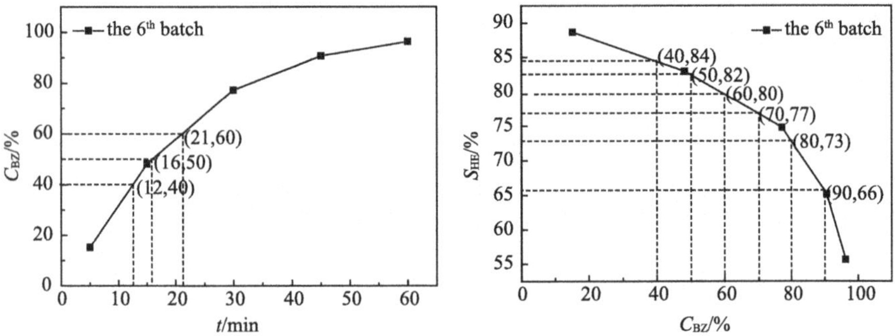

Fig. 5.46 C_{BZ}-t curve and S_{HE}-C_{BZ} curve of the sixth batch of Ru–Zn catalyst

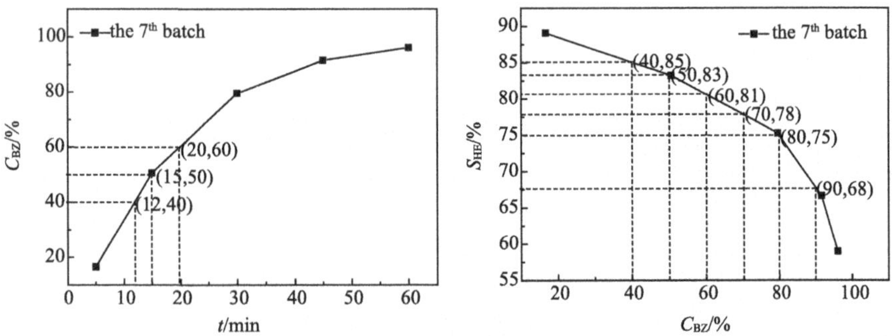

Fig. 5.47 C_{BZ}-t curve and S_{HE}-C_{BZ} curve of the seventh batch of Ru–Zn catalyst

The main technical indicators of the third batch of Ru–Zn catalyst:

$$t_{40} = 16\ \text{min},\ \gamma_{40} = 94,\ S_{40} = 85\%$$
$$t_{50} = 21\ \text{min},\ \gamma_{50} = 90,\ S_{50} = 83\%$$

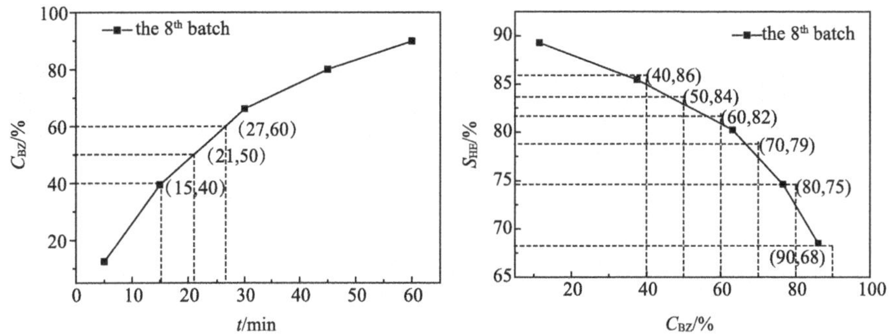

Fig. 5.48 C_{BZ}-t curve and S_{HE}-C_{BZ} curve of the eighth batch of Ru–Zn catalyst

$$t_{60} = 26\text{ min}, \gamma_{60} = 87, S_{60} = 81\%$$

The main technical indicators of the fourth batch of Ru–Zn catalyst:

$$t_{40} = 14\text{ min}, \gamma_{40} = 108, S_{40} = 86\%$$
$$t_{50} = 19\text{ min}, \gamma_{50} = 99, S_{50} = 84\%$$
$$t_{60} = 24\text{ min}, \gamma_{60} = 94, S_{60} = 82\%$$

The main technical indicators of the fifth batch of Ru–Zn catalyst:

$$t_{40} = 16\text{ min}, \gamma_{40} = 94, S_{40} = 86\%$$
$$t_{50} = 22\text{ min}, \gamma_{50} = 86, S_{50} = 83\%$$
$$t_{60} = 27\text{ min}, \gamma_{60} = 84, S_{60} = 81\%$$

The main technical indicators of the sixth batch of Ru–Zn catalyst:

$$t_{40} = 12\text{ min}, \gamma_{40} = 126, S_{40} = 84\%$$
$$t_{50} = 16\text{ min}, \gamma_{50} = 118, S_{50} = 82\%$$
$$t_{60} = 21\text{ min}, \gamma_{60} = 108, S_{60} = 80\%$$

The main technical indicators of the seventh batch of Ru–Zn catalyst:

$$t_{40} = 12\text{ min}, \gamma_{40} = 126, S_{40} = 84\%$$
$$t_{50} = 16\text{ min}, \gamma_{50} = 118, S_{50} = 82\%$$
$$t_{60} = 21\text{ min}, \gamma_{60} = 108, S_{60} = 80\%$$

The main technical indicators of the eighth batch of Ru–Zn catalyst:

$$t_{40} = 15\text{ min}, \gamma_{40} = 101, S_{40} = 86\%$$

$$t_{50} = 21\text{ min}, \gamma_{50} = 90, S_{50} = 84\%$$
$$t_{60} = 27\text{ min}, \gamma_{60} = 84, S_{60} = 82\%$$

In summary, the cyclohexene selectivity was above 80% and the yield was higher than 48% at the benzene conversion of 60%. Compared with the second generation catalyst, the benzene conversion has increased by 10 percentage points, and the cyclohexene selectivity has increased by 8 percentage points. The catalyst crystallite sizes are mainly distributed in the range of 3–5 nm, and the Fe content was ensured to be less than 1000 ppm due to the use of Hastelloy autoclave.

References

1. Nagahara, H., Konishi, M.: Process for Producing Cycloolefine. Japanese Patent, JP 981241395 (1998)
2. Nagahara, H., Konishi M.: Process for Producing Cyclohexene. USA Patent, US 4734536 (1988)
3. Wang, J.Q., Wang, Y.Z., Xie, S.H., et al.: Partial hydrogenation of benzene to cyclohexene on a Ru-Zn/*m*-ZrO_2 nanocomposite catalyst. Appl. Catal. A **272**, 29–36 (2004)
4. Hu, S.C., Chen, Y.W.: Partial hydrogenation of benzene on Ru-Zn/SiO_2 catalysts. Ind. Eng. Chem. Res. **40**, 6099–6104 (2001)
5. Liu, H.Z., Liang, S.G., Wang, W.T., et al.: The partial hydrogenation of benzene to cyclohexene over Ru-Cu catalyst supported on ZnO. J. Mol. Catal. A **341**, 35–41 (2011)
6. Zhang, P., Wu, T.B., Jiang, T., et al.: Ru-Zn supported on hydroxyapatite as an effective catalyst for partial hydrogenation of benzene. Green Chem. **15**, 152–159 (2013)
7. Liu, J.L., Zhu, L.J., Pei, Y., et al.: Ce-promoted Ru/SBA-15 catalyst prepared by a "two solvents" impregnation method for selective hydrogenation of benzene to cyclohexene. Appl. Catal. A **353**, 282–287 (2009)
8. Liu, J.L., Zhu, Y., Liu, J., et al.: Discrimination of the roles of $CdSO_4$ and $ZnSO_4$ in liquid phase hydrogenation of benzene to cyclohexene. J. Catal. **268**, 100–105 (2009)
9. Bu, J., Liu, J.L., Chen, X.Y., et al.: Ru/SBA-15 catalyst for partial hydrogenation of benzene to cyclohexene: tuning the Ru crystallite size of Ba. Catal. Commun. **9**, 2612–2615 (2008)
10. Niwa, S.I., Mizukami, F., Kuno, M., et al.: Selective hydrogenation of benzene to cyclohexene with new ruthenium catalysis prepared by a chemical mixing procedure. J. Mol. Catal. A **34**, 247–249 (1986)
11. Struijk, J., Moene, R., Kamp, T.V.D., et al.: Partial liquid phase hydrogenation of benzene to cyclohexene over ruthenium catalysts in the presence of an aqueous salt solution II. Influence of various salts on the performance of the catalyst. Appl. Catal. A **89**, 77–102 (1992)
12. Yuan, P.Q., Wang, B.Q., Ma, Y.M., et al.: Hydrogenation of cyclohexene over Ru-Zn/Ru(0001) surface alloy: a first principles density functional study. J. Mol. Catal. A **301**, 140–145 (2009)
13. He, H.M., Yuan, P.Q., Ma, Y.M., et al.: Theoretical and experimental study on the partial hydrogenation of benzene over Ru-Zn/ZrO_2 catalyst. Chin. J. Catal. **30**, 312–318 (2009)
14. Ramos-Fernándze, E.V., Ferreira, A.E.P., Sepúlveda-Escribano, A., et al.: Enhancing the catalytic performance of Pt/ZnO in the selective hydrogenation of cinnamaldehyde by Cr addition to the support. J. Catal. **258**, 52–60 (2008)
15. Sun, H.J.: Selective hydrogenation of benzene to cyclohexene over the Ru catalyst modified by $(Zn(OH)_2)_3(ZnSO_4)(H_2O)_x$. Ph.D. thesis, Zhengzhou University, Zhengzhou (2013)
16. Pan, Y.J.: Selective hydrogenation of benzene to cyclohexene over the Ru catalyst modified by transition-metals of the 4th period. Master thesis, Zhengzhou University, Zhengzhou (2013)

17. Dong, Y.Y.: Selective hydrogenation of benzene to cyclohexene over the Ru catalyst modified by rare earth elements of La, Ce. Master thesis, Zhengzhou University, Zhengzhou (2013)
18. Wang, H.X.: Study on the supported Ru catalyst for selective hydrogenation of benzene to cyclohexene. Master thesis, Zhengzhou University, Zhengzhou (2013)
19. Guo, H.B., Li, H.X., Xu, Y.P., et al.: Liquid phase glucose hydrogenation over Cr-promoted Ru-B amorphous alloy catalysts. Mater. Lett. **57**, 392–398 (2002)
20. Fan, G.Y., Li, R.X., Li, X.J., et al.: Effect of organic additives on partial hydrogenation of benzene. Catal. Commun. **9**, 1394–1397 (2008)
21. Morales, F., de Groot, F.M.F., Glatzel, P., et al.: *In Situ* X-ray absorption of Co/Mn/TiO_2 catalysts for Fischer-Tropsch synthesis. J. Phys. Chem. B **108**, 162071–162074 (2004)
22. Jia, Y.J., Cao, S., Tang, Y.W., et al.: Electrocatalytic performance of carbon supported Ru-Fe catalyst for oxygen reduction in direct formic acid fuel cell. Chin. J. Inorg. Chem. **25**, 674–678 (2009)
23. Wang, J.Q., Wang, Y.Z., Xie, S.H., et al.: Partial hydrogenation of benzene to cyclohexene on a Ru-Zn/*m*-ZrO_2 nanocomposite catalyst. Appl. Catal. A Gen. **272**, 29–36 (2004)
24. Zhou, X.L., Sun, H.J., Guo, W., et al.: Selective hydrogenation of benzene to cyclohexene on Ru-based catalysts promoted with Mn and Zn. J. Nat. Gas Chem. **20**, 53–59 (2011)
25. Sun, H.J., Wang, H.X., Jiang, H.B., et al.: Effect of $(Zn(OH)_2)_3(ZnSO_4)(H_2O)_5$ on the performance of Ru-Zn catalyst for benzene selective hydrogenation to cyclohexene. Appl. Catal. A **450**, 160–168 (2013)
26. Xue, W., Song, Y., Wang, Y.J., et al.: Effect of hydrazine hydrate on the Ru-Zn/SiO_2 catalysts performance for partial hydrogenation of benzene. Catal. Commun. **11**, 29–33 (2009)
27. Dai, W.L., Sun, Q., Deng, J.F., et al.: XPS studies of Cu/ZnO/Al_2O_3 ultra-fine catalysts derived by a novel gel oxalate co-precipitation for methanol synthesis by $CO_2 + H_2$. Appl. Surf. Sci. **177**, 172–179 (2001)
28. Silvestre-Albero, J., Serrano-Ruiz, J.C., Sepúlveda-Escribano, A., et al.: Modification of the catalytic behavior of platinum by zinc in crotonaldehyde hydrogenation and iso-butane dehydrogenation. Appl. Catal. A Gen. **292**, 244–251 (2005)
29. Nagahara, H., Ono, M., Konishi, M., et al.: Partial hydrogenation of benzene to cyclohexene. Appl. Surf. Sci. **121**(122), 448–451 (1997)
30. Struijk, J., d'Angremond, M., Regt, W.J.M., et al.: Partial liquid phase hydrogenation of benzene to cyclohexene over ruthenium catalysts in the presence of an aqueous salt solution: I. Preparation, characterization of the catalyst and study of a number of process variables. Appl. Catal. A Gen. **83**, 263–295 (1992)
31. Liu, S.C., Liu, Z.Y., Wang, Z., et al.: A novel amorphous alloy Ru-La-B/ZrO_2 catalyst with high activity and selectivity for benzene selective hydrogenation. Appl. Catal. A **313**, 49–57 (2006)
32. Schwab, F., Lucas, M., Claus, P.: Ruthenium-catalyzed selective hydrogenation of benzene to cyclohexene in the presence of an ionic liquid. Angew. Chem. Ind. Ed. **50**, 1–5 (2011)
33. Zonetti, P.D.C., Landers, R., Cobo, A.J.G.: Thermal treatment effects on the Ru/CeO_2 catalysts performance for partial hydrogenation of benzene. Appl. Surf. Sci. **254**, 6849–6853 (2008)
34. Ghiaci, M., Aghabarari, B., Rego, A.M.B.D., et al.: Efficient allylic oxidation of cyclohexene catalyzed by trimetallic hybrid nano-mixed oxide (Ru/Co/Ce). Appl. Catal. A **393**, 225–230 (2011)
35. Siriwardane, R.V., Poston, J.A.: Interaction of H_2S with zinc titanate in the presence of H_2 and CO. Appl. Surf. Sci. **45**, 131–139 (1990)
36. Eo, I., Hwangbo, S., Kim, J., et al.: Photoluminescence of chemical solution-derived amorphous ZnO layers prepared by low-temperature process. Curr. Appl. Phys. **10**, 1–4 (2010)
37. Xie, Y.C., Zhu, Y.X., Zhao, B.Y., et al.: An important principle for catalyst preparation spontaneous monolayer dispersion of solid compounds onto surfaces of supports. Stud. Surf. Sci. Catal. **118**, 441–449 (1998)
38. Xie, Y.C., Wang, C.B., Tang, Y.Q., et al.: Study on spontaneous dispersion of salts on carrier surface. Sci. China (Series B) **23**, 113–119 (1993)

39. Zhang, Y.P., Wang, X.L., Shen, K., et al.: WO_3 modification of MnO_x/TiO_2 catalysts for low temperature selective catalytic reduction of NO with ammonia. Chin. J. Catal. **33**, 1523–1531 (2012)
40. Sun, H.J., Pan, Y.J., Wang, H.X., et al.: Selective hydrogenation of benzene to cyclohexene over a Ru-Zn catalyst with diethanolamine as an additive. Chin. J. Catal. **33**, 610–620 (2012)
41. Sun, H.J., Chen, Z.H., Guo, W., et al.: Effect of organic additives on the performance of nano-sized Ru-Zn catalyst. Chin. J. Chem. **29**, 369–373 (2011)
42. Peplinski, B., Unger, W.E.S., Grohmann, I.: Characterization of Cu-Zn-Al oxide catalysts in the precipitaed, calcined and reduced state by means of XPS with the help of a finger-print data base. Appl. Surf. Sci. **62**, 115–129 (1992)
43. Sun, H.J., Zhang, X.D., Chen, Z.H., et al.: Monolayer dispersed Ru-Zn catalyst and its performance in the selective hydrogenation of benzene to cyclohexene. Chin. J. Catal. **32**, 224–230 (2011)
44. Xu, X.S., Chen, A.A., Zhou, L., et al.: Catalytic stability of *othro*-chloronitrobenzene hydrogenation on Ru-Fe/C catalyst. Chin. J. Catal. **34**, 391–396 (2013)
45. Chen, A.A., Xu, X.S., Hua, Y.X., et al.: Fe_3O_4 modified alumina supported ruthenium catalyst for novel *in-situ* liquid phase catalytic hydrogenation. Acta Phys. Chim. Sin. **29**(4), 799–805 (2013)
46. Liu, S.C., Liu, Z.Y., Zhao, S.H., et al.: Study on the nanosized amorphous Ru-Fe-B/ZrO_2 alloy catalyst for benzene selective hydrogenation to cyclohexene. J. Nat. Gas Chem. **15**, 319–326 (2006)
47. Da-Silva, J.W., Cobo, A.J.G.: The role of the titania and silica supports in Ru-Fe catalysts to partial hydrogenation of benzene. Appl. Catal. A Gen. **252**, 9–16 (2003)
48. Wei, S.Q., Huang, Z.X., Liu, S.C.: Preparation and properties of Ru-Fe/ZrO_2 catalysts for partial hydrogenation of benzene to cyclohexene. J. Henan Univer. Sci. Technol. Nat. Sci. **32**, 102–104 (2011)
49. Wang, W.T., Liu, H.Z., Wu, T.B., et al.: Ru catalyst supported on bentonite for partial hydrogenation of benzene to cyclohexene. J. Molec. Catal. A Chem. **355**, 174–179 (2012)
50. Wang, W.T., Liu, H.Z., Ding, G.D., et al.: Ru-Cd/Bentonite for the partial hydrogenation of benzene: a catalyst without additives. Chem. Cat. Chem. **4**, 1836–1843 (2012)
51. Sun, H.J., Jiang, H.B., Li S.H., et al.: Effect of alcohols as additives on the performance of a nano-sized Ru-Zn(2.8%) catalyst for selective hydrogenation of benzene to cyclohexene. Chem. Eng. J. **218**, 415–424 (2013)

Chapter 6
Fourth-Generation Catalyst of Benzene-Selective Hydrogenation to Cyclohexene—Ru–Zn@BZSS Core-Shell Catalyst

Compared with the preparation technology of the third-generation catalyst for benzene-selective hydrogenation, the fourth-generation Ru–Zn@BZSS core-shell catalyst overcomes the technical bottleneck of in situ preparation method of the BZSS salt. The surface modification of the catalyst by using the non-in situ prepared BZSS improves the selectivity and yield of cyclohexene, and the controllable preparation of the catalyst can be achieved. The main technical indicators of the catalyst are as follows: benzene conversion of 70%, cyclohexene selectivity of more than 80%, cyclohexene yield of more than 56%. Compared with the third-generation catalysts, the benzene conversion of the fourth-generation catalysts increased by 10% points, and the cyclohexene yield increased by 8% points.

The Ru–Zn catalysts with the same theoretical Zn content are prepared with different Zn precursors. As long as Zn content is the same, the catalyst activity and cyclohexene selectivity can be maintained at a certain level, the structural and textural properties of the catalyst are roughly the same. Zn exists as ZnO in the Ru–Zn catalyst, and ZnO rich on the surface of the catalyst can react with $ZnSO_4$ in the slurry to form an insoluble basic zinc sulfate complex salt. The surface adsorption of BZSS can reduce the Zn content in the catalyst, keeping the total Zn content in the surface and interior of the catalyst unchanged, and then the catalysts with high activity and selectivity can be prepared.

The main contents of this chapter are as follows: the effect of Zn precursor and Zn content on the properties of the Ru–Zn catalyst, and the research and development as well as industrial applications of the fourth-generation Ru–Zn@BZSS catalyst.

Z. Liu et al., *Catalytic Technology for Selective Hydrogenation of Benzene to Cyclohexene*,
https://doi.org/10.1007/978-981-15-6411-6_6

6.1 Effect of Zn Precursor on the Properties of the Ru–Zn Catalyst

According to the literature, in the Ru–Zn catalyst preparation process, compared with the precursor without Cl^-, the Ru/γ-Al_2O_3 catalyst prepared by the impregnation method with the Ru precursor where $RuCl_3$ acting as active component showed higher cyclohexene selectivity due to the residual Cl^- [1]. In the Ru-based catalysts and catalytic systems for benzene selective hydrogenation, Zn is found to be the best promoter, and $ZnSO_4$ is the best additive. At the optimum value, the catalyst not only has a very good activity, but also shows a high cyclohexene selectivity, and thus the maximum cyclohexene yield can be achieved [2, 3]. The properties of the catalyst can be modified through surface modification, thereby improving the catalytic activity and selectivity [4]. The determination of the optimum Zn content is the core issue of the preparation of the Ru–Zn catalyst. The residual negative ions have a significant effect on the performance of the catalyst with different precursors of Zn promoter, and thus need to be further studied through experiments.

6.1.1 *Effect of Zn Precursor*

The Ru–Zn catalysts with the same theoretical Zn content were prepared with the Ru precursor where $RuCl_3{\cdot}3H_2O$ acting as an active component, and $ZnSO_4{\cdot}7H_2O$, $ZnCl_2$, $5ZnO{\cdot}2CO_2{\cdot}4H_2O$, Zn powder, ZnO, $Zn(OH)_2$, $Zn(NO_3)_2$, and $Zn(Ac)_2$ as the Zn precursor, respectively. The activity and selectivity of the catalyst are tested through direct hydrogenation and hydrogenation after the pretreatment, and the effect of Zn precursor on the performance of the Ru–Zn catalyst is then investigated.

(1) $ZnSO_4{\cdot}7H_2O$ as the Zn precursor

Table 6.1 lists the activity and selectivity of direct hydrogenation over the Ru–Zn catalyst prepared with $ZnSO_4{\cdot}7H_2O$ as the Zn precursor.

Figure 6.1 presents the curves of C_{BZ}-t and S_{HE}-C_{BZ} of direct hydrogenation over the Ru–Zn catalyst prepared with $ZnSO_4{\cdot}7H_2O$ as the Zn precursor.

As can be seen in Fig. 6.1, in the direct hydrogenation over the Ru–Zn catalyst with $ZnSO_4{\cdot}7H_2O$ as the Zn precursor, the values of time t_{40}, t_{50}, and t_{60} and cyclohexene selectivity S_{40}, S_{50}, and S_{60} at the benzene conversion of 40, 50, and 60% are as follows:

$$t_{40} = 6\,\text{min},\ S_{40} = 81\%$$
$$t_{50} = 9\,\text{min},\ S_{50} = 79\%$$
$$t_{60} = 11\,\text{min},\ S_{60} = 77\%$$

Table 6.1 Activity and selectivity of direct hydrogenation over the Ru–Zn catalyst prepared with $ZnSO_4 \cdot 7H_2O$ as the Zn precursor

t/min	C_{BZ}/%[a]	S_{HE}/%[b]	Y_{HE}/%[c]
5	33.64	81.96	27.57
10	56.35	77.51	43.68
15	71.08	74.21	52.75
20	82.76	69.13	57.21
25	88.49	64.40	56.99

[a]Benzene conversion (molar fraction); [b]cyclohexene selectivity (molar fraction); [c]cyclohexene yield (molar fraction), the following are the same. Reaction conditions: temperature of 150 °C, H_2 pressure of 5.0 MPa, stirring rate of 1400 r/min, 140 mL of benzene, 280 mL of H_2O, 45.7 g of $ZnSO_4 \cdot 7H_2O$, and 9.8 g of ZrO_2

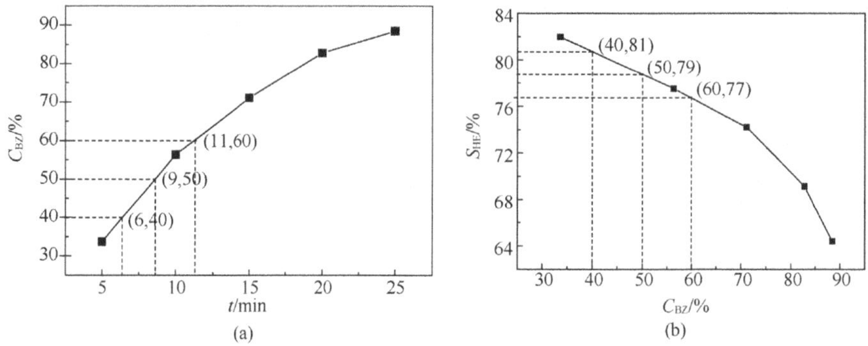

Fig. 6.1 **a** C_{BZ}-t curve and **b** S_{HE}-C_{BZ} curve of direct hydrogenation over the Ru–Zn catalyst prepared with $ZnSO_4 \cdot 7H_2O$ as the Zn precursor

The catalyst shows high activity with the benzene conversion of 60% in 11 min, and the cyclohexene selectivity is above 80% at a benzene conversion of 40%. The cyclohexene selectivity can be further improved at the sacrifice of the catalyst activity.

Table 6.2 lists the hydrogenation activity and selectivity of the Ru–Zn catalyst prepared with $ZnSO_4 \cdot 7H_2O$ as the Zn precursor after pretreatment for 22 h.

Figure 6.2 presents the curves of C_{BZ}-t and S_{HE}-C_{BZ} of hydrogenation over the Ru–Zn catalyst prepared from $ZnSO_4 \cdot 7H_2O$ as the Zn precursor after pretreatment for 22 h.

As can be seen in Fig. 6.2, the hydrogenation characters over the Ru–Zn catalyst prepared with $ZnSO_4 \cdot 7H_2O$ as the Zn precursor after pretreatment for 22 h are as follows:

$$t_{40} = 15\,\text{min},\ S_{40} = 84\%$$
$$t_{50} = 20\,\text{min},\ S_{50} = 81\%$$
$$t_{60} = 26\,\text{min},\ S_{60} = 78\%$$

Table 6.2 Hydrogenation activity and selectivity of the Ru–Zn catalyst prepared with $ZnSO_4 \cdot 7H_2O$ as Zn precursor after pretreatment for 22 h

t/min	C_{BZ}/%	S_{HE}/%	Y_{HE}/%
5	12.90	85.81	11.07
10	27.35	86.33	23.61
15	39.65	83.78	33.22
20	50.33	81.08	41.21
25	58.51	78.68	46.03

Pretreatment conditions: temperature of 140 °C, H_2 pressure of 5.0 MPa, stirring rate of 800 r/min, and time of 22 h; Reaction conditions: temperature of 150 °C, H_2 pressure of 5.0 MPa, and stirring rate of 1400 r/min, 140 mL of benzene, 280 mL of H_2O, 45.7 g of $ZnSO_4 \cdot 7H_2O$, and 9.8 g of ZrO_2

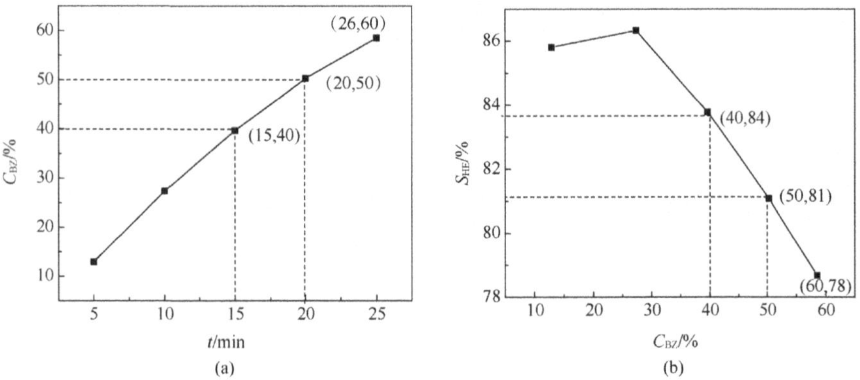

Fig. 6.2 **a** C_{BZ}-t curve and **b** S_{HE}-C_{BZ} curve of hydrogenation over the Ru–Zn catalyst prepared with $ZnSO_4 \cdot 7H_2O$ as the Zn precursor after pretreatment for 22 h

The activity of the pretreated catalyst decreased, and the reaction time increased to 26 min with the benzene conversion of 60%; while the cyclohexene selectivity increased to 84% at a benzene conversion of 40%.

(2) $ZnCl_2$ as the Zn precursor

Table 6.3 lists the activity and selectivity of direct hydrogenation over the Ru–Zn catalyst prepared with $ZnCl_2$ as the Zn precursor.

Figure 6.3 presents the curves of C_{BZ}-t and S_{HE}-C_{BZ} of direct hydrogenation over the Ru–Zn catalyst prepared with $ZnCl_2$ as the Zn precursor.

As can be seen in Fig. 6.3, in the direct hydrogenation over the Ru–Zn catalyst with $ZnCl_2$ as the Zn precursor, the values of time t_{40}, t_{50}, and t_{60} and cyclohexene selectivity S_{40}, S_{50}, and S_{60} at the benzene conversion of 40, 50, and 60% are as follows:

Table 6.3 Activity and selectivity of direct hydrogenation over the Ru–Zn catalyst prepared with $ZnCl_2$ as the Zn precursor

t/min	C_{BZ}/%	S_{HE}/%	Y_{HE}/%
5	27.95	84.11	23.51
10	53.45	77.60	41.48
15	72.10	73.19	52.77
20	83.97	66.65	55.97
25	90.19	61.04	55.05

Reaction conditions: temperature of 150 °C, H_2 pressure of 5.0 MPa, and stirring rate of 1400 r/min, 140 mL of benzene, 280 mL of H_2O, 45.7 g of $ZnSO_4{\cdot}7H_2O$, and 9.8 g of ZrO_2

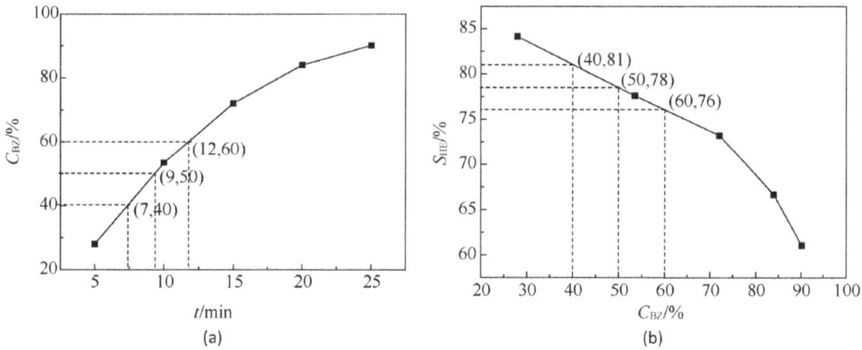

Fig. 6.3 **a** C_{BZ}-t curve and **b** S_{HE}-C_{BZ} curve of direct hydrogenation over the Ru–Zn catalyst prepared from $ZnCl_2$ as the Zn precursor

$$t_{40} = 7\,\text{min},\ S_{40} = 81\%$$
$$t_{50} = 9\,\text{min},\ S_{50} = 78\%$$
$$t_{60} = 12\,\text{min},\ S_{60} = 76\%$$

Table 6.4 lists the hydrogenation activity and selectivity of the Ru–Zn catalyst prepared from $ZnCl_2$ as the Zn precursor after pretreatment for 22 h.

Figure 6.4 presents the curves of C_{BZ}-t and S_{HE}-C_{BZ} of hydrogenation over the Ru–Zn catalyst prepared from $ZnCl_2$ as the Zn precursor after pretreatment for 22 h.

As can be seen in Fig. 6.4, the hydrogenation characters of Ru–Zn catalyst prepared from $ZnCl_2$ as the Zn precursor after pretreatment for 22 h are as follows:

$$t_{40} = 12\,\text{min},\ S_{40} = 80\%$$
$$t_{50} = 15\,\text{min},\ S_{50} = 78\%$$
$$t_{60} = 18\,\text{min},\ S_{60} = 75\%$$

Compared with the Ru–Zn catalyst prepared with $ZnSO_4{\cdot}7H_2O$ as the Zn precursor, in the direct hydrogenation and hydrogenation after pretreatment for 22 h,

Table 6.4 Hydrogenation activity and selectivity of Ru–Zn catalyst prepared from $ZnCl_2$ as the Zn precursor after pretreatment for 22 h

t/min	C_{BZ}/%	S_{HE}/%	Y_{HE}/%
5	18.91	86.14	16.29
10	35.50	83.76	29.57
15	51.77	80.99	41.93
20	64.77	76.79	49.74
25	76.86	72.16	55.46

Pretreatment conditions: temperature of 140 °C, H_2 pressure of 5.0 MPa, stirring rate of 800 r/min, and time of 22 h; Reaction conditions: temperature of 150 °C, H_2 pressure of 5.0 MPa, and stirring rate of 1400 r/min, 140 mL of benzene, 280 mL of H_2O, 45.7 g of $ZnSO_4{\cdot}7H_2O$, and 9.8 g of ZrO_2

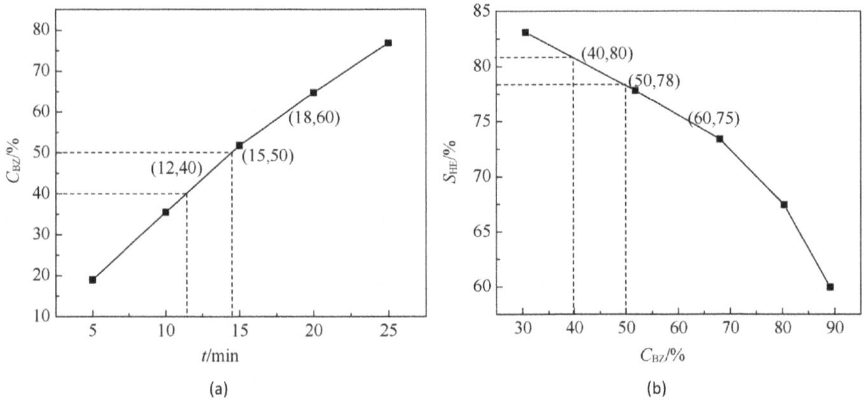

Fig. 6.4 **a** C_{BZ}-t curve and **b** S_{HE}-C_{BZ} curve of hydrogenation over the Ru–Zn catalyst prepared from $ZnCl_2$ as the Zn precursor after pretreatment for 22 h

the Ru–Zn catalyst prepared from $ZnCl_2$ as the Zn precursor showed relatively higher activity and lower cyclohexene selectivity.

(3) $5ZnO{\cdot}2CO_2{\cdot}4H_2O$ as the Zn precursor

Table 6.5 lists the activity and selectivity of direct hydrogenation over the Ru–Zn catalyst prepared from $5ZnO{\cdot}2CO_2{\cdot}4H_2O$ as the Zn precursor.

Figure 6.5 presents the curves of C_{BZ}-t and S_{HE}-C_{BZ} of direct hydrogenation over the Ru–Zn catalyst prepared from $5ZnO{\cdot}2CO_2{\cdot}4H_2O$ as the Zn precursor.

As can be seen in Fig. 6.5, in the direct hydrogenation over the Ru–Zn catalyst with $5ZnO{\cdot}2CO_2{\cdot}4H_2O$ as the Zn precursor, the values of time t_{40}, t_{50}, and t_{60} and cyclohexene selectivity S_{40}, S_{50}, and S_{60} at the benzene conversion of 40, 50, and 60% are as follows:

Table 6.5 Activity and selectivity of direct hydrogenation over the Ru–Zn catalyst prepared from $5ZnO{\cdot}2CO_2{\cdot}4H_2O$ as the Zn precursor

t/min	C_{BZ}/%	S_{HE}/%	Y_{HE}/%
5	30.64	83.09	25.46
10	51.76	77.82	40.28
15	67.98	73.40	49.90
20	80.35	67.45	54.20
25	89.12	59.95	53.43

Reaction conditions: temperature of 150 °C, H_2 pressure of 5.0 MPa, and stirring rate of 1400 r/min, 140 mL of benzene, 280 mL of H_2O, 45.7 g of $ZnSO_4{\cdot}7H_2O$, and 9.8 g of ZrO_2

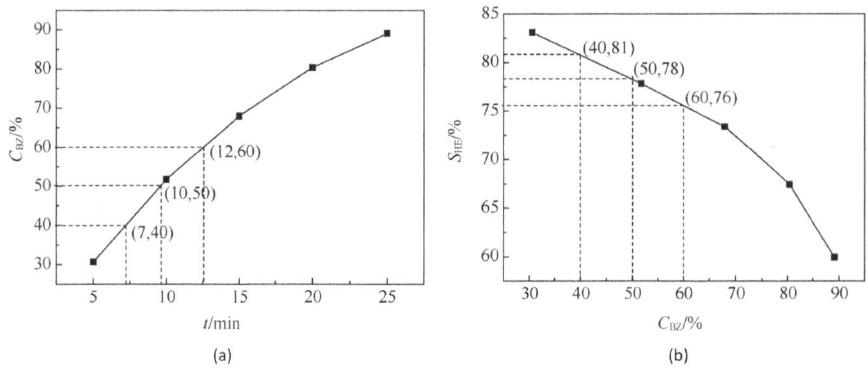

Fig. 6.5 **a** C_{BZ}-t curve and **b** S_{HE}-C_{BZ} curve of direct hydrogenation over the Ru–Zn catalyst prepared from $5ZnO{\cdot}2CO_2{\cdot}4H_2O$ as the Zn precursor

$$t_{40} = 7\,\text{min},\ S_{40} = 81\%$$
$$t_{50} = 10\,\text{min},\ S_{50} = 78\%$$
$$t_{60} = 12\,\text{min},\ S_{60} = 76\%$$

Table 6.6 lists the hydrogenation activity and selectivity over Ru–Zn catalyst prepared from $5ZnO{\cdot}2CO_2{\cdot}4H_2O$ as the Zn precursor after pretreatment for 22 h.

Figure 6.6 presents the curves of C_{BZ}-t and S_{HE}-C_{BZ} of hydrogenation over the Ru–Zn catalyst prepared with $ZnCl_2$ as the Zn precursor after pretreatment for 22 h.

As can be seen in Fig. 6.6, the hydrogenation characters of the Ru–Zn catalyst prepared with $5ZnO{\cdot}2CO_2{\cdot}4H_2O$ as the Zn precursor after pretreatment for 22 h are as follows:

$$t_{40} = 16\,\text{min},\ S_{40} = 80\%$$
$$t_{50} = 20\,\text{min},\ S_{50} = 78\%$$
$$t_{60} = 25\,\text{min},\ S_{60} = 75\%$$

Table 6.6 Hydrogenation activity and selectivity of the Ru–Zn catalyst prepared from $5ZnO{\cdot}2CO_2{\cdot}4H_2O$ as the Zn precursor after pretreatment for 22 h

t/min	C_{BZ}/%	S_{HE}/%	Y_{HE}/%
5	14.70	84.63	12.44
10	23.70	83.53	19.85
15	37.26	81.11	30.22
20	48.78	78.00	38.05
25	60.53	74.34	45.01

Pretreatment conditions: temperature of 140 °C, H_2 pressure of 5.0 MPa, stirring rate of 800 r/min, and time of 22 h; Reaction conditions: temperature of 150 °C, H_2 pressure of 5.0 MPa, stirring rate of 1400 r/min, 140 mL of benzene, 280 mL of H_2O, 45.7 g of $ZnSO_4{\cdot}7H_2O$, and 9.8 g of ZrO_2

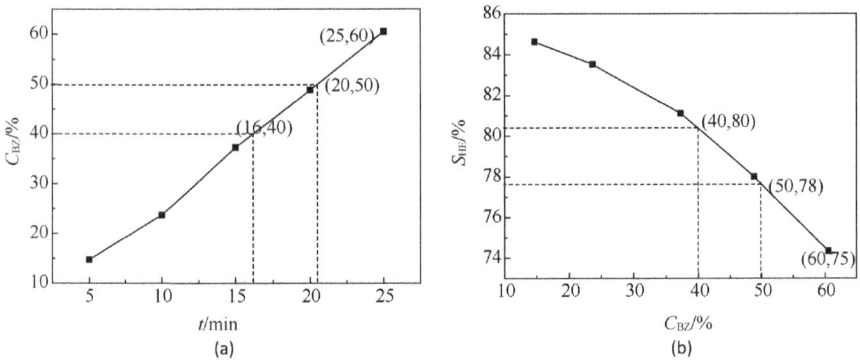

Fig. 6.6 **a** C_{BZ}-t curve and **b** S_{HE}-C_{BZ} curve of hydrogenation over the Ru–Zn catalyst prepared with $ZnCl_2$ as the Zn precursor after pretreatment for 22 h

Compared with the Ru–Zn catalysts prepared with $ZnSO_4{\cdot}7H_2O$ and $ZnCl_2$ as the Zn precursor, respectively, the Ru–Zn catalysts prepared with $5ZnO{\cdot}2CO_2{\cdot}4H_2O$ has shown the same activity and selectivity for direct hydrogenation as that catalyzes with $ZnCl_2$ as the Zn precursor and have the same trend after pretreatment.

(4) Zn powder as the Zn precursor

Table 6.7 lists the activity and selectivity of direct hydrogenation over the Ru–Zn catalyst prepared from Zn powder as the Zn precursor.

Table 6.8 lists the hydrogenation activity and selectivity of the Ru–Zn catalyst prepared with Zn powder as the Zn precursor after pretreatment for 22 h.

By using the same data processing methods with the Ru–Zn catalyst prepared with $ZnSO_4{\cdot}7H_2O$, $ZnCl_2$, and $4ZnO{\cdot}2CO_2{\cdot}4H_2O$ as the Zn precursors, the values of t_{40}, t_{50}, t_{60} and S_{40}, S_{50}, S_{60} of direct hydrogenation over the Ru–Zn catalyst prepared with Zn powder as the Zn precursor are as follows:

Table 6.7 Activity and selectivity of direct hydrogenation over the Ru–Zn catalyst prepared from Zn powder as the Zn precursor

t/min	C_{BZ}/%	S_{HE}/%	Y_{HE}/%
5	27.95	80.14	22.40
10	52.66	76.30	40.18
15	64.91	71.75	46.57
20	79.50	65.91	52.40
25	86.84	61.08	53.04

Reaction conditions: temperature of 150 °C, H_2 pressure of 5.0 MPa, and stirring rate of 1400 r/min, 140 mL of benzene, 280 mL of H_2O, 45.7 g of $ZnSO_4 \cdot 7H_2O$, and 9.8 g of ZrO_2

Table 6.8 Hydrogenation activity and selectivity of the Ru–Zn catalyst prepared from Zn powder as the Zn precursor after pretreatment for 22 h

t/min	C_{BZ}/%	S_{HE}/%	Y_{HE}/%
5	16.59	90.11	14.95
10	31.35	87.02	27.28
15	46.37	83.05	38.51
20	58.83	79.79	46.94
25	67.85	75.18	51.01

Pretreatment conditions: temperature of 140 °C, H_2 pressure of 5.0 MPa, stirring rate of 800 r/min, and time of 22 h; Reaction conditions: temperature of 150 °C, H_2 pressure of 5.0 MPa, and stirring rate of 1400 r/min, 140 mL of benzene, 280 mL of H_2O, 45.7 g of $ZnSO_4 \cdot 7H_2O$, and 9.8 g of ZrO_2

$$t_{40} = 7\,\text{min},\ S_{40} = 78\%$$
$$t_{50} = 9\,\text{min},\ S_{50} = 76\%$$
$$t_{60} = 13\,\text{min},\ S_{60} = 74\%$$

and the values of t_{40}, t_{50}, t_{60} and S_{40}, S_{50}, S_{60} of hydrogenation over the catalyst after pretreatment for 22 h are as follows:

$$t_{40} = 13\,\text{min},\ S_{40} = 85\%$$
$$t_{50} = 17\,\text{min},\ S_{50} = 82\%$$
$$t_{60} = 21\,\text{min},\ S_{60} = 79\%$$

(5) ZnO as the Zn precursor

Table 6.9 lists the activity and selectivity of direct hydrogenation over the Ru–Zn catalyst prepared with ZnO as the Zn precursor.

Table 6.10 lists the hydrogenation activity and selectivity of the Ru–Zn catalyst prepared with ZnO as the Zn precursor after pretreatment for 22 h.

By using the same data processing method, the values of t_{40}, t_{50}, t_{60} and S_{40}, S_{50}, S_{60} of direct hydrogenation over the Ru–Zn catalyst prepared with ZnO as the Zn

Table 6.9 Activity and selectivity of direct hydrogenation over the Ru–Zn catalyst prepared from ZnO as the Zn precursor

t/min	C_{BZ}/%	S_{HE}/%	Y_{HE}/%
5	32.32	84.68	27.37
10	59.87	78.72	47.13
15	75.26	74.30	55.92
20	84.11	69.72	58.64
25	89.15	65.05	57.99

Reaction conditions: temperature of 150 °C, H_2 pressure of 5.0 MPa, and stirring rate of 1400 r/min, 140 mL of benzene, 280 mL of H_2O, 45.7 g of $ZnSO_4 \cdot 7H_2O$, and 9.8 g of ZrO_2

Table 6.10 Hydrogenation activity and selectivity of the Ru–Zn catalyst prepared with ZnO as the Zn precursor after pretreatment for 22 h

t/min	C_{BZ}/%	S_{HE}/%	Y_{HE}/%
5	34.86	83.45	29.09
10	52.55	80.93	42.53
15	63.01	80.69	50.84
20	71.09	77.65	55.20
25	77.01	76.28	58.72

Pretreatment conditions: temperature of 140 °C, H_2 pressure of 5.0 MPa, stirring rate of 800 r/min, and time of 22 h; Reaction conditions: temperature of 150 °C, H_2 pressure of 5.0 MPa, and stirring rate of 1400 r/min, 140 mL of benzene, 280 mL of H_2O, 45.7 g of $ZnSO_4 \cdot 7H_2O$, and 9.8 g of ZrO_2

precursor can be derived as follows:

$$t_{40} = 6\,\text{min},\ S_{40} = 83\%$$
$$t_{50} = 9\,\text{min},\ S_{50} = 81\%$$
$$t_{60} = 14\,\text{min},\ S_{60} = 79\%$$

and the values of t_{40}, t_{50}, t_{60} and S_{40}, S_{50}, S_{60} of hydrogenation over the catalyst after pretreatment for 22 h are as follows:

$$t_{40} = 6\,\text{min},\ S_{40} = 83\%$$
$$t_{50} = 9\,\text{min},\ S_{50} = 81\%$$
$$t_{60} = 14\,\text{min},\ S_{60} = 80\%$$

(6) $Zn(Ac)_2$ as the Zn precursor

Table 6.11 lists the activity and selectivity of direct hydrogenation over the Ru–Zn catalyst prepared with $Zn(Ac)_2$ as the Zn precursor.

By using the same data processing method, the values of t_{40}, t_{50}, t_{60} and S_{40}, S_{50}, S_{60} of direct hydrogenation over the Ru–Zn catalyst prepared with $Zn(Ac)_2$ as the

Table 6.11 Activity and selectivity of direct hydrogenation over the Ru–Zn catalyst prepared with $Zn(Ac)_2$ as the Zn precursor

t/min	C_{BZ}/%	S_{HE}/%	Y_{HE}/%
5	24.38	86.96	21.20
10	42.24	83.74	35.37
15	59.32	80.93	48.01
20	71.36	77.66	55.42
25	77.80	74.31	57.81

Reaction conditions: temperature of 150 °C, H_2 pressure of 5.0 MPa, and stirring rate of 1400 r/min, 140 mL of benzene, 280 mL of H_2O, 45.7 g of $ZnSO_4{\cdot}7H_2O$, and 9.8 g of ZrO_2

Zn precursor can be derived as follows:

$$t_{40} = 10\,\text{min},\ S_{40} = 84\%$$
$$t_{50} = 12\,\text{min},\ S_{50} = 82\%$$
$$t_{60} = 15\,\text{min},\ S_{60} = 80\%$$

(7) $Zn(OH)_2$ as the Zn precursor

Table 6.12 lists the activity and selectivity of direct hydrogenation over the Ru–Zn catalyst prepared with $Zn(OH)_2$ as the Zn precursor.

By using the same data processing method, the values of t_{40}, t_{50}, t_{60} and S_{40}, S_{50}, S_{60} for direct hydrogenation over the Ru–Zn catalyst prepared with $Zn(OH)_2$ as the Zn precursor can be derived as follows:

$$t_{40} = 7\,\text{min},\ S_{40} = 88\%$$
$$t_{50} = 9\,\text{min},\ S_{50} = 85\%$$
$$t_{60} = 11\,\text{min},\ S_{60} = 82\%$$

Table 6.12 Activity and selectivity of direct hydrogenation over the Ru–Zn catalyst prepared with $Zn(OH)_2$ as the Zn precursor

t/min	C_{BZ}/%	S_{HE}/%	Y_{HE}/%
5	30.46	90.94	27.70
10	55.99	83.71	46.87
15	74.71	79.79	59.61
20	85.00	72.91	61.97
25	91.81	68.09	62.51

Reaction conditions: temperature of 150 °C, H_2 pressure of 5.0 MPa, and stirring rate of 1400 r/min, 140 mL of benzene, 280 mL of H_2O, 45.7 g of $ZnSO_4{\cdot}7H_2O$, and 9.8 g of ZrO_2

Table 6.13 Activity and selectivity of direct hydrogenation over the Ru–Zn catalyst prepared with $Zn(NO_3)_2$ as the Zn precursor

t/min	C_{BZ}/%	S_{HE}/%	Y_{HE}/%
5	42.59	84.01	35.78
10	64.70	77.26	49.99
15	82.89	70.35	58.31
20	90.57	64.66	58.56
25	93.95	59.29	55.70

Reaction conditions: temperature of 150 °C, H_2 pressure of 5.0 MPa, and stirring rate of 1400 r/min, 140 mL of benzene, 280 mL of H_2O, 45.7 g of $ZnSO_4 \cdot 7H_2O$, and 9.8 g of ZrO_2

(8) $Zn(NO_3)_2$ as the Zn precursor

Table 6.13 lists the activity and selectivity of direct hydrogenation over the Ru–Zn catalyst prepared from $Zn(NO_3)_2$ as the Zn precursor.

By using the same data processing method, the values of t_{40}, t_{50}, t_{60} and S_{40}, S_{50}, S_{60} of direct hydrogenation over the Ru–Zn catalyst prepared with $Zn(NO_3)_2$ as the Zn precursor can be derived as follows:

$$t_{40} = 5\ \text{min},\ S_{40} = 85\%$$
$$t_{50} = 7\ \text{min},\ S_{50} = 82\%$$
$$t_{60} = 9\ \text{min},\ S_{60} = 79\%$$

Table 6.14 lists the values of t_{40}, t_{50}, t_{60} and S_{40}, S_{50}, S_{60} of direct hydrogenation over the Ru–Zn catalyst prepared with eight different Zn precursors with the same theoretical Zn content.

From Table 6.14, the following conclusions can be drawn.

The Ru–Zn catalyst with the same theoretical Zn content was prepared with different Zn precursors, and the result of direct hydrogenation showed that the Ru–Zn

Table 6.14 Values of t_{40}, t_{50}, t_{60} and S_{40}, S_{50}, S_{60} of direct hydrogenation over the Ru–Zn catalyst prepared with eight different Zn precursors

Zn precursor	t_{40}/min	S_{40}/%	t_{50}/min	S_{50}/%	t_{60}/min	S_{60}/%
$ZnSO_4 \cdot 7H_2O$	6	81	9	79	11	77
$ZnCl_2$	7	81	9	78	12	76
$5ZnO \cdot 2CO_2 \cdot 4H_2O$	7	81	10	78	12	76
Zn powder	7	78	9	76	13	74
ZnO	6	83	9	81	14	79
$Zn(Ac)_2$	10	84	12	82	15	80
$Zn(OH)_2$	7	88	9	85	11	82
$Zn(NO_3)_2$	5	85	7	82	9	79

Table 6.15 Values of t_{40}, t_{50}, t_{60} and S_{40}, S_{50}, S_{60} of hydrogenation over the Ru–Zn catalyst after pretreatment for 22 h prepared with five different Zn precursors

Zn precursor	t_{40}/min	S_{40}/%	t_{50}/min	S_{50}/%
$ZnSO_4 \cdot 7H_2O$	15	84	20	81
$ZnCl_2$	12	80	15	78
$5ZnO \cdot 2CO_2 \cdot 4H_2O$	16	80	20	78
Zn powder	13	85	17	82
ZnO	6	83	9	81

catalyst showed not only good activity, but also high cyclohexene selectivity. The benzene conversion reached 40% in 10 min, 50% in 12 min, and 60% in 15 min. Moreover, the cyclohexene selectivity could reach above 78% at a benzene conversion of 40%, above 76% at a benzene conversion of 50%, and above 74% at a benzene conversion of 60%. The benzene conversion was increased from 40 to 60%, and the cyclohexene selectivity decreased by 2% points when the benzene conversion increased by 10% points.

Table 6.15 lists the values of t_{40}, t_{50}, t_{60} and S_{40}, S_{50}, S_{60} of hydrogenation over the Ru–Zn catalyst after pretreatment for 22 h prepared with five different Zn precursors.

From Table 6.15, the following conclusions can be drawn.

The Ru–Zn catalyst with same theoretical Zn content was prepared with different Zn precursors, and the hydrogenation performance result of the catalyst after pretreatment for 22 h showed that the reaction time was 6–16 min when reaching a benzene conversion of 40%, and it was 9–20 min when reaching a benzene conversion of 50%. The cyclohexene selectivity could reach above 80% at a benzene conversion of 40%, and it could reach above 78% at a benzene conversion of 50%. The activity and cyclohexene selectivity of the Ru–Zn catalysts with the same theoretical Zn content prepared with different Zn precursors could be maintained at a certain level, as long as Zn content was the same.

6.1.2 Characterizations of the Ru–Zn Catalyst Prepared with Different Zn Precursors

Figure 6.7 presents the XRD patterns of the Ru–Zn catalyst prepared with different Zn precursors.

As shown in Fig. 6.7, all the catalysts showed the diffraction peaks of metallic Ru (JCPDS 01-070-0274) at $2\theta = 38.4°$, 44.0°, 58.3°, 69.4°, 78.4°, and 84.7°, but the diffraction peaks of Zn or ZnO were not detected, indicating that the metallic Ru played a catalytic role under the hydrogenation conditions, providing the active sites for benzene hydrogenation, additionally, Zn and Ru form a solid solution. Since the actual Zn content is more than 7%, Ru phase is the only detectable phase when the solid solution is formed, and Zn was randomly distributed in the Ru lattice. According

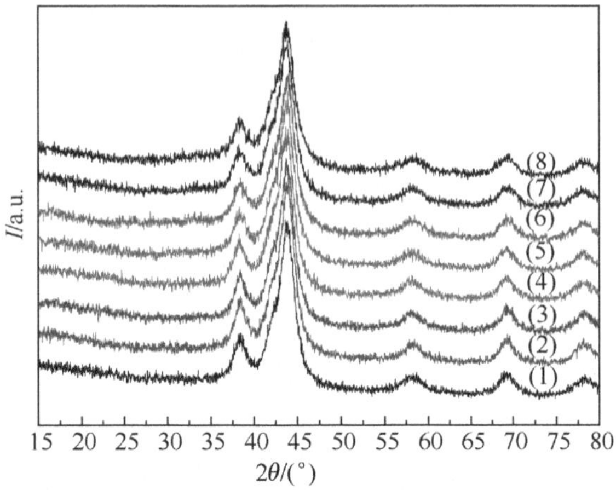

Fig. 6.7 XRD patterns of the Ru–Zn catalyst prepared with different Zn precursors. (1) $ZnSO_4$; (2) $Zn(Ac)_2$; (3) $Zn(OH)_2$; (4) ZnO; (5) $ZnCl_2$; (6) Zn powder; (7) $5ZnO{\cdot}2CO_2{\cdot}4H_2O$; (8) $Zn(NO_3)_2$

to the knowledge of structural chemistry, the tendency to form solid solution alloys depends on three factors: the position of two metal elements in the element periodic table and the proximity degree of their chemical and physical properties, the proximity degree of atomic radii; as well as the structural forms of the elements. The transition metal elements are most likely to form a solid solution phase, when the radius difference between two transition metal atoms is less than 15%, together with the same structure of the element, and often can form a complete solid solution system. The positions of Ru and Zn in the periodic table are close, with the atomic radii difference of less than 3%, and the structural forms are the same (A3-type), so the two elements are easy to form a solid solution. Zn randomly occupies the lattice position in the Ru lattice, and the properties of the Ru active center are modified by the electronic effect and geometrical effect, thus affecting the activity and cyclohexene selectivity of the catalyst. The Ru catalyst showed high activity and low cyclohexene selectivity, which can be understood as the strong ability to absorb benzene and desorb hydrogen. Each benzene molecule is surrounded by a large amount of hydrogen atoms, thus, benzene is easy to be converted to cyclohexane through direct hydrogenation. If Ru atoms are partially occupied by the Zn Species, and Zn itself has no ability to dissociate hydrogen, but it plugs the active centers of Ru, and thus reduces the hydrogen atoms around the benzene molecules. This is conducive to the step-by-step hydrogenation of benzene to cyclohexene, which can explain why the Zn promoter can improve the selectivity of cyclohexene. The reason why high cyclohexene selectivity can be achieved with appropriate Zn content is that the selectivity of the Ru–Zn catalyst is related to the number and distribution of Zn atoms around the Ru atoms.

The selective hydrogenation of benzene requires the active centers of more than one metals and it is a structurally sensitive reaction [5], which is closely related to the crystallite size and shape of the metal catalyst.

Table 6.16 shows the Ru crystallite sizes calculated from the strongest peak broadening at $2\theta = 44.0°$ in XRD patterns using the Scherrer equation.

Table 6.16 Zn/Ru atomic ratio and Ru crystallite size *d* of the Ru–Zn catalyst prepared with different Zn precursors

Zn precursor	Zn/Ru (mol/mol)[a]	*d*/nm[b]
$ZnSO_4$	0.31	4.6
$Zn(Ac)_2$	0.30	4.8
$Zn(OH)_2$	0.30	4.7
ZnO	0.29	4.8
$ZnCl_2$	0.28	5.1
Zn powder	0.31	4.7
$5ZnO{\cdot}2CO_2{\cdot}4H_2O$	0.32	4.9
$Zn(NO_3)_2$	0.28	5.0

[a]Measured by XRF; [b]Measured by XRD

As shown in Table 6.16, the Ru crystallite sizes of the Ru–Zn catalysts prepared with $ZnSO_4$, $Zn(Ac)_2$, $Zn(OH)_2$, ZnO, $ZnCl_2$, Zn powder, $5ZnO{\cdot}2CO_2{\cdot}4H_2O$, and $Zn(NO_3)_2$ as Zn precursors are mainly distributed in the narrow range of 4.6–5.1 nm, the Zn/Ru atomic ratios are 0.28–0.32, indicating that the Ru–Zn catalysts prepared with different Zn precursors have similar compositions and Ru crystallite sizes, and thus show similar activity and selectivity.

It is reported that for the Ru-Ba/SBA-15 catalyst for benzene selective hydrogenation, when the Ba/Ru is 0.1–1.0, the Ru crystallite size increases from 3.6 to 7.5 nm, and the maximum cyclohexene yield is obtained at the Ru crystallite size of 5.6 nm, because the concentration of active sites which are beneficial for the formation of cyclohexene is highest with the Ru crystallite size of 5.6 nm [6].

It is required that the Ru–Zn crystallite size of the industrial catalyst for benzene selective hydrogenation is no more than 7.0 nm. With the extension of the catalyst service time, the crystallite size increases gradually, while the selectivity of cyclohexene decreases, and the selectivity of cyclohexene is obviously reduced when the crystallite size increases to 15–20 nm, and this change is irreversible. The crystallite size is an important control index of the Ru–Zn catalyst.

Cl^- is not detected by X-ray fluorescence spectroscopy (XRF), indicating that Cl^- in the precursor has been washed away, and thus the impact on the catalyst performance can be ignored. For all the anions in different Zn precursors, since Cl^- is easy to anchor on the carrier or Ru surface, it is most difficult to be washed [1]. Additionally, it means that the anions in other precursors have been washed away.

Figure 6.8 shows the TEM images of the Ru–Zn catalysts prepared with different Zn precursors.

From Fig. 6.8, it can be seen that the crystallite morphology of the Ru–Zn catalyst is similar, and it is generally composed of spherical or ellipsoidal particles. Therefore, there is little difference in the catalyst performance due to the particle morphology.

Figure 6.9 shows the HTEM image of the Ru–Zn catalyst prepared with $ZnSO_4{\cdot}7H_2O$ as the Zn precursor, and the particle size distribution of Ru–Zn catalyst is given by the statistical method.

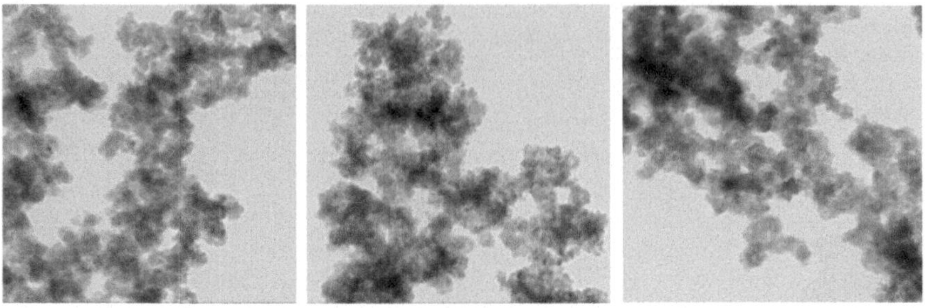

(a) (b) (c)

Fig. 6.8 TEM images of the Ru–Zn catalyst prepared with different Zn precursors. **a** $ZnSO_4$; **b** $Zn(OH)_2$; **c** Zn powder

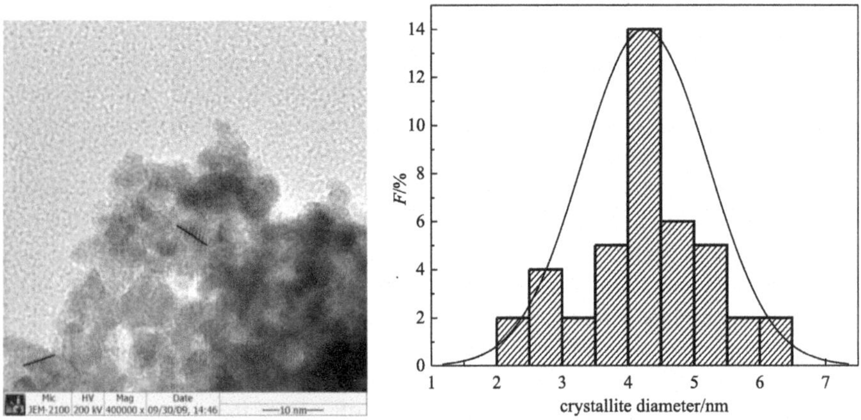

(a) (b)

Fig. 6.9 HTEM image (**a**) and particle size distribution (**b**) of the Ru–Zn catalyst prepared from $ZnSO_4$ as the Zn precursor

As can be seen from Fig. 6.9, the Ru–Zn catalyst is composed of the particles that are tightly bonded together with ellipsoidal crystallites. The crystallite particle size distribution curve shows that the crystallite size of the Ru–Zn catalyst is concentrated at around 4.3 nm, which is consistent with the crystallite size obtained from the XRD results. It is proved that the Ru–Zn catalyst belongs to the nanoparticle. Nanoparticles are in the transition zone between the cluster and the macroscopic object, and thus have some special properties, often showing high catalytic activity and selectivity. The concept of "interface confinement" of nanoparticles is proposed [7, 8], and the catalytic "scale control method" based on the confinement effect is further proposed on the basis of electronic control. The core goal is to control the properties of catalyst active sites through the nanosize effect and interface confinement effect, achieving high conversion and selectivity of almost 100% as well as high stability at low

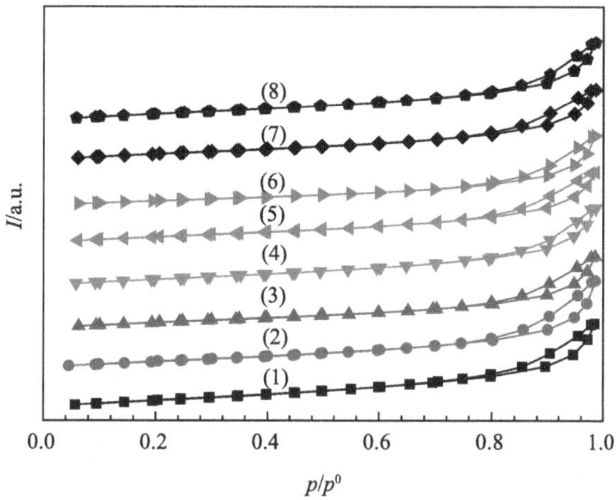

Fig. 6.10 Adsorption-desorption isotherms of the Ru–Zn catalyst prepared with different Zn precursors. (1) $ZnSO_4$; (2) $Zn(Ac)_2$; (3) $Zn(OH)_2$; (4) ZnO; (5) $ZnCl_2$; (6) Zn powder; (7) $5ZnO{\cdot}2CO_2{\cdot}4H_2O$; (8) $Zn(NO_3)_2$

temperatures. Moreover, the complete monolayer-dispersed Pt_1/FeO_x catalyst has been successfully obtained [9, 10], which greatly improves the catalyst selectivity and the single atom is considered to be the real active center. The uniform chemical environment around the single atom drives the reaction to be carried out in a certain direction. The research and scientific concepts in the nanofield provide theoretical support for the high activity and selectivity of the Ru–Zn catalyst.

The adsorption-desorption isotherms of Ru–Zn catalysts prepared with different Zn precursors are presented in Fig. 6.10.

Figure 6.10 shows that the adsorption-desorption isotherms of the Ru–Zn catalysts prepared with different Zn precursors all have hysteresis loops. The hysteresis loops are an important sign of the mesopores (2 nm < r < 50 nm), indicating that the Ru–Zn catalyst has a typical mesoporous structure. In general, the greater separation degree of the absorption-desorption isotherms means the higher content of the corresponding pores at a certain relative pressure p/p_0. For the Ru–Zn catalysts prepared with different Zn precursors, the adsorption-desorption isotherms show a large separation at $p/p_0 > 0.75$, indicating the existence of abundant mesopores. The adsorption isotherm is between type-IV and type-V according to the BDDT classification, indicating that the interaction between the adsorbate and the adsorbent is neither too strong nor too weak. The hysteresis loop belongs to H1 type according to the IUPAC classification. Combined with the TEM image of the Ru–Zn catalyst, the pore structure corresponding to the H1-type hysteresis loop can be understood as the pores formed by the accumulation of spherical particles with uniform size. The adsorption-desorption isotherms of Ru–Zn catalyst prepared with different Zn precursors coincide with each other at the relative pressure $(p/p_0) < 0.8$, indicating that the adsorption and desorption are completely reversible, which is related to the monolayer adsorption. The monolayer adsorption is reversible without hysteresis. The adsorption isotherms reflect the surface properties and pore distribution of the catalyst, as well as the information about the interaction between the catalyst and

Table 6.17 BET surface area S_{BET}, average pore diameter d_p, and pore volume V_p of the Ru–Zn catalysts prepared with different Zn precursors

Zn precursor	$S_{BET}/(m^2/g)$	d_p/nm	$V_p/(cm^3/g)$
$ZnSO_4$	56	0.13	10.2
$Zn(Ac)_2$	52	0.12	9.5
$Zn(OH)_2$	53	0.11	9.8
ZnO	62	0.13	11.2
$ZnCl_2$	53	0.14	10.4
Zn powder	54	0.13	9.3
$5ZnO{\cdot}2CO_2{\cdot}4H_2O$	58	0.15	13.3
$Zn(NO_3)_2$	57	0.14	10.6

the adsorbed gas. The adsorption-desorption isotherms of Ru–Zn catalysts prepared with different Zn precursors are similar, indicating the similar textural properties.

Table 6.17 gives the textural parameters of Ru–Zn catalysts prepared with different Zn precursors.

As can be seen from Table 6.17, the Ru–Zn catalysts prepared by using $ZnSO_4$, $Zn(Ac)_2$, $Zn(OH)_2$, ZnO, $ZnCl_2$, Zn powder, $5ZnO{\cdot}2CO_2{\cdot}4H_2O$, and $Zn(NO_3)_2$ as Zn precursors showed BET specific surface areas of 52–62 m^2/g, the average pore sizes of 0.11–0.15 nm, and the average pore volumes are concentrated in the range of 9.3–11.2 cm^3/g. It is shown that the Ru–Zn catalysts with the same theoretical Zn content prepared by the same method show similar textural properties.

Table 6.18 gives the maximum yield of cyclohexene in 20–25 min of Ru–Zn catalysts prepared with different Zn precursors.

Table 6.18 shows that the yield of cyclohexene reached 57.0% in 20–25 min, indicating that the Ru–Zn catalysts prepared with different Zn precursors show similar performances. The Zn content in the catalyst plays a key role in the performance of catalyst.

Table 6.18 Maximum yield of cyclohexene in 20–25 min of the Ru–Zn catalysts prepared with different Zn precursors

Zn precursor	C_{BZ}/%	S_{HE}/%	Y_{HE}/%[a]	t/min
$ZnSO_4$	79.4	75.1	59.6	25
$Zn(Ac)_2$	79.4	75.1	59.6	25
$Zn(OH)_2$	91.8	68.1	62.5	25
ZnO	89.2	63.9	57.0	25
$ZnCl_2$	78.3	76.3	59.7	25
Zn powder	96.6	62.5	60.4	20
$5ZnO{\cdot}2CO_2{\cdot}4H_2O$	88.0	67.5	59.4	25
$Zn(NO_3)_2$	90.6	64.7	58.6	20

[a]Maximum yield of cyclohexene

Based on the above characterization results, the essential reasons for the high activity and high selectivity of Ru–Zn catalysts prepared with different Zn precursors can be summarized as the following three aspects. The electronic properties, geometrical structure, and surface properties of the Ru-active centers can be modified by the Zn promoter, increasing the hydrophilicity of the catalyst. The suitable crystallite size is beneficial to improve the selectivity of cyclohexene. Moreover, the abundant mesopores and the cylindrical porous structure with both sides open are conducive to the evolution of cyclohexene from the surface and channel of the catalyst, and inhibit the further hydrogenation to cyclohexane.

In summary, the activity and selectivity of the Ru–Zn catalyst depend on the Zn content, and the Zn precursors have little effect. In contrast, $ZnSO_4$ and $ZnCl_2$ are cheap and easy to be obtained, and the Ru–Zn catalyst prepared by using $ZnSO_4$ and $ZnCl_2$ as the Zn precursor shows relatively good activity and selectivity, thus, the optimal Zn content in the catalyst is determined by using $ZnSO_4$ as the Zn precursor.

6.2 Effect of Zn Content on the Performance of Ru–Zn Catalyst

6.2.1 Effect of Zn Content

Figure 6.11 presents the effect of Zn content on the activity and cyclohexene selectivity of the Ru–Zn catalyst.

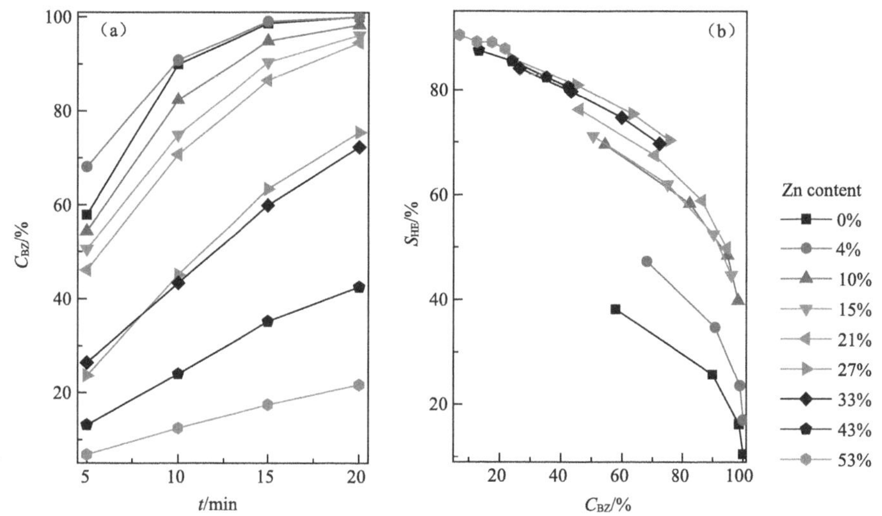

Fig. 6.11 Effect of Zn content on the activity and cyclohexene selectivity of Ru–Zn catalyst. **a** C_{BZ}-t curve; **b** S_{HE}-C_{BZ} curve

Figure 6.11a, b show that with the Zn content in the Ru–Zn catalyst being increased from 0 to 53% (mass fraction), the benzene conversion gradually decreases and the cyclohexene selectivity gradually increases. Zn plays an important role in reducing the catalyst activity and improving the cyclohexene selectivity. It is reported that the activity of the Ru–Zn/SiO_2 catalyst prepared by co-impregnation method decreases, and the cyclohexene selectivity increases first and then decreases with the increase of Zn content [11]. For the Ru–Zn/ZrO_2 catalyst prepared by Ru/ZrO_2 reduction in $ZnSO_4{\cdot}7H_2O$ solution, the activity and selectivity of the catalyst both increased first and then decreased with the increase of Zn content [12], and the Ru–Zn catalyst prepared by co-precipitation method was found to have the same influence law of Zn content [13]. Although the preparation method and hydrogenation conditions of Ru–Zn catalysts are different, and the influence of Zn is not exactly the same, it is clearly proven that Zn plays an important role in reducing the catalyst activity and improving the selectivity of cyclohexene.

Figure 6.11 also shows that the Ru–Zn catalyst with a Zn content of 15–27% showed a high benzene conversion within the same reaction time, and high cyclohexene selectivity at the same benzene conversion, and thus the maximum yield of cyclohexene could be achieved. Therefore, the effects of Zn content on the activity and cyclohexene selectivity of the catalyst are investigated by using the pretreated catalysts with a Zn content of 15%, 21%, and 27%, respectively.

Figure 6.12 gives the hydrogenation activity, cyclohexene selectivity, and yield of the pretreated Ru–Zn catalyst with a Zn content of 15, 21, and 27%.

From Fig. 6.12, it can be seen that Ru–Zn (21%) catalyst not only has a high activity, but also shows the highest cyclohexene selectivity at the same benzene conversion, and the maximum yield of cyclohexene reaches 59%.

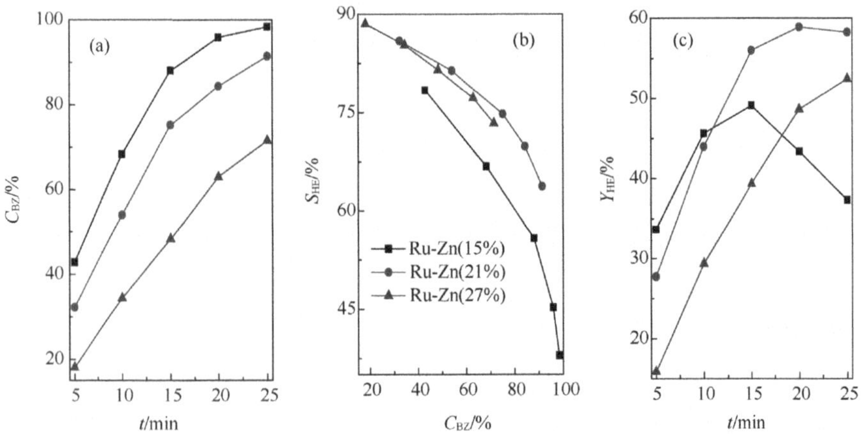

Fig. 6.12 Hydrogenation activity, cyclohexene selectivity, and yield of the pretreated Ru–Zn catalyst with a Zn content of 15, 21, and 27%. **a** C_{BZ}-t curve; **b** S_{HE}-C_{BZ} curve; **c** Y_{HE}-t curve

Table 6.19 Actual Zn contents in the Ru–Zn catalysts (mass fraction)

Theoretical Zn content of Ru–Zn catalyst	Actual Zn content measured by atomic absorption spectroscopy/%
Ru–Zn (0)	0
Ru–Zn (4%)	2.55
Ru–Zn (10%)	5.17
Ru–Zn (15%)	7.66
Ru–Zn (21%)	8.56
Ru–Zn (27%)	9.55
Ru–Zn (33%)	12.42
Ru–Zn (43%)	14.85
Ru–Zn (53%)	29.09

The actual Zn contents measured by atomic absorption spectroscopy in the Ru–Zn catalysts prepared with various theoretical Zn contents are listed in Table 6.19.

As can be seen from Table 6.19, the actual Zn content increases with the increase of theoretical Zn content in the catalyst, but it is much lower than the theoretical value. This is due to the high concentration of NaOH during the precipitation and reduction process. $Zn(OH)_2$ and ZnO are acidic, and the following reactions occurred:

$$Zn(OH)_2 + 2NaOH = Na_2ZnO_2 + 2H_2O$$
$$ZnO + 2NaOH = Na_2ZnO_2 + H_2O$$

Therefore, Zn may partially remain in the form of zincate in the solution. Since Zn and Ru are easy to form a solid solution, Zn in the Ru lattice of solid solution is difficult to be dissolved. It is confirmed by XPS that a small amount of Zn^{2+} (about 1.1%) can be reduced to the metallic Zn. Thus, Zn in the catalyst may be presented as both the metallic Zn and ZnO.

The amphoteric properties of ZnO have caused great difficulties in the preparation of Ru–Zn catalysts. The amount of NaOH must be taken into account during the preparation of the Ru–Zn catalyst. A big difference is found in the theoretical and actual Zn contents due to the different concentrations of NaOH. It is reported in the foreign patents and literature that the use of 30% NaOH solution is clearly not desirable. The usage of NaOH was varied from more than the theoretical amount of 1/3, 2/3, 3/3, 4/3, 5/3, 6/3, 7/3, and until 8/3. The actual Zn content in the catalyst was simultaneously determined by both atomic absorption spectrum (AAS) and inductively coupled plasma atomic emission spectrometry (ICP-AES), and the activity and selectivity of the catalyst were determined, so the optimal Zn content was ultimately determined.

6.2.2 Characterizations of the Ru–Zn Catalyst

Figure 6.13 displays the XRD patterns of catalysts with different Zn contents.

Figure 6.13 shows that the diffraction peaks of Ru (JCPDS 01-070-0274) are detected at $2\theta = 38.4°, 43.9°, 58.3°, 69.4°$, and 78.1°. The actual Zn content increases from 2.55 to 14.85% when the theoretical content increases from 4 to 43%. The diffraction peaks of Zn or ZnO are not observed in the XRD patterns, which further confirmed the view of the formation of the solid solution by Zn and Ru. The Zn atoms randomly occupy the lattice sites of the Ru lattice, and since Ru is the main body, the point group symmetry of the Ru crystal hexagonal system is maintained [14]. Therefore, only the diffraction peaks of the elemental Ru can be detected in the XRD patterns, or ZnO is highly dispersed on the surface of Ru, with the grain size less than the detection limit of 4 nm [15]. According to the principle of solid surface segregation, the surface structure of the solid is different from the bulk and gradually approaches the bulk phase within 10 atomic layers. Therefore, it is possible that a small part of Zn is enriched on the catalyst surface, but a majority of Zn must enter into the bulk phase.

Figure 6.13 also shows that the diffraction peaks of ZnO (JCPDS 01-070-2551) are detected at 2θ = 31.9°, 34.5°, 36.3°, 47.6°, 56.6°, 62.9°, 68.0°, and 69.2° in the XRD pattern of the catalyst with Zn content of 53%. It is shown that the incomplete solid solution is formed by ZnO and Ru, and Zn will be precipitated separately from the Ru lattice when Zn content is higher than a certain value.

Figure 6.14 shows the XRD patterns of Ru–Zn (15%) (mass fraction) catalysts with different aging times.

It can be seen from Fig. 6.14 that the diffraction peaks of ZnO are also found on the Ru–Zn (15%) fresh catalyst, but disappear after aging, and this may because the substitution between ZnO and Ru lattice takes a certain time.

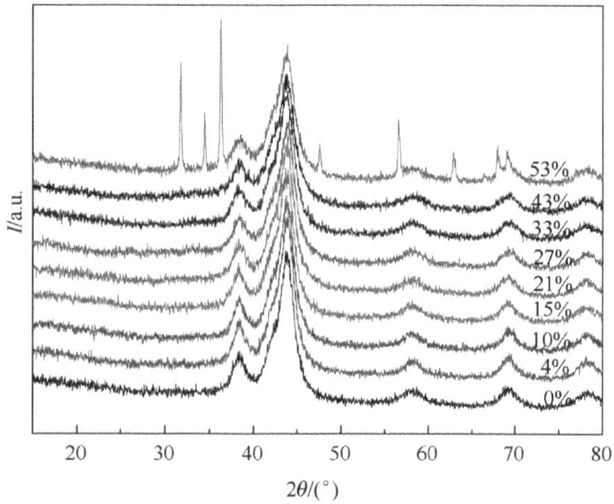

Fig. 6.13 XRD patterns of the Ru–Zn catalysts with different Zn contents

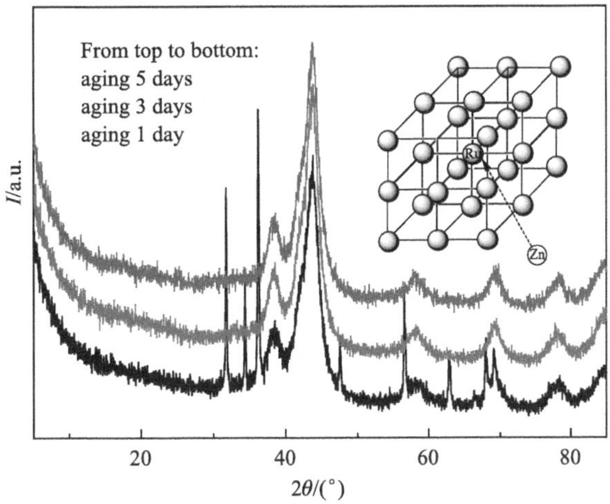

Fig. 6.14 XRD patterns of the Ru–Zn (15%) catalyst with different aging time

Figure 6.15 shows the XRD patterns of the Ru–Zn catalysts with different Zn contents after hydrogenation.

Figure 6.15 shows that only shorter and broaden diffraction peaks of Ru are detected at $2\theta = 43.8°$ in the Ru–Zn catalyst with different Zn contents after hydrogenation, indicating that the Ru crystallite is of small size and highly dispersed on ZrO_2. In the XRD patterns of the catalyst with Zn content of 27–53%, The ZnO enriched on the catalyst surface reacted with $ZnSO_4$ in the slurry to form the $(Zn(OH)_2)_3(ZnSO_4)(H_2O)_5$ salt. Two diffraction peaks of the $(Zn(OH)_2)_3(ZnSO_4)(H_2O)_5$ salt (JCPDS 01-078-0246) were detected at 2θ = 8.2° and 43.8°, and the intensity of the peaks gradually increased with the increase of Zn content, while the diffraction peaks of ZnO disappeared. The

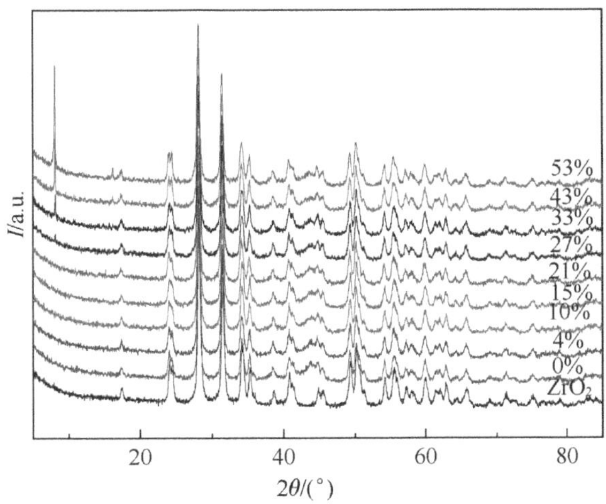

Fig. 6.15 XRD patterns of the Ru–Zn catalysts with different Zn contents after hydrogenation

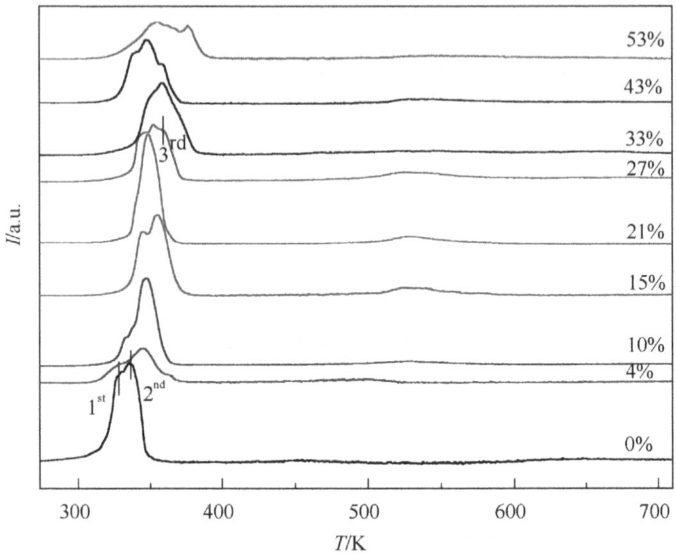

Fig. 6.16 H_2-TPR profiles of the Ru–Zn catalysts with different Zn contents before hydrogenation

$(Zn(OH)_2)_3(ZnSO_4)(H_2O)_5$ salt is monolayer dispersed on the surface of the catalyst when Zn content is less than 21%. When Zn content is higher than 27%, the $(Zn(OH)_2)_3(ZnSO_4)(H_2O)_3$ salt aggregates to form a new phase on the catalyst surface. The monolayer dispersion capacity of the $(Zn(OH)_2)_3(ZnSO_4)(H_2O)_3$ salt should be in the range of 21–27%. The actual Zn content is about 8.56% with the theoretical Zn content of 21% [16]. The diffraction peaks of the $(Zn(OH)_2)_3(ZnSO_4)(H_2O)_3$ salt are extremely weak, and thus the salt can be approximated to be monolayer dispersed.

Figure 6.16 illustrates the H_2-TPR profiles of the Ru–Zn catalyst with different Zn contents before hydrogenation.

The H_2-TPR profiles in Fig. 6.16 for the Ru–Zn catalysts at Zn contents of 0–15% before hydrogenation show two reduction peaks between 328–345 and 337–355 K, which is consistent with the result reported by Wang et al. [17]. The first reduction peak ascribes to the reduction peak of $RuCl_3$ which is not completely precipitated, and the second is attributed to the reduction peak of $Ru(OH)_3$ or Ru oxides. However, the temperatures of these two reduction peaks are much lower than the values reported in the literature (320–410 K and 410–510 K), which indicates that the precursor of Ru–Zn catalyst is highly dispersed and easy to be reduced. The H_2-TPR profile for the Ru–Zn catalyst with a Zn content of 21% shows only a symmetric reduction peak at 350 K, indicating the same coordination environment and chemical properties of Ru and Zn. The H_2-TPR profile for the catalyst with the Zn content of 33–53% shows a third peak between 359 and 378 K, in addition to the first two peaks, and it shifted to the high temperature with the increase of Zn content. For the Ru–Zn/SiO_2 catalyst, since the interaction of Zn and Ru delayed the reduction of Ru, the TPR

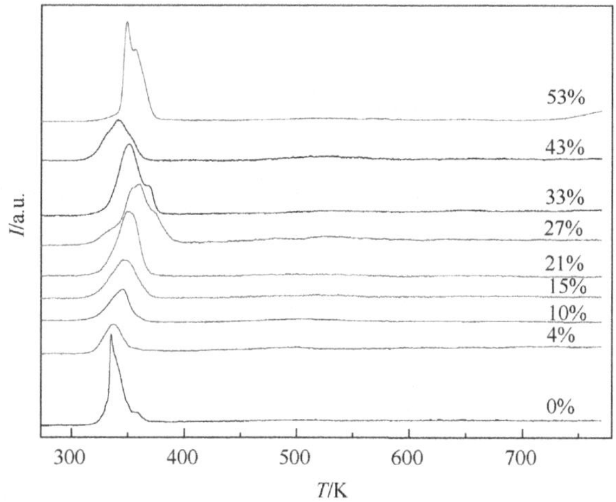

Fig. 6.17 H_2-TPR profiles of the Ru–Zn catalysts with different Zn contents after hydrogenation

reduction peak also shifted to the high temperature with the increase of Zn content [14]. According to Pauling electronegativity principle, the electronegativity of Ru and Zn are 2.2 and 1.65, respectively, and thus the electrons of Zn are partially shifted to Ru, which makes Ru difficult to be reduced. This may also be the reason why the reduction peak shifted to the high temperature and the third reduction peak appeared when the Zn content increased to 33%.

Figure 6.17 shows the H_2-TPR profiles of Ru–Zn catalysts with different Zn contents after hydrogenation.

The H_2-TPR profiles in Fig. 6.17 for the Ru–Zn catalysts at Zn contents of 0–21% after hydrogenation show that the reduction of the Ru species becomes easier due to the dispersion of ZrO_2, indicating the weak interaction between Ru and the dispersant ZrO_2. Only a reduction peak appears between 335 and 400 K. The H_2-TPR profiles for the Ru–Zn catalysts with Zn content of 27–53% after hydrogenation shows three reduction peaks, and the reduction temperature shifts to the high temperature due to the interaction between Ru and Zn. The catalyst with Zn content of 27% is also a turning point, it is further confirmed that the Ru–Zn catalyst which is close to monolayer dispersed performs different reduction behavior in the hydrogen atmosphere, and thus show different catalytic properties.

Table 6.20 gives the texture characters of the Ru–Zn catalysts with different Zn contents.

As can be seen from Table 6.20, the pure Ru catalyst has a larger specific surface area than any Zn-containing Ru–Zn catalyst. With the increase of Zn content, the specific surface area of the Ru–Zn catalyst increased first and then decreased, and there is a turning point between Ru–Zn (15%) and Ru–Zn (21%). Then the pore size decreased first and then increased, which was related to the dispersion state of Zn.

Figure 6.18 shows a schematic representation of the Zn species dispersed on the Ru surface.

Table 6.20 BET surface area S_{BET}, average pore diameter d_p, and pore volume V_p of the Ru–Zn catalysts with different Zn contents (mass fraction)

Sample	$S_{BET}/(m^2/g)$	$V_p/(cm^3/g)$	d_p/nm
Ru–Zn (0)	88.1	0.1819	4.1320
Ru–Zn (4%)	76.9	0.1844	4.7985
Ru–Zn (10%)	66.6	0.1522	4.5695
Ru–Zn (15%)	73.5	0.1653	4.5006
Ru–Zn (21%)	77.2	0.1550	4.0182
Ru–Zn (27%)	54.9	0.1453	5.2944
Ru–Zn (33%)	66.4	0.2737	8.2383
Ru–Zn (43%)	63.5	0.2664	8.3956
Ru–Zn (53%)	56.8	0.2213	7.7961

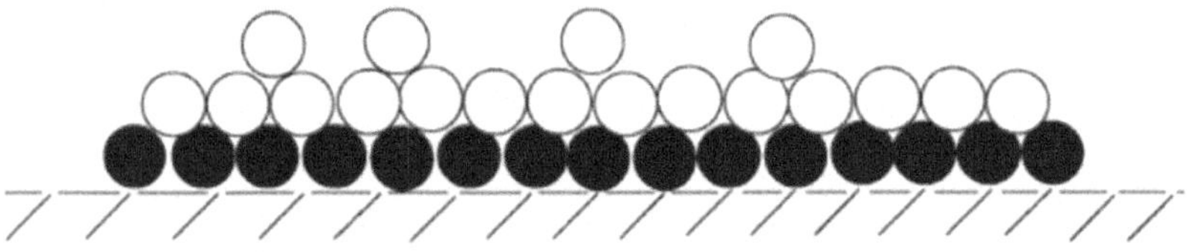

Dispersion state of Zn species

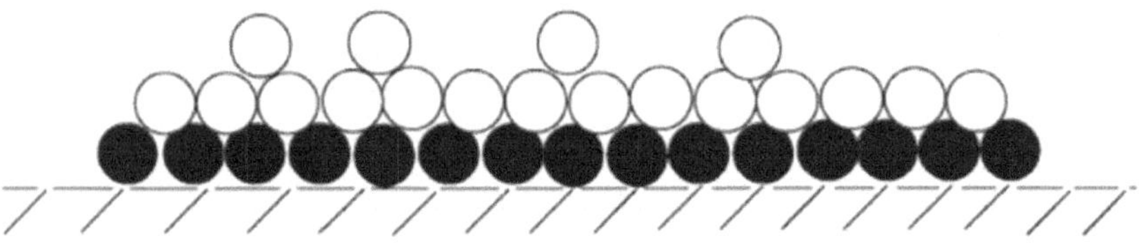

Monolayer dispersion state of Zn species

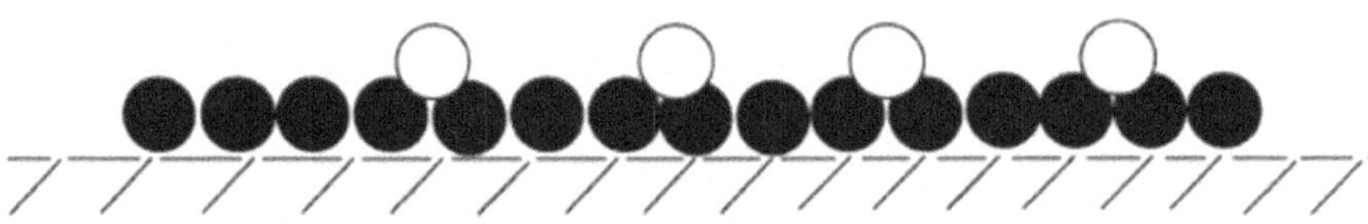

Non-monolayer dispersion state of Zn species

Fig. 6.18 Schematic representation of the dispersion of Zn species on Ru

In Fig. 6.18, the black solid sphere represents the Ru atom, and the hollow sphere represents the Zn species. The lowest layer indicates that the Zn species occupies the partial surface of the Ru before it reaches the monolayer dispersion state, and thus the specific surface area of the Ru–Zn catalyst is reduced. With the increase of Zn content, the monolayer dispersion of Zn species becomes severe and the specific surface area increases. Meanwhile, the Zn species may be dispersed in the pores of

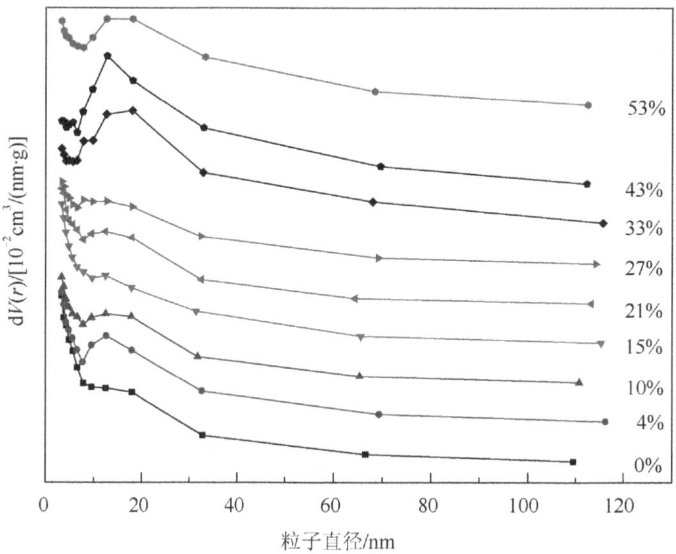

Fig. 6.19 Pore size distribution of Ru–Zn catalyst with different Zn contents

Ru, which is the main reason for the decrease of the pore size with the increase of Zn content. When Zn species reaches the monolayer dispersion state, the specific surface area is the largest and the pore size is the smallest. After the Zn species reaches the monolayer dispersion state, it starts to accumulate and the specific surface area decreases. Zn species itself or with Ru can form new stacking holes, so the pore size increases. It can also be seen from Table 6.20 that there is little change in the pore volume with increasing Zn content.

Figure 6.19 shows the pore size distribution of Ru–Zn catalysts with different Zn contents.

It can be seen from Fig. 6.19, the pore size distribution was very similar when the Zn content was from 0 to 15%. When Zn content reached 21%, the Zn species blocked some micropores, but most of the mesopores were retained due to the monolayer dispersion of Zn content. At Zn content of 27%, the pores with a size of 2–6 and 6–33 nm reduced possibly due to the Zn species blocked part of the micropores and mesopores. When Zn content was 33–53%, the Zn content blocked some microporous and the accumulation of Zn and Ru or Zn itself led to the formation of new pores so that the number of micropores reduced and the mesopores increased.

The results of low-temperature N_2 adsorption and H_2-TPR of catalysts before and after hydrogenation showed that the interaction between Zn species and Ru changed the reduction performance and texture properties of the Ru catalyst. XRD confirmed that the critical Zn content of the monolayer dispersion of Zn species on Ru and ZrO_2 was close to 21%. The monolayer dispersion of Zn species resulted in the similar coordination environment and the same chemical properties of the active component Ru and the auxiliary Zn, and the relative large specific surface area and suitable pore structure, therefore, the catalysts not only have higher catalytic activity, but also have higher selectivity to cyclohexene.

6.3 Development of the Fourth-Generation Ru–Zn@BZSS Catalyst

6.3.1 Preparation of Ru–Zn@BZSS Catalyst

Ru–Zn catalyst and Ru–Zn@BZSS catalyst were prepared by using Ru as active component, Zn as additive, and basic zinc sulfate BZSS as modifier. In the Ru–Zn@BZSS catalyst, Ru–Zn is the nucleus and BZSS is the shell, which is a core-shell Ru–Zn catalyst covered with basic zinc sulfate.

Ru–Zn catalyst: Firstly, based on the molar amounts of active compound Ru and the additive Zn, the amount of Ru is set as 1, the amount of Zn is 0.1–1, and the mixed solution of $RuCl_3 \cdot H_2O$ and $ZnSO_4 \cdot 7H_2O$ and the alkali solution with a mass fraction of 5–30% are prepared, respectively. Then, at 70–100 °C, the alkali solution and the mixed solution were simultaneously introduced into the reaction tank with stirring. After the completion of the precipitation, the mixture was stirred for 10–30 min and the pH of the mother liquor was controlled at about 12. Then the precipitate together with the mother liquor was reduced at 100–150 °C under 3–5 MPa H_2 for 1–12 h, and finally the resulting solid was washed with distilled water until the filtrate did not contain Cl^-.

Preparation of BZSS: The first method: take a certain amount of zinc sulfate heptahydrate and sodium hydroxide in two beakers, then 20 and 30 mL of steamed water were added, respectively. The beaker with zinc sulfate was equipped with transferred to the magnetic stirrer, and the sodium hydroxide solution was added dropwise along the glass rod after zinc sulfate was fully dissolved, then the reaction lasted for about 5 min. The precipitate was washed three times with steamed water and then filtered and vacuum dried. The second method: the sodium hydroxide solution was poured into the zinc sulfate solution in one-time, and the rest of the method was the same as the first method. In each preparation method, $ZnSO_4$ can be over 50% and 100%, respectively, and BZSS can be obtained under four conditions. From XRD analysis, their phase was all $(Zn(OH)_2)_3(ZnSO_4)(H_2O)_3$.

Preparation of Ru–Zn@BZSS catalyst. The Ru–Zn catalyst and BZSS were placed in a certain proportion in the 1 L Hastelloy autoclave, then 280 mL water was added and the air was replaced with hydrogen. Pretreatment was carried out at 140 °C, 4 MPa H_2, and 800 r/min for 1 h.

The fourth-generation Ru–Zn@BZSS catalyst is different from the third generation, which breaks through the technical bottleneck of the formation of alkali salt in situ, and uses the alkali salt prepared by non-in situ for the chemical adsorption on the catalyst surface.

6.3.2 Reduction of Ru–Zn Catalyst

Catalyst reduction is an important part of the preparation process, because the real active compound is the metal state of Ru in the hydrogenation reaction conditions, the reduction is a catalyst activation process.

Table 6.21 shows the catalyst activity and selectivity for reduction with hydrogen at 3–10 h.

It can be seen from Table 6.21 that the activity of the direct hydrogenated catalyst was higher and the selectivity of cyclohexene was lower when comparing the reduction time of 3 h to that of 6 and 10 h, which indicated that the reduction time should not be lower than 3 h. However, compared with the reduction of 6 and 10 h, the reaction time for the same benzene conversion was prolonged with the increase of reduction time, that was, the activity of the catalyst was reduced, but the selectivity of cyclohexene was unchanged, indicating that there was a best reduction time.

From Table 6.21, the results of hydrogenation after 22 h pretreatment showed that the catalyst with 3 h reduction time had not only high activity, but also the best selectivity. This is because the catalyst has undergone another reduction in the hydrogen atmosphere during the pretreatment process. It can be concluded that the catalyst reduction time can be shortened.

Figure 6.20 shows the industrial equipment for the preparation of the Ru–Zn catalyst.

In Fig. 6.20, $RuCl_3{\cdot}3H_2O$ was added to the salt box 8 at room temperature and dissolved in deionized water with stirring. For the base tank 9, deionized water was added first and then NaOH was added with stirring with the controlled temperature at 50–60 °C.

The base solution was introduced into the high-pressure reactor 1 via the liquid feed pipe by starting the vacuum system, and then the salt solution was introduced under stirring at control reaction temperature for 30 min before aging for 3–6 h. Before catalyst reduction, N_2 was introduced through the intake pipe 2 to the aged high-pressure reactor to replace the air, and then through the high purity H2 was introduced with the pressure of 4–5 MPa before increasing the temperature. The

Table 6.21 The catalyst activity and selectivity of catalysts under hydrogen reduction in 3–10 h, t_{40}, S_{40}, t_{50}, S_{50}, and t_{60}, S_{60}

Reduction time	Hydrogenation conditions	t_{40}/min	t_{50}/min	t_{60}/min	S_{40}/%	S_{50}/%	S_{60}/%
3	Direct hydrogenation	5.3	6.9	8.5	78	76	73
6		6.9	9.2	12.0	80	78	75
10		9.6	12.6	15.6	80	78	75
3	Pretreated hydrogenation	8.4	11.1	13.8	85	83	80
6		9.0	11.7	14.7	82	80	78
10		12.0	15.0	18.8	85	82	79

Reduction condition: 140 °C, 4.5 MPa, 800 r/min

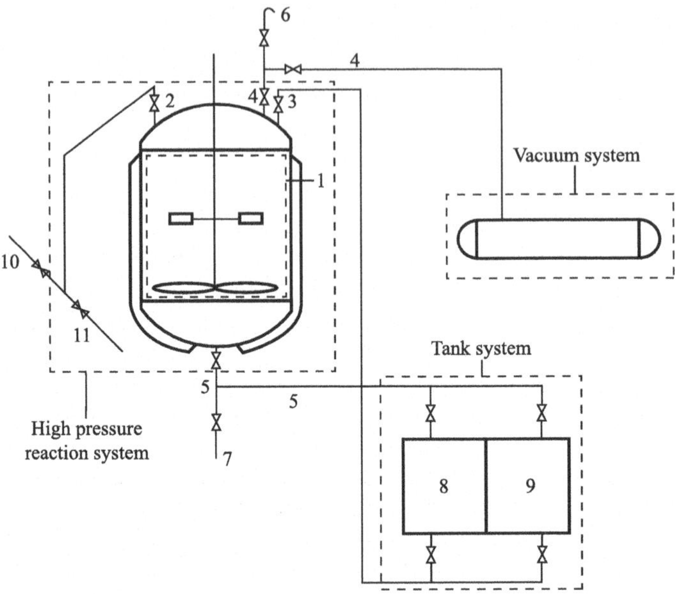

Fig. 6.20 The industrial equipment for the preparation of the Ru–Zn catalyst. 1. Autoclave; 2. Intake pipe; 3. Liquid feed pipe; 4. Exhaust pipe; 5. Discharge pipe; 6. Exhaust pipe; 7. Discharge port; 8. Salt box; 9. Base tank; 10. N_2 pipeline; 11. H_2 pipeline

reactor pressure was kept at 4.5 MPa during the heating process. When the temperature became 150 °C, the reaction time was calculated. The reduced catalysts were introduced into the salt box and the base tank through the discharge pipe 5, washed to pH $\approx$ 7, and there is no Cl^- in the solution.

6.3.3 Evaluation of Catalyst Activity and Selectivity

The Ru–Zn and Ru–Zn@BZSS catalysts were prepared according to the procedure in Sect. 6.3.1. The activity and selectivity were evaluated for hydrogenation reaction before and after pretreatment.

(1) Direct hydrogenation of Ru–Zn catalyst

Table 6.22 gave the results about the activity and selectivity of the direct hydrogenation using the Ru–Zn catalyst (140320). The number in parentheses represents the catalyst number.

Figure 6.21 shows the C_{BZ}-t curve obtained by plotting the benzene conversion C_{BZ} versus time t and the S_{HE}-C_{BZ} curve obtained by plotting the cyclohexene selectivity S_{HE} versus benzene conversion C_{BZ}.

Table 6.22 The activity and selectivity of Ru–Zn catalyst (140320) in the direct hydrogenation

t/min[a]	T/°C[b]	C_{BZ}/%[c]	S_{HE}/%[d]	Y_{HE}/%[e]
5	158	54.14	72.40	39.20
10	162	81.90	62.42	51.12
15	157	94.35	52.36	49.40
20	151	97.91	43.75	42.84
25	149	99.13	37.35	37.03

Slurry composition: 1.96 g catalyst, 280 mL water, 9.8 g ZrO_2, 45.7 g $ZnSO_4{\cdot}7H_2O$; hydrogenation conditions: 150 °C, 5 MPa H_2, 1400 r/min. After adding 140 mL benzene, the reaction started. The samples were obtained every 5 min

[a]Sampling time, [b]sampling temperature, [c]C_{BZ} conversion of benzene (mol%), [d]S_{HE} selectivity of cyclohexene, [e]Y_{HE} yield of cyclohexene

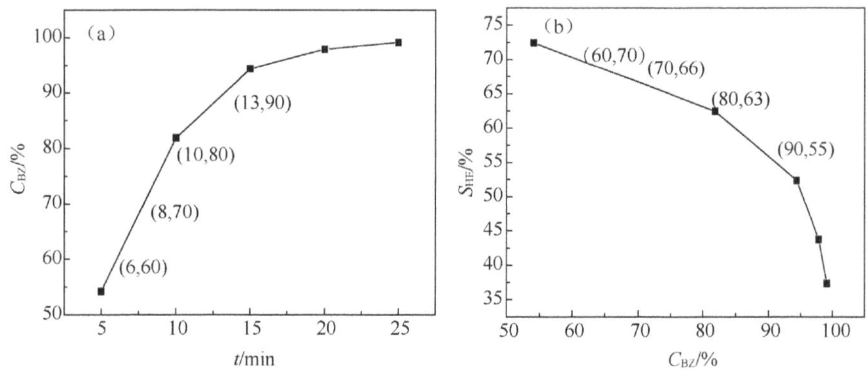

Fig. 6.21 The performance of Ru–Zn catalyst for direct hydrogenation. **a** C_{BZ}-t curve; **b** S_{HE}-C_{BZ} curve

In Fig. 6.21a, the time (min) was determined by interpolation method for 40%, 50%, 60%, and 70% conversion of benzene, respectively, which was recorded as t_{40}, t_{50}, t_{60}, t_{70}. The shorter the reaction time for the same benzene conversion was, the higher the activity of the catalyst was.

In Fig. 6.21b, the selectivity of cyclohexene was determined by interpolation method for 40%, 50%, 60%, 70% conversion of benzene, respectively, which was recorded as S_{40}, S_{50}, S_{60}, and S_{70}. The comparison of cyclohexene selectivity was carried out at the same benzene conversion.

From Fig. 6.21, the results can be obtained for the direct hydrogenation by Ru–Zn catalyst (140320):

$$t_{60} = 6\,\text{min},\ S_{60} = 70\%;\ t_{70} = 8\,\text{min},\ S_{70} = 66\%$$
$$t_{80} = 10\,\text{min},\ S_{80} = 63\%;\ t_{90} = 13\,\text{min},\ S_{90} = 55\%$$

It can be seen that the catalyst activity is very high for direct hydrogenation, the conversion of benzene was more than 50% at 5 min, thus t_{40} and t_{50} cannot be obtained. At 6 min, 8 min, 10 min, 13 min, the benzene conversions were 60%, 70%, 80%, and 90%, respectively. When Benzene conversions were 60%, 70%, 80%, and 90%, the cyclohexene selectivities were 70%, 66%, 63%, and 55%, and the yields of cyclohexene were 42%, 46%, 50%, and 49.5%, respectively. The catalyst activity was too high and the cyclohexene selectivity was relatively low. Although the yield of cyclohexene was quite high, the by-product cyclohexane was greatly produced, thus the system was not suitable for industrial implementation. The industrially ideal catalyst should achieve a higher benzene conversion and cyclohexene selectivity at a shorter reaction time, resulting in higher yields of cyclohexene. The residence time of benzene is normally 15–20 min, and the benzene conversion can be 40%, the cyclohexene selectivity can be above 80% and the cyclohexene yield can be above 32%. Although the benzene conversion is low, the benzene can be recycled and the by-product cyclohexane is less, leading to an overall higher economic benefit.

For selective hydrogenation of benzene, cyclohexene selectivity is the most important factor, the benzene conversion should be as high as possible when the cyclohexene selectivity is above 80%. Due to the dynamic characteristics of the continuous reaction, it is a critical condition to achieve 80% cyclohexene selectivity. Only when benzene conversion is low (5% or less), cyclohexene selectivity may be as high as 90%, however, this shows the poor possibility for industrial application.

(2) The hydrogenation by Ru–Zn catalyst with the pretreatment for 22 h

Table 6.23 gives the hydrogenation activity and selectivity of the Ru–Zn catalyst (140320) after pretreatment for 22 h.

Figure 6.22 shows the C_{BZ}-t curve and the S_{HE}-C_{BZ} curve of the hydrogenation by Ru–Zn catalyst with 22 h pretreatment.

From Fig. 6.22, the results of hydrogenation by Ru–Zn catalyst with pretreatment for 22 h can be obtained.

Table 6.23 Hydrogenation activity and selectivity by Ru–Zn catalyst (140320) with pretreatment for 22 h

t/min	T/°C	C_{BZ}/%	S_{HE}/%	Y_{HE}/%
5	155	36.81	81.83	30.12
10	160	58.23	77.71	45.25
15	160	76.79	71.61	54.99
20	156	88.26	65.15	57.50
25	152	93.58	59.24	55.44

Slurry composition: 1.96 g catalyst, 280 mL water, 9.8 g ZrO_2, 45.7 g $ZnSO_4 \cdot 7H_2O$; pretreatment conditions: 140 °C, 4 MPa H_2, 800 r/min, 22 h; hydrogenation conditions: 150 °C, 5 MPa H_2, 1400 r/min, 140 mL

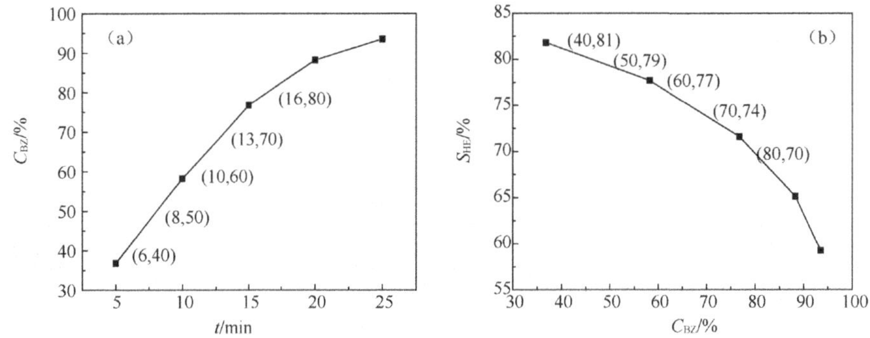

Fig. 6.22 Hydrogenation by Ru–Zn catalyst with 22 h pretreatment. **a** C_{BZ}-t curve; **b** S_{HE}-C_{BZ} curve

$$t_{40} = 6\,\text{min},\ S_{40} = 81\%,\ Y_{HE} = 32\%$$
$$t_{50} = 8\,\text{min},\ S_{50} = 79\%,\ Y_{HE} = 39\%$$
$$t_{60} = 10\,\text{min},\ S_{60} = 77\%,\ Y_{HE} = 46\%$$
$$t_{70} = 13\,\text{min},\ S_{70} = 74\%,\ Y_{HE} = 52\%$$
$$t_{80} = 16\,\text{min},\ S_{80} = 70\%,\ Y_{HE} = 56\%$$

Compared with the results in Table 6.22 and Fig. 6.21, it can be seen that the catalysts after pretreatment required longer reaction time for the same benzene conversion and the activity of the catalysts decreased: the selectivity of cyclohexene increased at the same benzene conversion. Higher selectivity and yield of cyclohexene can be achieved at a suitably high benzene conversion.

(3) Direct hydrogenation by Ru–Zn@BZSS catalyst

Table 6.24 shows the activity and selectivity of direct hydrogenation by the Ru–Zn@BZSS catalyst prepared by adding 300 mg BZSS to 1.96 g Ru–Zn catalyst (140320).

Table 6.24 The activity and selectivity of direct hydrogenation after addition of 300 mg BZSS to 1.96 g Ru–Zn catalyst (140320)

t/min	T/°C	C_{BZ}/%	S_{HE}/%	Y_{HE}/%
5	153	13.87	88.97	12.34
10	158	22.74	88.70	20.17
15	163	34.87	86.49	30.16
20	168	47.69	83.94	40.03
25	173	62.62	79.67	49.89
30	178	70.88	76.55	54.26

Slurry composition: 1.96 g catalyst, 280 mL water, 300 mg BZSS, 9.8 g ZrO_2, 45.7 g $ZnSO_4 \cdot 7H_2O$; hydrogenation conditions: 150 °C, 5 MPa H_2, 1400 r/min, 140 mL benzene

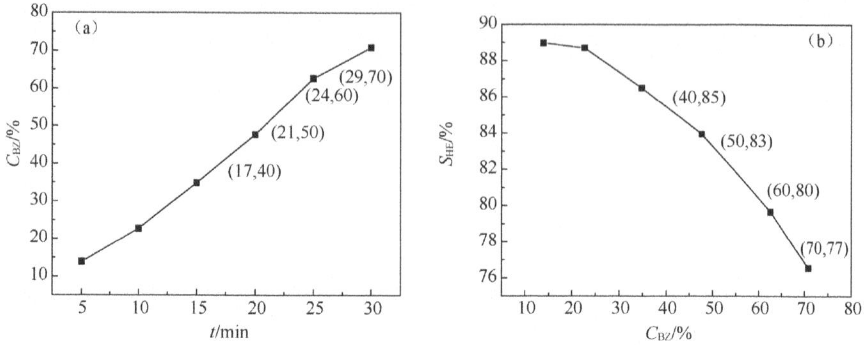

Fig. 6.23 Direct hydrogenation by Ru–Zn@BZSS catalyst. **a** C_{BZ}-t curve; **b** S_{HE}-C_{BZ} curve

Figure 6.23 shows the C_{BZ}-t curve and the S_{HE}-C_{BZ} curve of the hydrogenation by Ru–Zn@BZSS catalyst.

From Fig. 6.23, the results of hydrogenation by Ru–Zn@BZSS catalyst can be obtained.

$$\mathrm{t}_{40} = 17\ \mathrm{min},\ \mathrm{S}_{40} = 85\%;\ \mathrm{t}_{50} = 21\ \mathrm{min},\ \mathrm{S}_{50} = 83\%$$
$$\mathrm{t}_{60} = 24\ \mathrm{min},\ \mathrm{S}_{60} = 80\%;\ \mathrm{t}_{70} = 29\ \mathrm{min},\ \mathrm{S}_{70} = 77\%$$

The results show that the activity of the catalyst obviously decreased while the selectivity of cyclohexene obviously improved after adding BZSS. The performance of Ru–Zn@BZSS for direct hydrogenation showed higher cyclohexene selectivity compared with Ru–Zn catalyst with 22 h pretreatment. The chemical adsorption of BZSS on the Ru–Zn catalyst played a role as pretreatment.

(4) The hydrogenation by Ru–Zn@BZSS catalyst with pretreatment

Table 6.25 gives the hydrogenation activity and selectivity by 1.96 g Ru–Zn catalyst (140320) with the addition of 300 mg BZSS after pretreatment for 22 h.

Figure 6.24 shows the C_{BZ}-t curve and the S_{HE}-C_{BZ} curve of the hydrogenation by Ru–Zn catalyst (140320) with the addition of BZSS and 22 h pretreatment.

From Fig. 6.24, the results of hydrogenation by Ru–Zn@BZSS catalyst with 22 h pretreatment can be obtained.

$$t_{40} = 25\,\mathrm{min},\ S_{40} = 87\%;\ t_{50} = 32\,\mathrm{min},\ S_{50} = 85\%$$
$$t_{60} = 40\,\mathrm{min},\ S_{60} = 82\%;\ t_{70} = 54\,\mathrm{min},\ S_{70} = 79\%$$

The results showed that the activity of the catalyst was further reduced after pretreatment, and the selectivity of cyclohexene was further improved.

For the Ru–Zn catalyst with the addition of BZSS and pretreatment, although the selectivity of cyclohexene increased, the time to achieve the same benzene conversion was prolonged. Too long reaction time will lead to lower production capacity,

Table 6.25 The hydrogenation activity and selectivity by 1.96 g Ru–Zn catalyst (140320) with the addition of 300 mg BZSS after pretreatment for 22 h

t/min	T/°C	C_{BZ}/%	S_{HE}/%	Y_{HE}/%
5	147	6.92	88.00	6.09
10	150	15.46	90.75	14.03
15	155	24.78	89.83	22.26
20	155	33.64	88.29	29.70
25	155	56.08	83.08	46.59
30	151	51.16	84.75	43.36
40	146	59.27	82.76	49.05
50	146	66.46	80.33	53.39
60	152	73.29	77.49	56.79

Slurry composition: 1.96 g catalyst, 280 mL water, 300 mg BZSS, 9.8 g ZrO_2, 45.7 g $ZnSO_4 \cdot 7H_2O$; hydrogenation conditions: 150 °C, 5 MPa, 1400 r/min, 140 mL

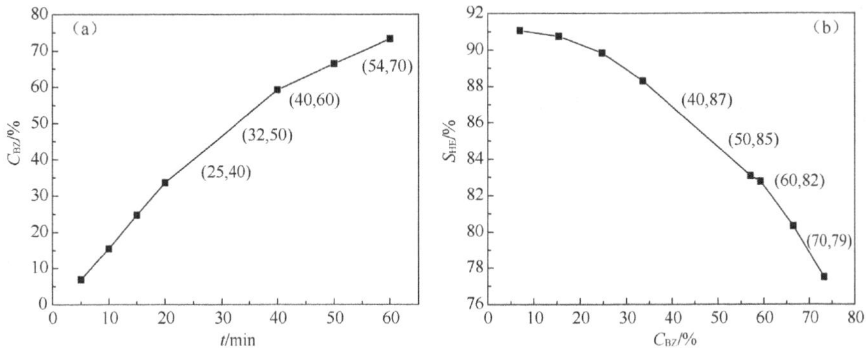

Fig. 6.24 The hydrogenation of 1.96 g Ru–Zn catalyst (140320) with the addition of 300 mg BZSS and 22 h pretreatment. **a** C_{BZ}-t curve; **b** S_{HE}-C_{BZ} curve

indicating that the hydrogenation can be directly carried out using a catalyst with the addition of BZSS and without pretreatment, thus it can save the reaction time and improve production efficiency.

Table 6.26 shows the comparison of hydrogenation activity and selectivity for the hydrogenation by Ru–Zn and Ru–Zn@BZSS catalysts with pretreatment.

As can be seen from Table 6.26, the Ru–Zn catalyst can improve the selectivity of cyclohexene after pretreatment. When the Ru–Zn@BZSS was used for direct hydrogenation, the cyclohexene selectivity increased significantly because the addition of BZSS played a role as pretreatment. When using Ru–Zn@BZSS with pretreatment, although the selectivity of cyclohexene can be improved, the reaction time to achieve the same benzene conversion was significantly longer, and the catalyst activity was significantly reduced, making it not conducive to industrial applications. Ru–Zn@BZSS exhibited a higher selectivity to cyclohexene without pretreatment

Table 6.26 The results of hydrogenation activity and selectivity by Ru–Zn and Ru–Zn@BZSS catalysts with and without pretreatment of hydrogenation

Hydrogenation conditions	t_{40}/min	t_{50}/min	t_{60}/min	S_{40}/%	S_{50}/%	S_{60}/%
Ru–Zn Direct hydrogenation	–	–	6	–	–	70
Ru–Zn Pretreated hydrogenation	6	8	10	81	79	77
Ru–Zn@BZSS Direct hydrogenation	17	21	24	85	83	80
Ru–Zn@BZSS Pretreated hydrogenation	25	32	40	87	85	82

and avoided the fluctuations of selectivity and activity caused by the addition of new catalysts, which was important for maintaining the stability of industrial processes.

6.3.4 *Catalyst Characterization*

(1) X-ray diffraction (XRD)

Figure 6.25 shows the XRD patterns of BZSS prepared using different methods. Different methods included using $ZnSO_4$ with an excess of 50 and 100%, and adding the sodium hydroxide solution to the zinc sulfate solution by dropwise and at once.

It can be seen from Fig. 6.25 that the BZSS's formula was $(Zn(OH)_2)_3(ZnSO_4)(H_2O)_x$ ($x = 3$, 4 or 5), and it was a kind of insoluble alkaline salt of zinc sulfate. There is a remarkable strongest peak when 2θ was around 10°.

Figure 6.26 shows the particle size of the microcrystals by the XRD and half-width method of Ru–Zn and Ru–Zn@BZSS catalysts.

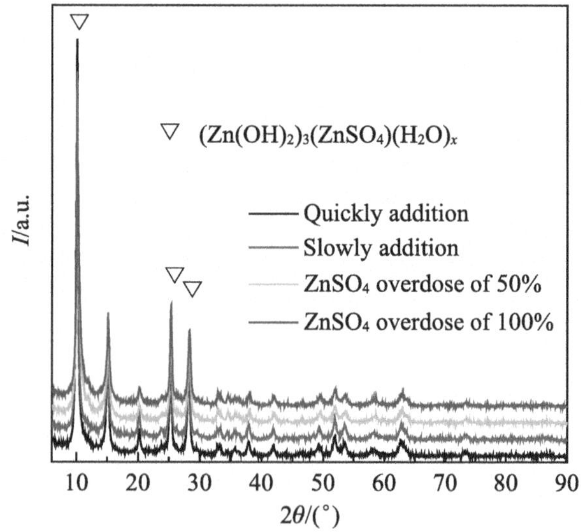

Fig. 6.25 The XRD patterns of BZSS prepared using different methods

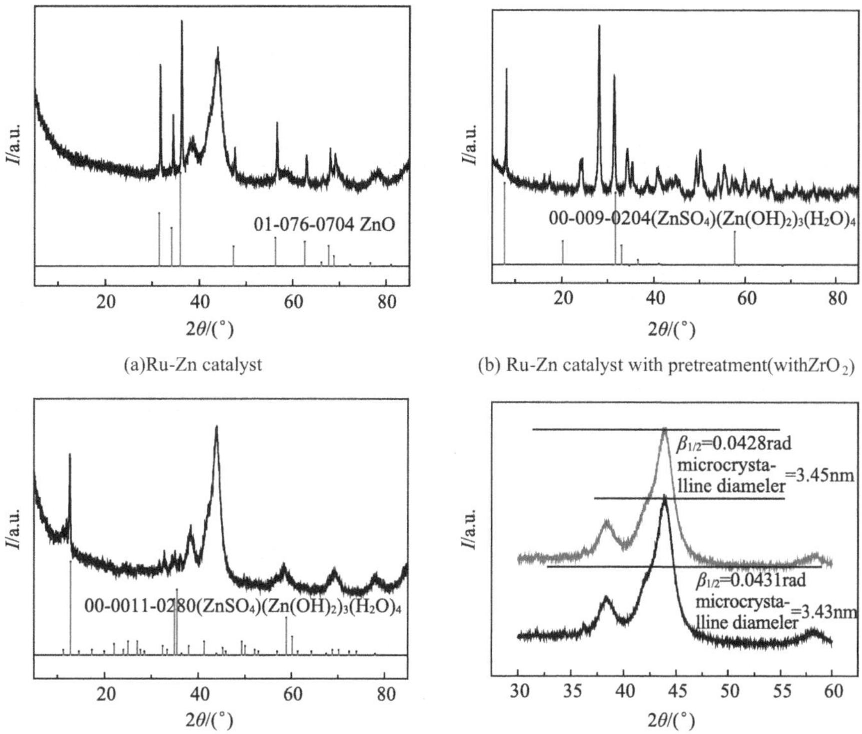

(a)Ru-Zn catalyst

(b) Ru-Zn catalyst with pretreatment(withZrO_2)

(c)Ru-Zn@BZSScatalysts

(d) Particle size of the microcrystals of catalysts

Fig. 6.26 XRD patterns and microcrystalline particle sizes of Ru–Zn and Ru–Zn@BZSS catalysts

It can be seen from Fig. 6.26a that Zn was present as ZnO in the Ru–Zn catalyst. From (b), with the pretreatment in a reaction slurry containing $ZnSO_4$ for 22 h, the characteristic peaks of BZSS[$(ZnSO_4)(Zn(OH)_2)_3(H_2O)_4$] appeared when 2θ was around 10° in the pattern of the Ru–Zn catalyst. It showed that the ZnO on the surface of the catalyst and the $ZnSO_4$ in the slurry formed insoluble BZSS in situ in the pretreatment process. In (c), BZSS was first prepared and then underwent heat treatment with a certain ratio of the Ru–Zn catalyst under certain conditions. Then the characteristic peak of BZSS appeared in the XRD pattern of samples. Figure 6.26d gave the half-height and width of the strongest peak of Ru (100) plane and the microcrystalline particle size (around 3.4 nm) was obtained according to the Scherrer formula.

(2) X-ray photoelectron spectroscopy (XPS)

Figure 6.27 shows the X-ray photoelectron spectroscopy of the catalyst before and after the addition of BZSS.

It can be seen from Fig. 6.27 that the intensity of the photoelectron spectrum of Ru and Zn decreased and the displacement occurred when BZSS was added to the

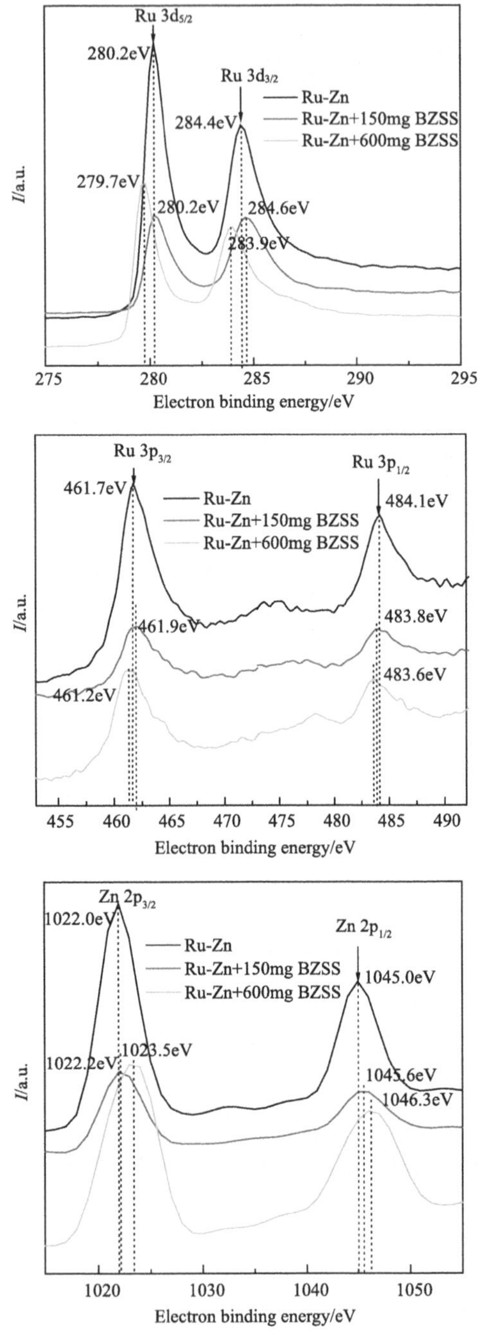

Fig. 6.27 XPS spectra of Ru 3p, Ru 3d, Zn 2p of Ru–Zn catalysts with different amounts of BZSS

Ru–Zn catalyst. This was due to the BZSS adsorption on the surface of the Ru–Zn catalyst to form a solid stagnant water film, which had a shielding effect on the escape of internal electrons. In the Ru–Zn and Ru–Zn@BZSS catalysts with pretreatment, the electron binding energy of Ru decreased and the electron binding energy of Zn increased, indicating that Zn electrons shifted to Ru and regulated the electrical properties of Ru-active center.

6.3.5 Surface Modification and Reaction Mechanism

The results of activity and selectivity showed that the pretreatment reduced the activity of the Ru–Zn catalyst and improved the selectivity of cyclohexene. It can be seen from the XRD pattern that the BZSS peak appeared after the pretreatment of the Ru–Zn catalyst. This indicated that ZnO on the surface of the catalyst and the $ZnSO_4$ in the slurry formed insoluble BZSS in situ during the pretreatment. It can be presumed that the Zn in the catalyst and Zn of BZSS played important roles in the modification of the space arrangement and the electronic properties of Ru-active center due to the geometric effect and electron effect from the XPS spectrum. And due to the formation of core-shell structure of Ru–Zn@BZSS, BZSS formed a layer of stagnant water film on the surface of the Ru–Zn catalyst, which had a hydrophilic modification effect on the catalyst surface.

The Zn in the Ru–Zn catalyst or BZSS was preferentially adsorbed on the Ru strongest active center and weakened the hydrogen coverage on the surface of the Ru–Zn catalyst, which reduced the direct hydrogenation of benzene to cyclohexane. The electron effect between Zn and Ru made Ru a rich electron center, which can weaken the adsorption of π-bond of the benzene ring and double-bond electrons of cyclohexene. This was beneficial to avoid the deep hydrogenation of benzene and to promote the cyclohexene desorption.

Therefore, the BZSS-modified Ru-active center and catalyst surface properties were helpful to reduce the catalyst activity and improve the selectivity of cyclohexene.

The results of the Ru–Zn catalyst with 22 h pretreatment and Ru–Zn@BZSS for direct hydrogenation both proved that the modification of Ru catalyst by BZSS and the mechanism of the highly selective production cyclohexene.

In 2015, the comparison of catalytic performance among 59 hydrogenation catalysts was carried out in the world and Ru–Zn@BZSS catalyst was the best one [5].

6.4 Industrial Applications for Fourth-Generation Catalysts

In the preparation of 58 batches of industrial catalysts, the influences of the impurity ions on the furnace wall, the Zn promoter, the Zn content in the catalyst, and the BZSS adsorption on the surface were studied. The optimization of the precipitation temperature, the reduction time, the aging time, the catalyst washing, the storage conditions, and other industrial preparation parameters improved the fourth-generation catalyst preparation technology.

6.4.1 Effect of Impurity Ions on Wall

For the 1st, 2nd, 3rd batch of catalyst, the precipitation temperature was 95 °C with a reaction time of 2 h, and the reduction was conducted at 150 °C for 10 h, including the catalyst aging, washing, and other processes, thus preparation cycle for each batch of catalyst needs about 24 h.

Figure 6.28 shows the activity and selectivity of hydrogenation with and without pretreatment of the 1st, 2nd, and 3rd batch of catalysts.

Table 6.27 gives the t_{40}, t_{50}, t_{60} and S_{40}, S_{50}, S_{60} of the 1st, 2nd, 3rd batch of catalyst for direct hydrogenation with and without pretreatment.

It can be seen from Table 6.27 that the direct hydrogenation was compared with the small test data. For the 1st, 2nd, and 3rd batch of the catalyst, t_{60} which was the time to achieve the 60% conversion of benzene increased from the 6 min in the small test to 16 min, and the selectivity of cyclohexene, that was S_{60}, increased from 70 to 76%, indicating that the catalyst activity decreased while the selectivity of cyclohexene increased under the conditions of industrial production.

According to the catalyst characterization and analysis of the content of impurity ions (Fe, Cr, Ni from the wall), the results showed that the structure and texture of the catalyst were basically the same as those catalysts in the small test, however, the Fe, Cr, Ni ions severely exceed the limit. The reason was that the ingredients tank and the material delivery pipe were made of 304 stainless steel and the pH of the ruthenium trichloride solution was 1–2 with a strong corrosive effect. For this, the industrial equipment was modified and the tanks were made by using polypropylene and the pipelines were made by polytetrafluoroethylene. All parts of the autoclave which contacted the liquid were made of Hastelloy alloy to solve the problem of excessive impurity ions.

6.4.2 The Precursor of Zn Promoter

Figure 6.29 and Fig. 6.30 give the hydrogenation activity and selectivity of 4th–7th batches of catalysts for hydrogenation without and with 22 h pretreatment using $ZnSO_4 \cdot 7H_2O$ and $ZnCl_2$ as Zn precursors, respectively.

Table 6.28 shows the t_{40}, S_{40}, t_{50}, S_{50}, t_{60}, S_{60} and t_{70}, S_{70} of hydrogenation by the 4th–7th batches of catalysts without and with 22 h pretreatment.

From Table 6.28, the results of the direct hydrogenation showed that the 4th–7th batches of catalyst exhibited high activity, and Benzene conversion can be 40% with 5–6 min reaction time. Meanwhile, a high selectivity was also achieved and

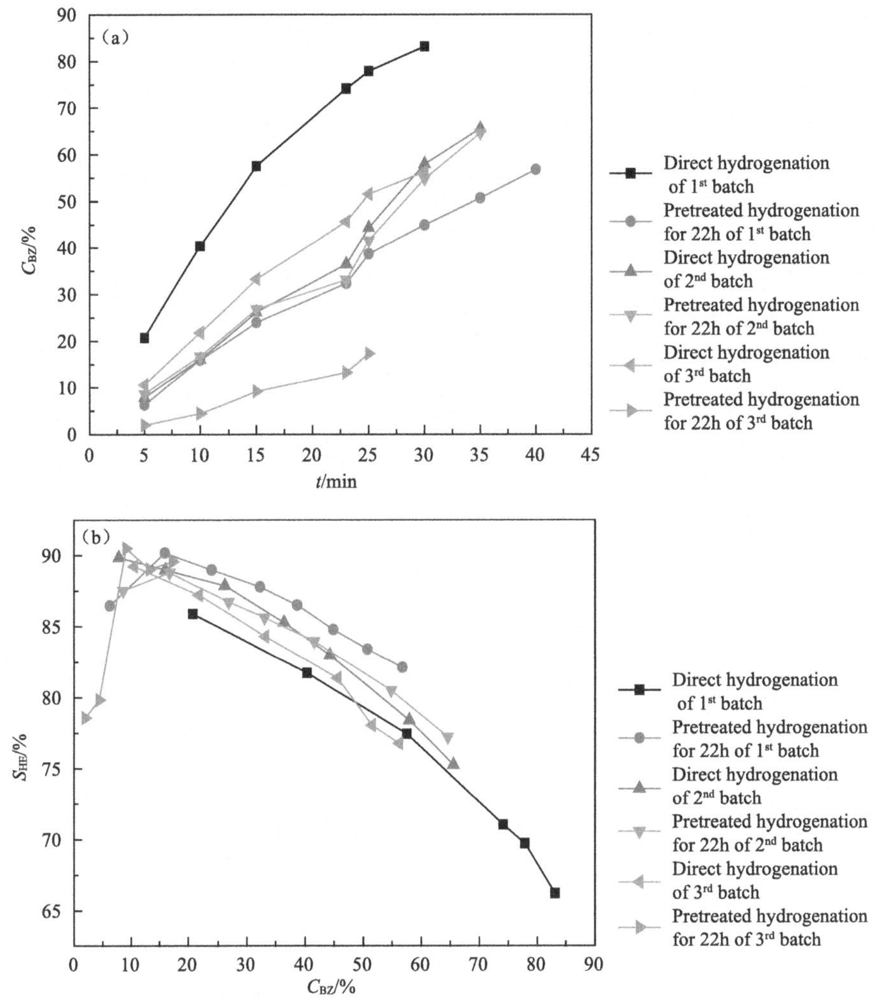

Fig. 6.28 1st, 2nd, 3rd batches catalyst direct hydrogenation and pretreatment hydrogenation. **a** C_{BZ}-t curve; **b** S_{HE}-C_{BZ} curve

the cyclohexene selectivity was 73–77% when benzene conversion was 40%. After pretreatment, the reaction time to achieve 40% benzene conversion extended to 12–16 min and the corresponding cyclohexene selectivity increased from 73–77% to 82–88%. The results were obtained using the 5th–7th batches of catalysts with the addition of BZSS and pretreatment. When using $ZnCl_2$ instead of $ZnSO_4 \cdot 7H_2O$ as Zn precursor, there was no significant difference in the selectivity of catalyst activity. According to the preparation conditions of the 4th and 5th batches of catalysts, the activity and selectivity using $RuCl_3$ as Ru precursor and using $ZnCl_2$ as Zn precursor were investigated by the preparation of 18 batches of catalysts.

Table 6.27 t_{40}, t_{50}, t_{60} and S_{40}, S_{50}, S_{60} of 1st, 2nd, 3rd batches catalyst direct hydrogenation and pretreatment hydrogen

Hydrogenation conditions	Batch	t_{40}/min	t_{50}/min	t_{60}/min	S_{40}/%	S_{50}/%	S_{60}/%
Direct hydrogenation	1	10	12	16	82	79	76
	2	19	24	33	83	79	75
	3	24	27	31	85	81	78
Lab-scale data in Fig. 6.21		–	–	6	–	–	70
Pretreated hydrogenation	1	26	34	43	86	83	81
	2	24	28	32	84	82	79
	3	–	–	–	–	–	–
Lab-scale data in Fig. 6.22		6	8	10	81	79	77

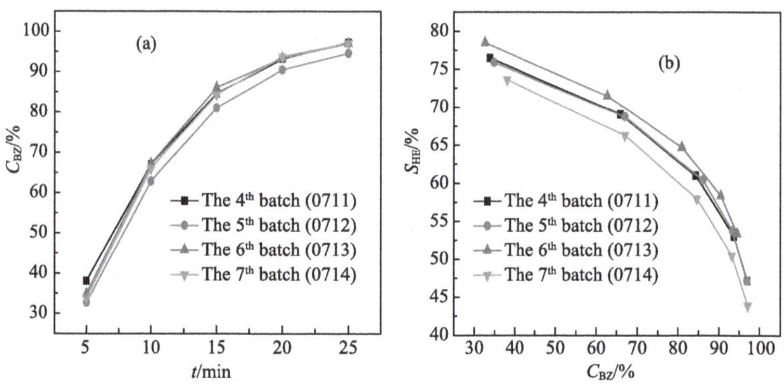

Fig. 6.29 The results of direct hydrogenation by the 4th–7th batches of catalysts. **a** C_{BZ}-t curve; **b** S_{HE}-C_{BZ} curve

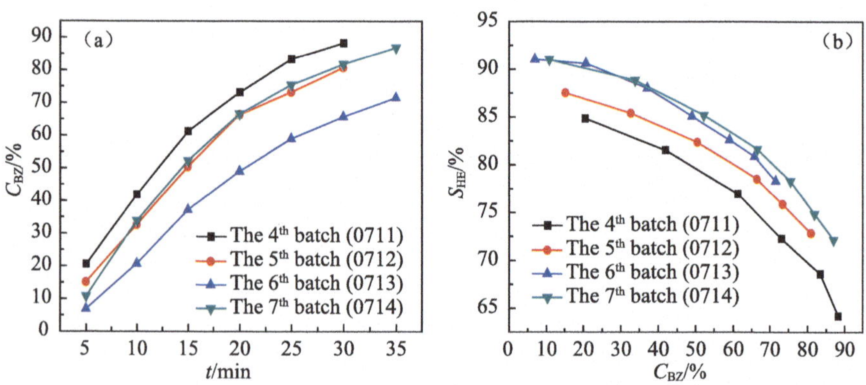

Fig. 6.30 The results of hydrogenation by 4th–7th batches of catalysts with 22 h pretreatment. **a** C_{BZ}-t curve; **b** S_{HE}-C_{BZ} curve

Table 6.28 The t_{40}, S_{40}, t_{50}, S_{50}, t_{60}, S_{60} and t_{70}, S_{70} of the hydrogenation by 4th–7th batches of catalysts without and with 22 h pretreatment

Condition	Batch	t_{40}/min	t_{50}/min	t_{60}/min	t_{70}/min	S_{40}/%	S_{50}/%	S_{60}/%	S_{70}/%
Direct hydrogenation	4[a]	6	8	9	–	75	73	70	–
	5[a]	6	7	9	–	75	73	70	–
	6[a]	6	8	10	–	77	74	72	–
	7 ($ZnCl_2$)	5	7	9	–	73	71	68	–
Pretreated hydrogenation	4[a]	10	12	15	–	82	80	77	–
	5[a]	12	15	18	23	84	83	80	77
	6[a]	16	21	26	–	87	85	82	–
	7 ($ZnCl_2$)	12	14	18	22	88	86	83	80

[a]Represented that $ZnSO_4 \cdot 7H_2O$ was the Zn precursor. The 5th, 6th, 7th batches of catalysts were prepared with the addition of BZSS and pretreatment

Figures 6.31 and 6.32 show the C_{BZ}-t curve and the S_{HE}-C_{BZ} curve of hydrogenation using catalysts with pretreatment and $ZnCl_2$ precursor.

Table 6.29 gives the main indicators of the first 8–19 batches catalyst pretreatment after hydrogenation.

From the results in Table 6.29, according to the preparation conditions of 4th–5th batches of catalysts, it could be seen that the catalyst was not only highly active and selective but also had good reproducibility and stability with $RuCl_3$ as Ru precursor and $ZnCl_2$ as Zn precursor. In the pretreatment of 22 h, the conversion of benzene was 40% and the selectivity of cyclohexene was 81–87% at 8–15 min. At 9–18 min, the conversion of benzene was 50% and the selectivity of cyclohexene was 77–85%.

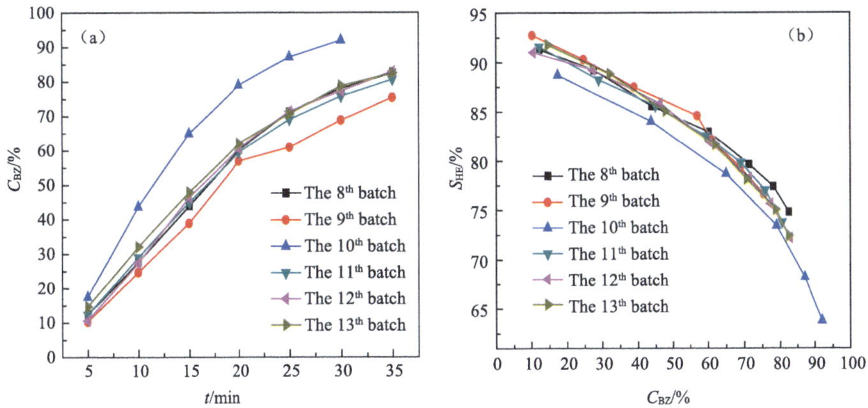

Fig. 6.31 The results of hydrogenation using 8th–13th batches of catalysts with 22 h pretreatment. **a** C_{BZ}-t curve; **b** S_{HE}-C_{BZ} curve

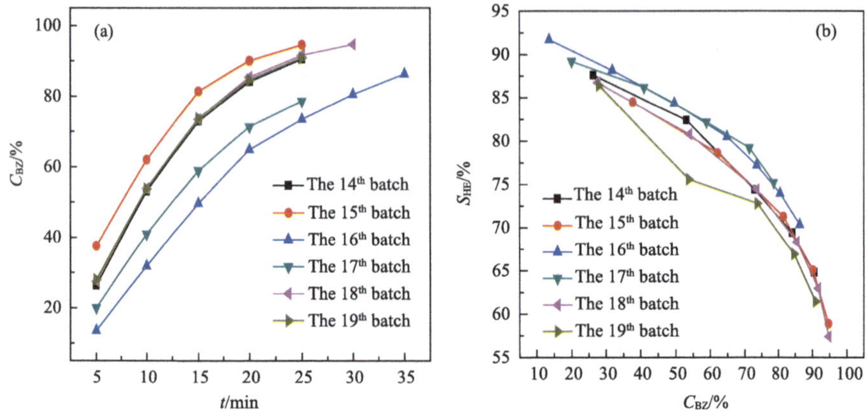

Fig. 6.32 The results of hydrogenation using 14th–19th batches of catalysts with 22 h pretreatment. **a** C_{BZ}-t curve; **b** S_{HE}-C_{BZ} curve

Table 6.29 The t_{40}, S_{40}, t_{50}, S_{50}, t_{60}, S_{60} and t_{70}, S_{70} of the hydrogenation by 8th–19th batches of catalysts with 22 h pretreatment

Batch	t_{40}/min	t_{50}/min	t_{60}/min	t_{70}/min	S_{40}/%	S_{50}/%	S_{60}/%	S_{70}/%
8	14	17	20	24	86	85	83	80
9	15	18	24	31	87	85	83	79
10	9	12	14	17	85	82	80	77
11	14	17	20	25	86	85	82	80
12	14	17	20	24	87	85	82	79
13	12	16	19	25	87	85	82	78
14	8	10	12	15	85	83	80	76
15	6	8	9	12	84	81	79	76
16	12	15	18	23	86	84	82	79
17	10	12	15	19	86	84	82	80
18	8	9	12	14	84	82	78	76
19	8	9	12	14	81	77	74	73

Note The 15th, 17th, and 19th batches were pretreated for 12 h; the other catalysts were pretreated for 22 h

At 12–24 min, the conversion of benzene was 60% and the selectivity of cyclohexene was 74–83%. At 14–25 min, the conversion of benzene was 70%, and the selectivity of cyclohexene was 73–80%. The selectivity of the catalysts was fully capable of meeting the industrial requirements.

6.4.3 Zn Content in the Catalyst and Absorption Amount of BZSS on Surface

The results showed that the chemical adsorption of BZSS on the surface of the catalysts by non-in situ method could overcome the bottleneck of BZSS in situ formation, which further improved the selectivity and yield of cyclohexene, and made the catalyst preparation more controllable and made the use of catalyst and the shift of activity and selectivity more convenient.

The key to the adsorption of BZSS on the surface of the catalyst by non-in situ preparation was to adjust the Zn content in the catalyst and determine the adsorption amount of BZSS on the catalyst surface. For catalysts, the activity and selectivity were highest at the optimum content of Zn and BZSS on the surface, thus the highest cyclohexene yield could be obtained. Due to the presence of BZSS on the surface, it was necessary to reduce the Zn content in the catalyst so that the Zn content on the catalyst surface and the total internal Zn content could remain unchanged.

The optimum Zn content in the Ru–Zn catalyst was about 10%. In the process of industrialization of the catalyst, the Zn content was reduced to about 9%, and the Zn content in the added BZSS was about 1%. The amount of $ZnCl_2$ and the amount of BZSS were determined on the basis of the amount of each batch of catalyst.

Figure 6.33 gives the C_{BZ}-t curve and the S_{HE}-C_{BZ} curve of hydrogenation using the catalysts prepared according to the method described above with 22 h pretreatment.

Table 6.30 shows the t_{40}, S_{40}, t_{50}, S_{50}, t_{60}, S_{60} and t_{70}, S_{70} of hydrogenation by the 7th, 8th, 11th, 27th, and 31st batches of catalysts with 22 h pretreatment.

As can be seen from Table 6.30, the main indicators of the performance of the catalysts with the reduction of the Zn content and the surface adsorption of BZSS and keeping the total Zn content constant are showed as follows:

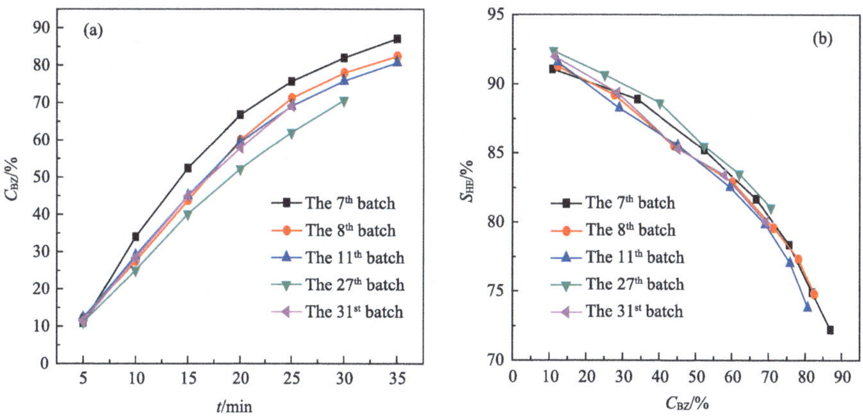

Fig. 6.33 The results of hydrogenation using 7th, 8th, 11th, 27th, and 31st batches of catalysts with 22 h pretreatment. **a** C_{BZ}-t curve; **b** S_{HE}-C_{BZ} curve

Table 6.30 The t_{40}, S_{40}, t_{50}, S_{50}, t_{60}, S_{60} and t_{70}, S_{70} of hydrogenation by the 7th, 8th, 11th, 27th, and 31st batches of catalysts with 22 h pretreatment

Batch	t_{40}/min	t_{50}/min	t_{60}/min	t_{70}/min	S_{40}/%	S_{50}/%	S_{60}/%	S_{70}/%
7	12	14	18	22	88	86	83	80
8	14	17	20	24	86	85	83	80
11	14	17	20	25	86	85	82	80
27	15	19	24	30	89	86	84	81
31	14	17	21	25	87	85	83	80

$$t_{40} = 12\text{–}15\,\text{min},\ S_{40} = 86\text{–}89\%$$
$$t_{50} = 14\text{–}19\,\text{min},\ S_{50} = 84\text{–}86\%$$
$$t_{60} = 18\text{–}28\,\text{min},\ S_{60} = 83\text{–}84\%$$
$$t_{70} = 22\text{–}30\,\text{min},\ S_{70} = 80\text{–}81\%$$

The breakthrough was made that the benzene conversion was 70% and the cyclohexene selectivity was more than 80% at 22 min.

6.4.4 *Optimization of Industrial Preparation Parameters*

(1) Precipitation temperature

The effects of precipitation temperature on the activity and selectivity of the catalysts were investigated at 35–95 °C for the preparation of four batches of catalysts.

Figure 6.34 and Table 6.31 show the activity and selectivity of the catalysts prepared at different precipitation temperatures, respectively.

As can be seen from Table 6.31, the activity and selectivity of the catalyst showed no apparent dependency on the generation temperature of hydroxide precipitate in the range of 35–95 °C. At 9–14 min, benzene conversion was 40% and cyclohexene selectivity was 86–87%. At 12–17 min, benzene conversion was 50% and cyclohexene selectivity was 84–85%. At 15–21 min, benzene conversion was 60% and cyclohexene selectivity was 80–83%. At 19–25 min, benzene conversion was 70% and cyclohexene selectivity was 76–80%. All these results proved that low temperature not only made the operation process convenient, but also reduced energy consumption for the precipitation reaction at 35–45 °C.

(2) Reduction time

Figure 6.35 shows the C_{BZ}-t curve and the S_{HE}-C_{BZ} curve of the hydrogenation with different reduction times using catalysts with 22 h pretreatment.

Table 6.32 shows the t_{40}, S_{40}, t_{50}, S_{50}, t_{60}, S_{60}, and t_{70}, S_{70} of hydrogenation by the catalysts with 22 h pretreatment at different reduction times.

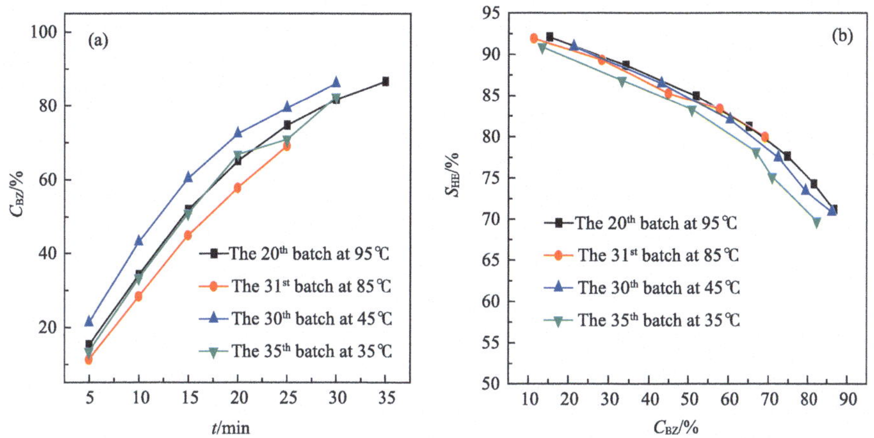

Fig. 6.34 The results of hydrogenation by catalysts with precipitation temperature of 35–95 °C. **a** C_{BZ}-t curve; **b** S_{HE}-C_{BZ} curve

Table 6.31 The t_{40}, S_{40}, t_{50}, S_{50}, t_{60}, S_{60} and t_{70}, S_{70} of hydrogenation by catalysts with precipitation temperature of 35–95 °C

Batch	T/°C	t_{40}/min	t_{50}/min	t_{60}/min	t_{70}/min	S_{40}/%	S_{50}/%	S_{60}/%	S_{70}/%
20	95	12	15	18	23	87	85	83	79
31	85	14	17	21	25	87	85	83	80
30	45	9	12	15	19	87	85	82	78
35	35	12	15	16	23	86	84	80	76

Note T refers to the temperature of the precipitation process

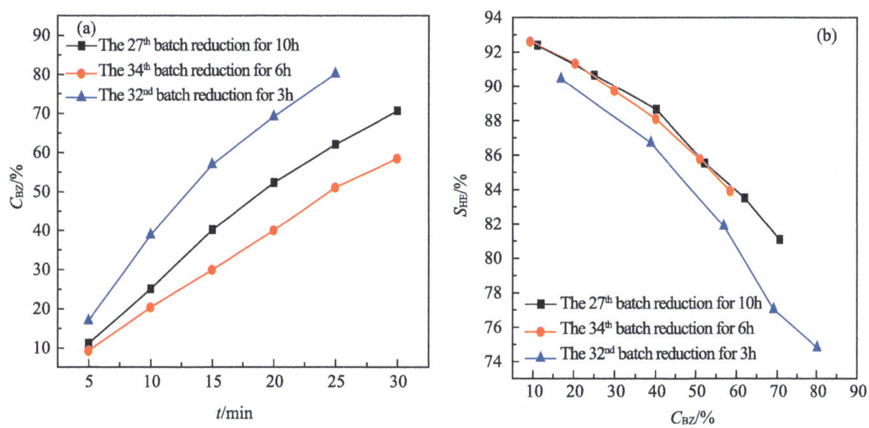

Fig. 6.35 The results of hydrogenation by catalysts with 22 h pretreatment at different reduction times. **a** C_{BZ}-t curve; **b** S_{HE}-C_{BZ} curve

Table 6.32 The t_{40}, S_{40}, t_{50}, S_{50}, t_{60}, S_{60} and t_{70}, S_{70} of hydrogenation by the catalysts with 22 h pretreatment at different reduction times

Batch	t/h	t_{40}/min	t_{50}/min	t_{60}/min	t_{70}/min	S_{40}/%	S_{50}/%	S_{60}/%	S_{70}/%
27	10	15	19	24	30	89	86	84	81
34	6	20	25	30	–	88	86	83	–
32	3	10	13	16	20	86	84	81	77

Note t refers to the reduction time (h); the first 27th, 34th batches of catalysts were prepared with 22 h pretreatment and the 32nd batch of the catalyst was prepared with 15 h pretreatment

From Table 6.32, the different conversion rates and the selectivity of cyclohexene of the 27th and 34th batches of catalysts showed that although the activity of the catalyst was different, the selectivity of cyclohexene was substantially the same for 10 and 6 h reduction. Moreover, high activity and high selectivity can be obtained with 3–10 h reaction of catalyst.

The catalyst reduction time was reduced from the beginning of 10 h to finally 3–4 h without reducing the catalyst performance.

(3) Aging time

Aging is a thermodynamic equilibrium process. According to the Kelvin equation, the aging process can make the small particles dissolve and the larger particles grow further, thus the catalyst particle size distribution and the structure could become more stable, which is beneficial to stabilize the catalyst performance. The process involves two aspects: one is the aging of the catalyst precursor after precipitation; the second is the aging of the catalyst after reduction. For the former, the same batch of prepared catalyst is divided into two (the 7th batch), respectively, and aged for 3 and 10 h, then their activity and selectivity are determined under the same conditions.

Figure 6.36 shows the catalyst activity and selectivity of catalysts prepared at different aging times.

Table 6.33 shows the t_{40}, S_{40}, t_{50}, S_{50}, t_{60}, S_{60} and t_{70}, S_{70} of hydrogenation by the catalysts prepared at different aging times (7th batch).

It can be seen from Table 6.33 that the activity and selectivity of the catalyst were basically the same for 10 and 3 h.

Figure 6.37 shows the hydrogenation activity and selectivity of catalysts by mixing the catalysts in Table 6.33 and adding BZSS with 22 h pretreatment.

From Fig. 6.37a, b, the reaction time to achieve different benzene conversions and cyclohexene selectivity were determined by interpolation method, and the activity index was calculated according to the amount of catalyst.

$$t_{40} = 12\,\text{min},\ \gamma_{40} = 126,\ S_{40} = 88\%$$
$$t_{50} = 14\,\text{min},\ \gamma_{50} = 135,\ S_{50} = 86\%$$
$$t_{60} = 18\,\text{min},\ \gamma_{60} = 126,\ S_{60} = 83\%$$
$$t_{70} = 22\,\text{min},\ \gamma_{70} = 120,\ S_{70} = 80\%$$

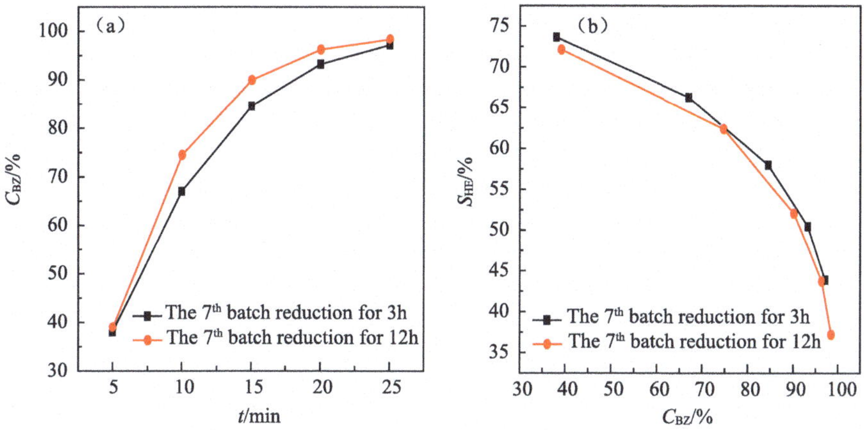

Fig. 6.36 The results of direct hydrogenation by catalysts prepared at different aging times. **a** C_{BZ}-t curve; **b** S_{HE}-C_{BZ} curve

Table 6.33 The t_{40}, S_{40}, t_{50}, S_{50}, t_{60}, S_{60} and t_{70}, S_{70} of hydrogenation by catalysts with prepared at different aging times (7th batch)

Batch	t/h	t_{40}/min	t_{50}/min	t_{60}/min	S_{40}/%	S_{50}/%	S_{60}/%
7	3	5	7	9	73	71	68
7	10	5	7	8	72	70	66

Note t refers to the aging time

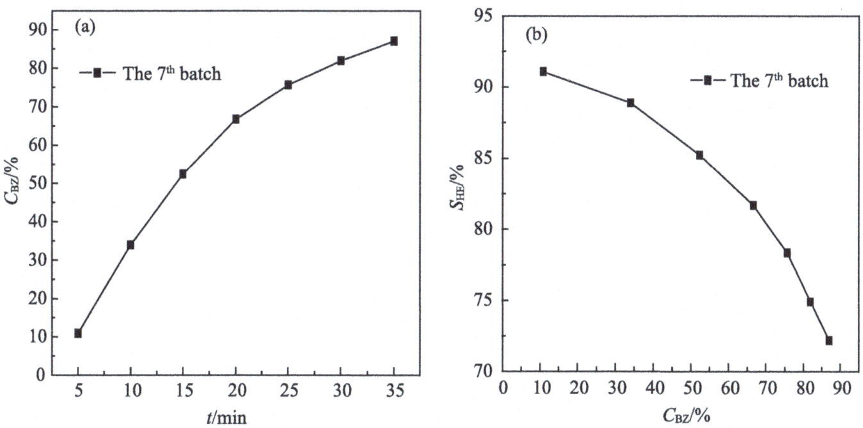

Fig. 6.37 The results of hydrogenation by catalysts prepared at different aging times with pretreatment. **a** C_{BZ}-t curve; **b** S_{HE}-C_{BZ} curve

After industrial amplification of Ru–Zn@BZSS catalyst, $\gamma_{50} = 135$, $S_{50} = 86\%$. This can fully meet the industrial production requirements.

The aging time of the catalyst precursor did not significantly affect the catalyst performance because it was not the final state. From the hydroxide precipitation to the reduced state, the catalyst is needed to go through the reduction process. The reduction process started from the temperature rise to the end of reduction, in fact, the catalyst itself underwent a thermodynamic spontaneous balance process, which could make up for the lack of precipitation time.

The studies have repeatedly shown that the reduction of the catalyst after aging was necessary for the selective hydrogenation of benzene to cyclohexane, but there was no special requirement for aging temperature. Whether at room temperature or at higher temperatures, their effect on the catalyst is not significantly different. In order to save the preparation time of each batch of catalysts, the reduced catalyst was put into the washing tank, the washing operation is carried out after 4 h to ensure the aging process after the reduction.

The reduced catalyst is thermodynamically unstable, and the irregularity of the catalyst surface results in many active centers, resulting in a high catalyst activity and a decrease in the selectivity of cyclohexene. After the aging process, the catalysts spontaneously tends to a thermodynamically stable state, and some point defects, line defects, and even surface defects (or called stackings) can be reduced to a certain extent, which not only helps to reduce the activity of the catalysts and improve the selectivity of cyclohexene, but also is conducive to the stability of the catalysts.

(4) Catalyst washing and storage conditions

Washing and storage is the final step in the preparation of the catalyst. The purpose of the washing is to remove the adsorbed impurity ions on the catalyst. These impurity ions are derived from the raw materials, e.g., $RuCl_3{\cdot}3H_2O$, $ZnCl_2$, NaOH, even H_2O, etc. Cr, Na^+, OH^-, and other adsorbed substrates on the catalyst will occupy the surface of the Ru-active center, affecting the performance of the catalyst. Using a multiple method, the catalyst was washed until the pH of the washing solution was about 7, while Cl^- cannot be detected by the 0.1 mol/L silver nitrate solution.

The studies shown that the floc appeared then the catalyst was stored in the water after two weeks, and these flocs were from the hydrolysis of ZnO from the catalyst. If the catalyst is stored in an alkaline aqueous solution, the appearance of the floc can be avoided and the structure of the catalyst can be maintained and stabilized.

Figure 6.38 shows the activity and selectivity of the catalyst for direct hydrogenation after keeping in an alkaline aqueous solution for 15 days.

Table 6.34 shows the t_{40}, S_{40}, t_{50}, S_{50}, t_{60}, S_{60}, and t_{70}, S_{70} of direct hydrogenation by the catalysts stored in alkaline solution for 15 days.

As can be seen from Table 6.34, the activity of the catalyst was slightly decreased after being stored in an alkaline aqueous solution for 15 days. The time for 40% conversion of benzene was increased from 6–7 to 8–10 min, and the selectivity of cyclohexene increased from 80–83 to 84%. The time for 50% conversion of benzene was increased from 7–9 to 10–12 min, and the selectivity of cyclohexene increased

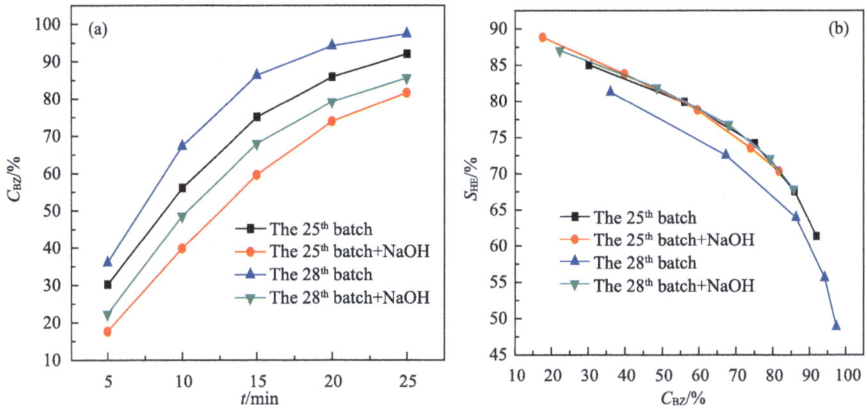

Fig. 6.38 The results of direct hydrogenation by catalysts stored in alkaline solution for 15 days. **a** C_{BZ}-t curve; **b** S_{HE}-C_{BZ} curve

Table 6.34 The t_{40}, S_{40}, t_{50}, S_{50}, t_{60}, S_{60}, and t_{70}, S_{70} of direct hydrogenation by the catalysts stored in alkaline solution for 15 days

Batch	t/d	t_{40}/min	t_{50}/min	t_{60}/min	t_{70}/min	S_{40}/%	S_{50}/%	S_{60}/%	S_{70}/%
25	0	7	9	11	14	83	80	77	74
25	15	10	12	15	18	84	81	79	75
28	0	6	7	9	11	80	78	75	72
28	15	8	10	13	15	84	82	80	76

Note t is the time stored in alkaline aqueous solution; 0 refers to the fresh catalyst; 15 refers to 15 days

from 78–80 to 81–82%. The time for 60% conversion of benzene was increased from 9–11 to 13–15 min, and the selectivity of cyclohexene increased from 75–77 to 79–80%. The time for 70% conversion of benzene was increased from 11–14 to 15–18 min, and the selectivity of cyclohexene increased from 72–74 to 75–76%. The main indicators of the performance of the catalyst did not reduce but improved.

6.4.5 Industrial Preparation of the Catalyst

Industrial preparation of the catalyst was carried out in a 500 L autoclave. The effective volume of the autoclave was 350–400 L. The key equipment and process flow as shown in Fig. 6.20. 6–8 kg of catalysts were prepared per batch according to the active ingredient precursors and lye concentration.

The preparation conditions of the catalysts mainly include: the concentration of alkali and salt in the precipitation process, the precipitation temperature, the reaction time, the reduction temperature, the reduction time, aging, washing, and so on. By optimizing the catalyst preparation parameters, the catalysts have high activity and high selectivity, and the preparation procedure is simplified. Through the preparation of 400 kg catalyst, the optimization of the preparation parameters was achieved and the industrial preparation technology for the preparation of the fourth generation of catalysts was developed.

The 1st batch catalyst used 7.5 kg trichloride and the amount of catalyst was finally 2.92 kg. For the 2nd and 3rd batches, the amount of each batch of trichloride was 15 kg and the catalyst was 5.84 kg. It could be found from the preparation of the 1st batch catalyst that the preparation equipment caused several problems. As the ingredients trough and the washing tank were 304 steel, the ingredients trough was corroded seriously and Fe content was up to 12% or more from detection, thus the catalyst activity was very low. Then the equipment was modified, and the material of ingredients trough and washing tank (304 steel) was replaced by polypropylene. For the 1st, 2nd, 3rd batches of catalysts, the preparation conditions included the precipitation temperature of 95 °C, the reaction time of 2 h, and the reduction time of 10 h. From the raw materials to the post-treatment, the preparation cycle for each batch was about 24 h and the catalyst microcrystalline particle size was around 2 nm.

For the 4th batch, the amount of trichloride was 15 kg and the amount of catalyst was 5.84 kg. For 5th–25th batches, the amount of trichloride was 17.5 kg and the amount of catalyst was 6.8 kg. The preparation cycle was about 24 h. From the 4th batch, BZSS was added to the catalyst, i.e., the BZSS slurry was formed using the strict stoichiometric amount of $ZnSO_4 \cdot 7H_2O$ and NaOH and the amount of BZSS was determined according to the amount of catalyst. The benzene conversion was 70% by the 7th batch of catalyst and the cyclohexene selectivity was 80%. With the optimization of process parameters, the reaction temperature was reduced from 95 °C to about 80 °C and the reaction time was reduced from 3 to 1 h. Moreover, the reduction time was reduced from 10 to 5 h. From the raw materials to the product, the production cycle was reduced from 24 to 12 h for each batch of catalyst.

For the 27th batch, the amount of trichloride was 7.5 kg and the amount of catalyst was 6.8 kg. The reaction time was firstly reduced to 0.5 h and the reduction time was reduced to 3 h. The main indicators of the performance of catalysts can be indicated that the cyclohexene selectivity was above 80% when benzene conversion was 70%. For 30th, 31st, and 32nd batches, each batch of trioxide increased from 17.5 to 20 kg and the amount of each batch of catalyst increased from 6.8 to 7.8 kg.

For 45th–56th batches, the amount of each batch of trioxide increased from 17.5 to 20 kg, and the amount of each batch of catalyst increased from 6.8 to 7.8 kg. For the 57th and 58th batches, the amount of each batch of trichloride increased from 20 to 22.5 kg and the amount of each batch of catalyst increased from 7.8 to 8.8 kg. The preparation time of a single batch catalyst was shortened from 24 to 8 h, and the yield of a single reactor was increased from 1000 to 3000 kg/a.

6.4.6 *Industrial Catalyst Characterization*

(1) XRD

Figure 6.39 shows the XRD patterns of the catalysts prepared at different precipitation temperatures in the industry.

It can be seen from Fig. 6.39 that the microcrystalline particles of the catalyst precipitated at 45 °C and 95 °C were 3.46 nm and 3.51 nm, respectively, and these two were basically the same.

(2) N_2 physical adsorption at low temperature

Figure 6.40 shows the N_2 adsorption-desorption isotherms and pore size distributions curve of industrial catalysts at different reduction times.

It can be seen from Fig. 6.40a that the catalysts with different reduction times had similar pore structures and thus had similar hysteresis loops with typical mesoporous structures.

Table 6.35 shows the texture parameters for catalysts with different reduction times.

As can be seen from Table 6.35, the average pore diameter of the catalyst was 4.2–5.1 nm and the BET specific surface area was 52–65 m^2/g. From Fig. 6.40 and Table 6.35, the catalysts with different reduction times had no obvious difference in texture, and they belonged to a mesoporous structure with similar hysteresis loop and texture parameters.

(3) TEM

Figure 6.41 shows the TEM photographs of industrial catalysts with different reduction times.

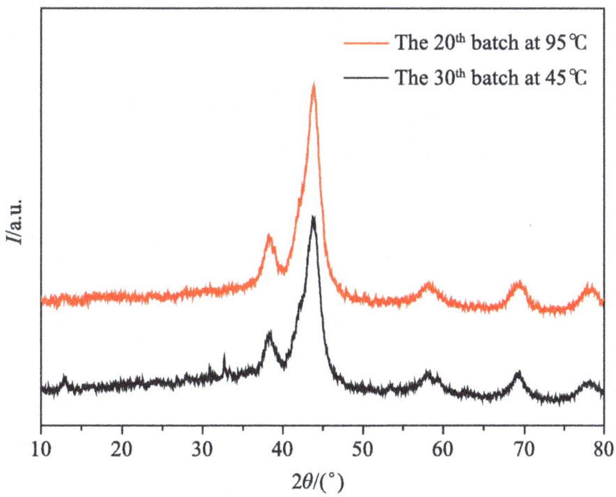

Fig. 6.39 The XRD patterns of the catalysts prepared at different precipitation temperatures in the industry

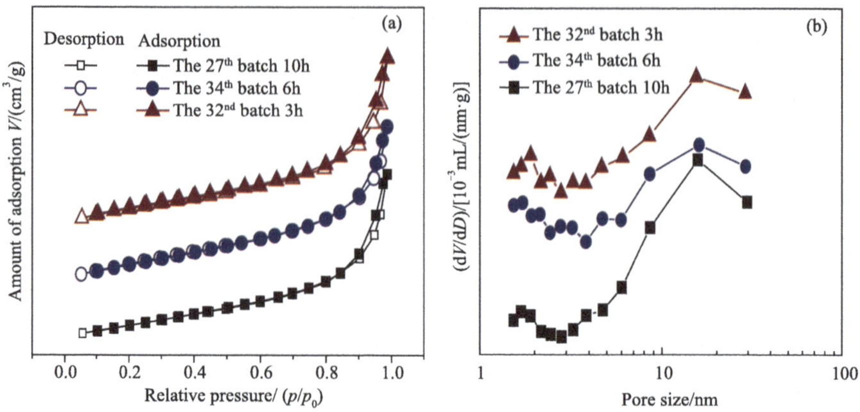

Fig. 6.40 The N_2 adsorption-desorption isotherms (**a**) and pore size distributions curve (**b**) of industrial catalysts at different reduction times

Table 6.35 The BET specific surface area S_{BET}, the pore volume V_p, and the average pore diameter d_p of the catalyst at different reduction times

Batch	t/h	S_{BET}/(m^2/g)	V_p/(cm^3/g)	d_p/nm
27	10	57	0.14	5.1
32	6	52	0.11	4.5
34	3	65	0.14	4.2

Note t refers to the reduction time

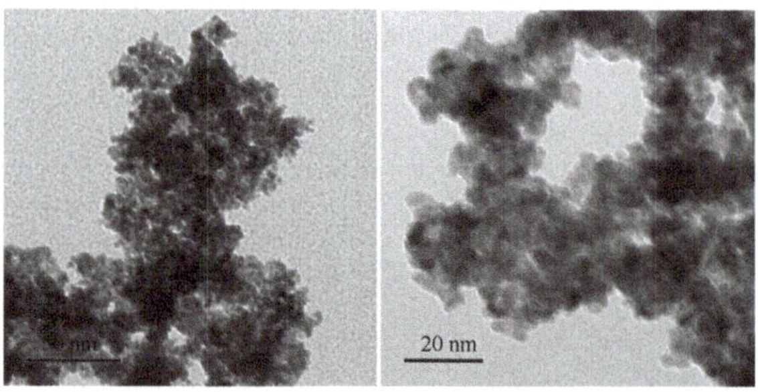

Fig. 6.41 TEM photographs of industrial catalysts at different reduction times

It can be seen from Fig. 6.41 that the catalyst was spherical particles that were randomly agglomerated by microcrystalline with a diameter of about 5 nm. This was in accordance with the results from XRD patterns.

Figure 6.42 shows the operating performance of the catalyst on the 100,000 t/a cyclohexene industrial plant for 280 h. The industrial design objective was to obtain 80% cyclohexene selectivity at 40% benzene conversion.

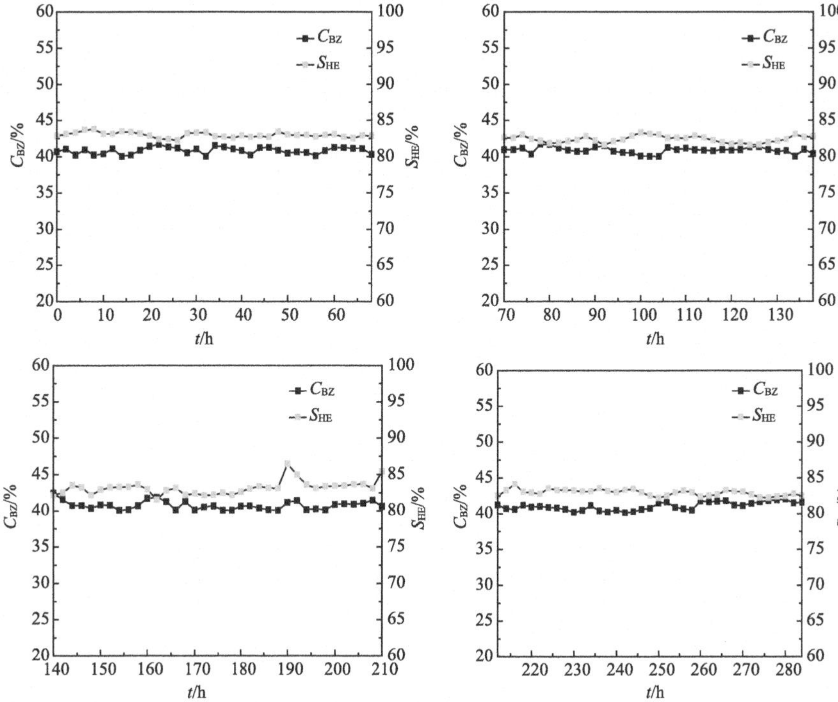

Fig. 6.42 The operating performance of the catalyst for 280 h

It can be seen from Fig. 6.42, benzene conversion rate remained at 40% and cyclohexene selectivity was stable at 80% or more, the catalysts had a very good stability.

The plant was implemented in December 2014. Figure 6.43 shows the changes in the microcrystalline particle size of the catalyst after 10 months.

It can be seen from Fig. 6.43 that the results of the test were 10.7 nm in September 2015, 12.4 nm in December 2015, 13.8 nm in January 2016, and 14.3 nm in March 2016 after 10 months of operation. The expected lifetime of the catalyst was 18 months with a microcrystalline diameter within 20 nm, and this will not have a significant impact on the selectivity of cyclohexene.

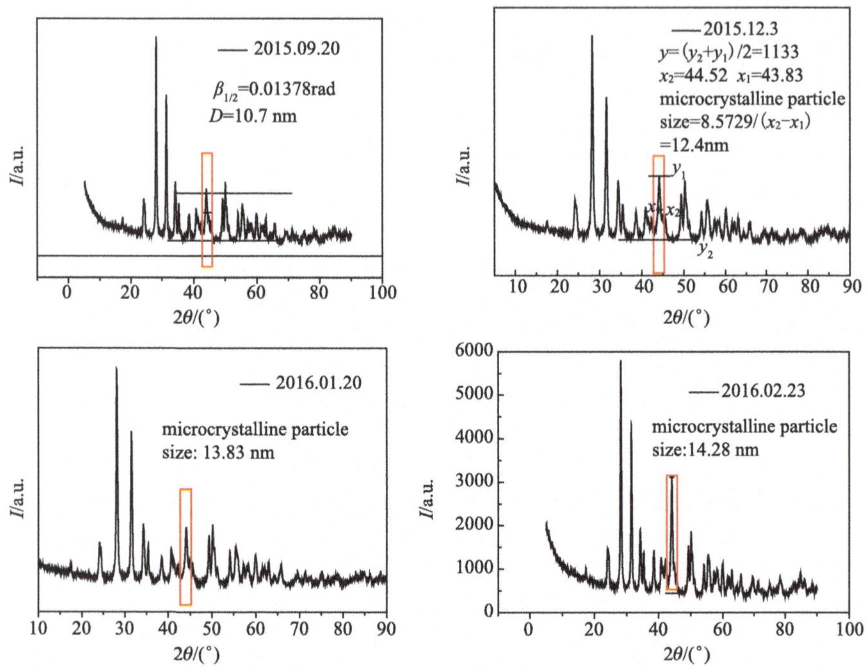

Fig. 6.43 The changes of microcrystalline particle size of the catalyst after 10 months

References

1. Milone, C., Neri, G., Donato, A., et al.: Selective hydrogenation of benzene to cyclohexene on Ru/γ-Al2O3. J. Catal. **159**, 253–258 (1996)
2. He, H.M., Yuan, P., Ma, Y.M., et al.: Theoretical and experimental study on the partial hydrogenation of benzene over Ru–Zn/ZrO_2 catalyst. Chin. J. Catal. **30**, 312–318 (2009)
3. Yuan, P.Q., Wang, B.Q., Ma, Y.M., et al.: Hydrogenation of cyclohexene over Ru–Zn/Ru (0001) surface alloy: a first principles density functional study. J. Molec. Catal. A Chem. **301**, 140–145 (2009)
4. Ye, D.Q., Pang, X.C., Huang, Z.T., et al.: A study on the selective hydrogenation of benzene to cyclohexene by catalytic surface modification. Chem. React. Eng. Technol. **8**(2), 210–213 (1992)
5. Foppa, L., Dupont, J.: Benzene particle hydrogenation: advances and perspectives. Chem. Soc. Rev. **44**, 1886–1897 (2015)
6. Liu, J.L., Zhu, L.J., Pei, Y.: Ce-promoted Ru/SBA-15 catalyst prepared by a "two solvents" impregnation method for selective hydrogenation of benzene to cyclohexene. Appl. Catal. A **252**, 9–16 (2003)
7. Guo, X., Fu, Q.: Ferrous centers confined on core-shell nanostructures for low-temperature CO oxidatio. J. Am. Chem. Soc. **134**, 12350–12353 (2012)
8. Bao, J., He, J.J., Zhang, Y., et al.: A core/shell catalyst produces a spatially confined effect and shape selectivity in a consecutive reaction. Angew. Chem. Int. Ed. **47**, 353–356 (2008)
9. Chen, W., Fan, Z.L., Pan, X.L., et al.: Effect of confined in carbon nanotubes on the activity of Fischer-Tropsch iron catalyst. J. Am. Chem. Soc. **130**, 9414–9419 (2008)
10. Fu, Q., Li, W.X., Yao, Y.X., et al.: Interface-confined ferrous centers for catalytic oxidation. Science **328**, 1141 (2010)

11. Bu, J., Liu, J.L., Chen, X.Y., et al.: Ru/SBA-15 catalysts for partial hydrogenation of benzene to cyclohexene: tuning the Ru crystalline size by Ba. Catal. Commun. **9**, 2612–2615 (2008)
12. Wang, J.Q., Wang, Y.Z., Xie, S.H., et al.: Partial hydrogenation of benzene to cyclohexene. Appl. Catal. A **272**, 29–36 (2004)
13. Sun, H.J., Pan, Y.J., Wang, H.X., et al.: Selective hydrogenation of benzene to cyclohexene over a Ru–Zn catalyst with diethanolamine as an additive. Chin. J. Catal. **33**, 610–620 (2012)
14. Liu, C.S., Luo, G., Xie, Y.L.: Study on Ru–Zn catalyst prepared by precipitation method for selective hydrogenation of benzene to cyclohexene. J. Mol. Catal. **16**(5), 349–354 (2002)
15. Liu, S.C., Liu, Z.Y., Luo, G.: Characterization of Ru–Zn catalysts prepared by precipitation method for selective hydrogenation of benzene to cyclohexene. Petrochem. Technol. **31**, 720–724 (2002)
16. Hu, S.C., Chen, Y.W.: Partial hydrogenation of benzene hydrogenation on Ru–Zn/SiO_2 catalyst. Ind. Eng. Chem. Res. **40**, 6099–6104 (2001)
17. Wang, S.R., Lin, W., Zhu, Y.X., et al.: Preparation and catalytic activity of monolayer dispersed Pt/Ni bimetallic catalyst for C=C and C=O hydrogenation. Chin. J. Catal. **27**(4), 301–303 (2006)

Chapter 7
Modulation of Activity and Selectivity of the Catalyst for Benzene Selective Hydrogenation

Benzene selective hydrogenation catalysts were applied in the industrial process. With the prolonging of reaction time and the fluctuations of reaction conditions, the activity and selectivity of catalysts may deviate from the normal value. Timely measures to restore the main technical indicators of the catalysts to the normal level are known as activity and selectivity modulation.

Activity and selectivity modulation is divided into physical processes and chemical processes. The physical process is to adjust the catalyst activity by regulating the temperature, pressure, and other factors. The chemical process is by adding acid, alkali, and other chemical substances to adjust the catalyst activity and the cyclohexene selectivity. In the industrial production process, the organic phase containing cyclohexene, cyclohexane, and benzene is continuously separated, and the water and the additives dissolved in the organic phase are subsequently discharged; or because the Fe, Cr, Ni, Mo, and other trace metals from the reactor wall and the utility processes deposit on the catalyst surface; or because the sulfur compounds in benzene and hydrogen exceed the standard level; or due to the introduction of nitrogen compounds (extractant: *N*, *N*-Dimethylacetamide, DMAC); and due to the introduction of CO and other chemical factors from hydrogen, these may all lead to deviation of activity and selectivity from normal levels.

In the specified benzene-selective hydrogenation catalyst evaluation system, when the reaction was within 10 min, benzene conversion rate was higher than 40% and the selectivity of cyclohexene was less than 80%. $Zn(OH)_2$, NaOH, or basic salt can be added to reduce the activity of the catalysts and improve the selectivity of cyclohexene. According to the reversibility principle of chemical reactions, adding H_2SO_4 to the reaction system can improve the catalyst activity. When the reaction time was within 20 min and the conversion of benzene was less than 40% as well as the selectivity of cyclohexene was about 90%, the activity and selectivity of the catalysts can be restored to normal level by H_2SO_4 modulation. The scientific nature of the modulation is to modulate the Zn content of the catalysts and the coverage of the base salt on the catalyst surface.

Z. Liu et al., *Catalytic Technology for Selective Hydrogenation of Benzene to Cyclohexene*,
https://doi.org/10.1007/978-981-15-6411-6_7

This chapter focuses on the most common fluctuations of activity and selectivity of catalysts in the industry, as well as the corresponding modulation methods and modulation mechanism.

7.1 Modulation of Activity and Selectivity of the Catalyst for Benzene Selective Hydrogenation

7.1.1 Modulation Method

In order to systematically study the modulation method of selective hydrogenation of Ru–Zn catalyst, a series of Ru–Zn catalysts with different Zn contents were prepared and their phase analysis, microcrystalline particle size determination and activity, and selectivity evaluation were carried out. The effects of Zn content, slurry pH, and pretreatment on the selectivity and activity of the catalyst were investigated.

Figure 7.1 shows the XRD patterns of Ru–Zn catalysts with different Zn contents.

It can be seen from Fig. 7.1, when the Zn content was more than 10%, there was a weak diffraction peak of ZnO in addition to the metal Ru diffraction peak in the XRD pattern.

Table 7.1 shows the microcrystalline particle sizes of Ru–Zn catalysts with different Zn contents.

It can be seen from Table 7.1 that the microcrystalline particle size distribution of Ru–Zn catalyst was between 4.2 and 6.2 nm, and the microcrystalline particle size decreased with the increase of Zn content. The results of activity and selectivity measurement showed that the activity of Ru–Zn catalyst decreased with the increase

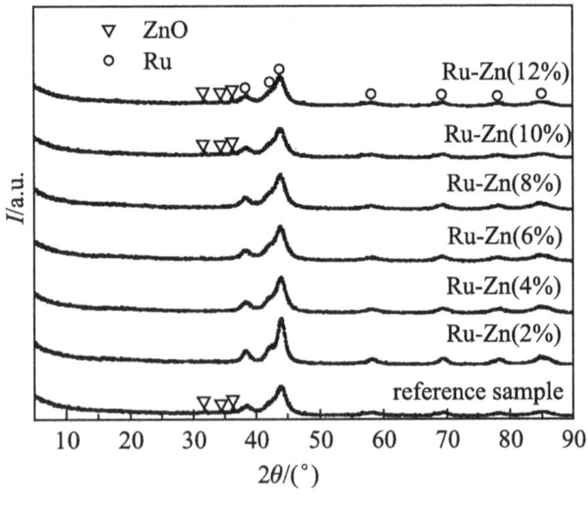

Fig. 7.1 XRD patterns of Ru–Zn catalysts with different Zn contents

Table 7.1 Microcrystalline particle size of Ru–Zn catalysts with different Zn Contents

Catalyst	$2\theta/(°)$	Full width at half maximum/rad	Microcrystalline particle size/nm
Reference sample	43.91	1.66	5.3
Ru–Zn(2%)	43.96	1.42	6.2
Ru–Zn(4%)	43.94	1.93	4.5
Ru–Zn(6%)	43.94	1.92	4.6
Ru–Zn(8%)	43.87	1.99	4.4
Ru–Zn(10%)	43.92	1.99	4.4
Ru–Zn(12%)	43.91	2.07	4.2

of Zn content, and the selectivity of cyclohexene was increasing at the same benzene conversion.

Tables 7.2 and 7.3 show the activity and selectivity of direct hydrogenation of the Ru–Zn (4%) catalysts when the slurry pH was 5.79 and 5.88, respectively.

Figure 7.2 shows the curves of the conversion of benzene C_{BZ} versus time t and the cyclohexene selectivity S_{HE} versus benzene conversion C_{BZ} at different pHs.

Table 7.2 The activity and selectivity of direct hydrogenation of the Ru–Zn (4%) catalysts at slurry pH of 5.79

t/min	T/°C	C_{BZ}/%	S_{HE}/%	Y_{HE}/%
5	157	56.65	77.56	43.94
10	159	86.98	64.67	56.25
15	152	96.03	54.37	52.21
20	145	98.59	47.22	46.55
25	143	99.34	42.14	41.86

Note C_{BZ} is benzene conversion, S_{HE} is cyclohexene selectivity, Y_{HE} is cyclohexene yield; reaction conditions: 150 °C, 5 MPa H_2, 1400 r/min, 1.96 g catalyst, 9.8 g ZrO_2, 45.7 g $ZnSO_4 \cdot 7H_2O$, 140 mL benzene, 280 mL water

Table 7.3 The activity and selectivity of direct hydrogenation of Ru–Zn (4%) catalyst at slurry pH of 5.88

t/min	T/°C	C_{BZ}/%	S_{HE}/%	Y_{HE}/%
5	149	38.51	77.80	29.96
10	152	61.83	72.17	44.62
15	153	79.48	64.95	51.62
20	150	89.48	58.75	52.57
25	146	94.70	53.53	50.69

Reaction conditions 150 °C, 5 MPa H_2, 1400 r/m, 1.96 g catalyst, 9.8 g ZrO_2, 45.7 g $ZnSO_4 \cdot 7H_2O$, 140 mL benzene, 280 mL water

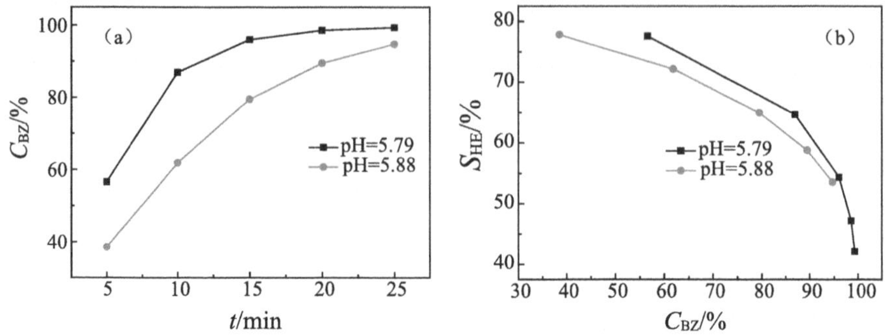

Fig. 7.2 The results of direct hydrogenation of Ru–Zn (4%) catalyst at different pHs. **a** C_{BZ}-t curve; **b** S_{HE}-C_{BZ} curve

It can be seen from Fig. 7.2 that the benzene conversion decreased at the same reaction time as the pH in the reaction system increased. And the selectivity of cyclohexene increased at the same benzene conversion. The small change of pH had a significant effect on the activity and selectivity of Ru–Zn catalysts. The pH affected the presence of the Zn species in the reaction system. Zn was present in the form of ionic form as Zn^{2+} at lower pH. At higher pH, Zn may be present as the basic complex salt of $Zn(OH)^+$, $Zn(OH)_2$, and zinc sulfate $3Zn(OH)_2 \cdot ZnSO_4 \cdot xH_2O$ (BZSS). The different Zn species were chemically adsorbed on the surface of the catalysts, which directly affected the surface properties of the catalysts and the diffusion and adsorption of benzene and hydrogen, and affected the formation and desorption of cyclohexene, thus affecting the activity of the catalyst and the selectivity of cyclohexene [1].

Table 7.4 gives the activity and selectivity of direct hydrogenation of Ru–Zn(6%) catalyst.

Table 7.5 gives the hydrogenation pretreatment of Ru–Zn (6%) catalyst after 22 h pretreatment.

Figure 7.3 shows the C_{BZ}-t curve and S_{HE}-C_{BZ} curve of direct hydrogenation by Ru–Zn (6%) catalyst after pretreatment for 22 h.

Table 7.4 The activity and selectivity of Ru–Zn (6%) catalysts for direct hydrogenation

t/min	T/°C	C_{BZ}/%	S_{HE}/%	Y_{HE}/%
5	149	33.95	82.44	27.99
10	154	58.27	76.32	44.47
15	156	80.35	68.84	55.31
20	154	89.98	62.53	56.26
25	149	94.37	56.84	53.64

Reaction conditions 150 °C, 5 MPa H_2, 1400 r/min, 1.96 g catalyst, 9.8 g ZrO_2, 45.7 g $ZnSO_4 \cdot 7H_2O$, 140 mL of benzene, 280 mL of water

Table 7.5 Hydrogenation activity and selectivity of Ru–Zn (6%) catalyst after 22 h pretreatment

t/min	T/°C	C_{BZ}/%	S_{HE}/%	Y_{HE}/%
5	140	10.10	89.90	9.08
10	145	20.59	89.51	18.43
15	151	32.91	88.51	29.13
20	159	47.59	88.33	40.61
25	163	61.91	81.62	50.53

Reaction conditions 150 °C, 5 MPa H_2, 1.96 g catalyst, 9.8 g ZrO_2, 45.7 g $ZnSO_4 \cdot 7H_2O$, 140 mL of benzene, 280 mL of water

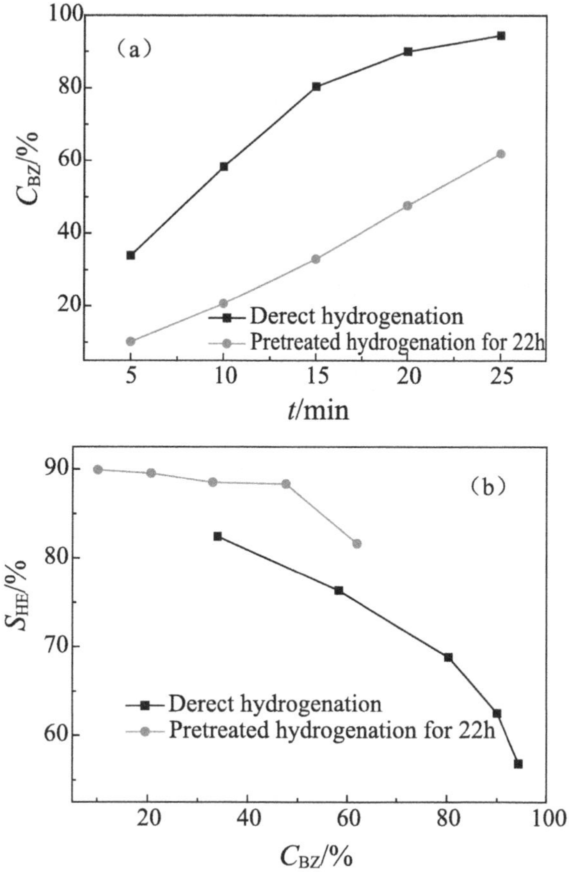

Fig. 7.3 The results of direct hydrogenation by Ru–Zn (6%) catalyst after pretreatment for 22 h. **a** C_{BZ}-t curve; **b** S_{HE}-C_{BZ} curve

It can be seen from Fig. 7.3 that the benzene conversion decreased at the same reaction time after pretreatment, and while the selectivity of cyclohexene increased at the same benzene conversion. By pretreatment, the ZnO on the surface of the catalyst interacted with $ZnSO_4$ in the slurry to produce insoluble base salt chemically adsorbed on the surface of the catalyst, which modified the electronic properties of the Ru active center and changed the geometrical arrangement of the Ru active center. Then hydrophilicity of the surface of the catalyst was improved. In the competitive adsorption of benzene and hydrogen in the same active site, the pretreatment process allowed the hydrogen to be preferentially adsorbed on the catalyst surface, reducing the number of adsorbed benzene molecules on the catalyst surface [2–5].

In summary, for Ru–Zn catalyst modulation method, the first is to adjust the Zn content of the catalyst, the second is to add acid or alkali to adjust the pH of reaction system, and the third is the catalyst pretreatment.

7.1.2 Modulation of $Zn(OH)_2$

The Ru–Zn (8%), Ru–Zn (9%) and Ru–Zn (10%) catalysts were used as modulated samples. The use of $Zn(OH)_2$ can reduce the activity of the catalyst and improve the selectivity of cyclohexene.

$Zn(OH)_2$ was prepared by $ZnSO_4 \cdot 7H_2O$ and NaOH. The same solution was prepared according to the stoichiometric amount of $ZnSO_4 \cdot 7H_2O$ and NaOH. NaOH solution was added dropwise to $ZnSO_4$ solution with stirring. After 5 min, white flocculent $Zn(OH)_2$ precipitation was formed. Then a certain amount of $Zn(OH)_2$ was added to the catalyst with stirring for 1 h to make it fully adsorbed on the catalyst surface, and then the catalyst was added to the reaction system. The temperature was raised to 140 °C, and 140 mL of benzene was directly hydrogenated by the catalyst, or after 22 h pretreatment of the catalyst.

Figure 7.4 shows the C_{BZ}-t curve and S_{HE}-C_{BZ} curve of direct hydrogenation by Ru–Zn (8%) catalysts with the addition of different amounts of $Zn(OH)_2$ to the reaction system.

Table 7.6 shows the t_{40}, t_{50}, t_{60} and S_{40}, S_{50}, S_{60} of direct hydrogenation by Ru–Zn (8%) catalysts with the addition of different amounts of $Zn(OH)_2$ to the reaction system.

As can be seen from Table 7.6, with the addition of 49 mg of $Zn(OH)_2$, the time t_{40} at which the benzene conversion was 40% was extended from 3 to 4 min, and the selectivity of cyclohexene S_{40} increased from 68 to 79%. Then the time t_{60} was extended from 6 to 8 min and the cyclohexene selectivity S_{60} increased from 62 to 73%. With the increase of the amount of $Zn(OH)_2$, the time to achieve 40% conversion benzene gradually prolonged, which meant that the activity of the catalyst gradually decreased and the selectivity of cyclohexene gradually increased. When the amount of $Zn(OH)_2$ increased from 49 mg to 245 mg, the time t_{40} increased from 4 min to 14 min and the selectivity of cyclohexene increased from 79 to 86%. The time of 60% benzene conversion was extended from 8 to 20 min and the cyclohexene

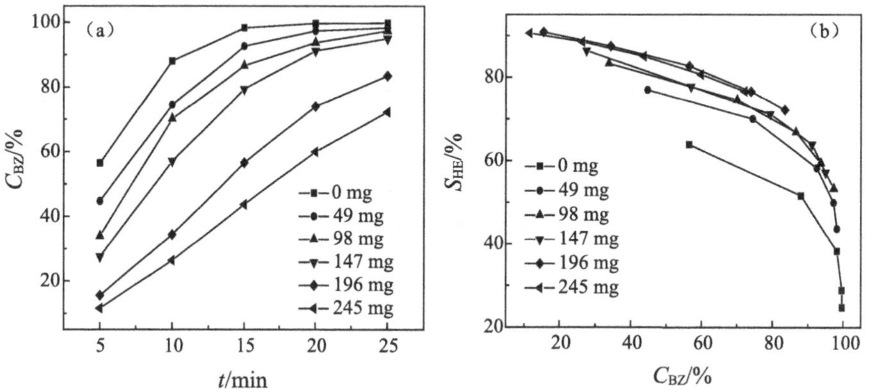

Fig. 7.4 The results of direct hydrogenation by Ru–Zn (8%) catalysts with the addition of different amounts of $Zn(OH)_2$ to the reaction system. **a** C_{BZ}-t curve; **b** S_{HE}-C_{BZ} curve

Table 7.6 The t_{40}, t_{50}, t_{60} and S_{40}, S_{50}, S_{60} of direct hydrogenation by Ru–Zn (8%) catalysts with the addition of different amounts of $Zn(OH)_2$ to the reaction system

$Zn(OH)_2$/mg	t_{40}/min	t_{50}/min	t_{60}/min	S_{40}/%	S_{50}/%	S_{60}/%
0	3	4	6	68	66	62
49	4	6	8	79	76	73
98	6	7	9	82	79	77
147	7	9	11	83	80	77
196	11	14	16	86	84	82
245	14	17	20	86	84	81

selectivity S_{60} increased from 73 to 81%. When the amount of $Zn(OH)_2$ was 196 mg, the selectivity of cyclohexene reached its maximum. When the amount of $Zn(OH)_2$ increased to 245 mg, the activity of the catalyst decreased and the selectivity of cyclohexene was unchanged. Thus the optimal amount of $Zn(OH)_2$ was 196 mg for Ru–Zn (8%) catalyst.

Figure 7.5 shows the C_{BZ}-t curve and S_{HE}-C_{BZ} curve of direct hydrogenation by Ru–Zn (9%) catalysts with the addition of different amounts of $Zn(OH)_2$ to the reaction system.

Table 7.7 shows the t_{40}, t_{50}, t_{60} and S_{40}, S_{50}, S_{60} of direct hydrogenation by Ru–Zn (9%) catalysts with the addition of different amounts of $Zn(OH)_2$ to the reaction system.

Figure 7.6 shows the C_{BZ}-t curve and S_{HE}-C_{BZ} curve of direct hydrogenation by Ru–Zn (10%) catalysts with the addition of different amounts of $Zn(OH)_2$ to the reaction system.

Table 7.8 shows the t_{40}, t_{50}, t_{60} and S_{40}, S_{50}, S_{60} of direct hydrogenation by Ru–Zn (10%) catalysts with the addition of different amounts of $Zn(OH)_2$ to the reaction system.

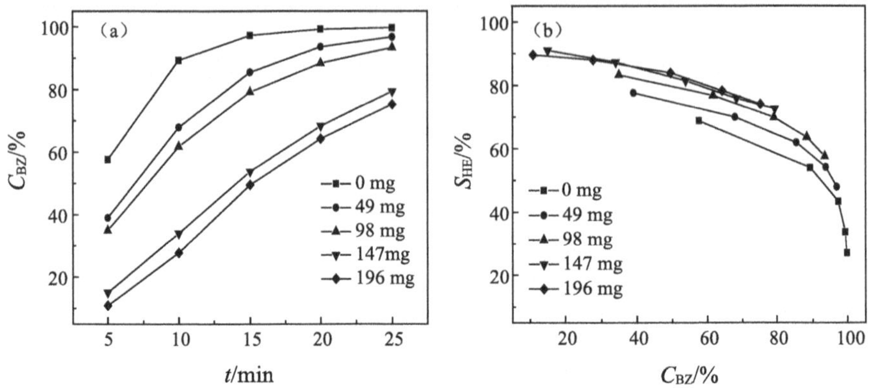

Fig. 7.5 The results of direct hydrogenation by Ru–Zn (9%) catalysts with the addition of different amounts of $Zn(OH)_2$ to the reaction system. **a** C_{BZ}-t curve; **b** S_{HE}-C_{BZ} curve

Table 7.7 The t_{40}, t_{50}, t_{60} and S_{40}, S_{50}, S_{60} of direct hydrogenation by Ru–Zn (9%) catalysts with the addition of different amounts of $Zn(OH)_2$ to the reaction system

$Zn(OH)_2$/mg	t_{40}/min	t_{50}/min	t_{60}/min	S_{40}/%	S_{50}/%	S_{60}/%
0	3	4	5	75	72	68
49	5	7	9	77	75	72
98	6	8	10	82	80	77
147	10	14	17	86	83	79
196	13	15	19	86	84	80

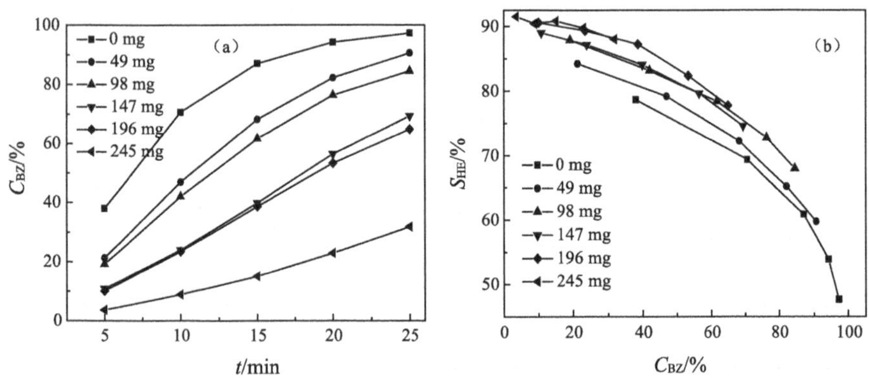

Fig. 7.6 The results of direct hydrogenation by Ru–Zn (10%) catalysts with the addition of different amounts of $Zn(OH)_2$ to the reaction system. **a** C_{BZ}-t curve and **b** S_{HE}-C_{BZ} curve

Table 7.8 The t_{40}, t_{50}, t_{60} and S_{40}, S_{50}, S_{60} of direct hydrogenation by Ru–Zn (10%) catalysts with the addition of different amounts of $Zn(OH)_2$ to the reaction system

$Zn(OH)_2$/mg	t_{40}/min	t_{50}/min	t_{60}/min	S_{40}/%	S_{50}/%	S_{60}/%
0	5	7	9	78	75	72
49	9	11	13	81	78	75
98	10	12	15	84	81	79
147	15	18	22	84	81	78
196	16	19	23	87	83	80
245	32	35	39	87	83	80

From Tables 7.7 and 7.8, it can be seen that $Zn(OH)_2$ had the same effects on Ru–Zn (9%), Ru–Zn (10%) catalysts with Ru–Zn (8%) catalysts. The catalyst activity decreased and the cyclohexene selectivity was improved. With the increase of the amount of $Zn(OH)_2$, the time for the same benzene conversion was gradually prolonged, and the selectivity of cyclohexene gradually increased at the same conversion of benzene. When $Zn(OH)_2$ reached a certain amount, the selectivity of cyclohexene was maximized. When the addition of $Zn(OH)_2$ was over this amount, the catalytic activity was reduced while the selectivity of cyclohexene cannot be improved. For Ru–Zn (8%), Ru–Zn (9%), and Ru–Zn (10%) catalysts, the optimal amount of the $Zn(OH)_2$ was 196 mg, and it was expected that this rule may be applied to Ru–Zn catalyst with other Zn contents.

7.1.3 Modification of NaOH

As the reaction system contains a large amount of $ZnSO_4$, the direct addition of NaOH to the reaction system will also lead to the in situ formation of $Zn(OH)_2$. It has a very small influence on the $ZnSO_4$ concentration in the slurry, but it can reduce the preparation procedures of $Zn(OH)_2$. In theory, the effect of the catalyst should be the same as long as the amount of $Zn(OH)_2$ is the same. In the presence of Ru–Zn (10%) catalyst, different amounts of NaOH were added to the slurry, stirred, and adsorbed for 30–60 min at 50 °C. Then the activity and selectively were evaluated by adding benzene to start the reaction directly.

Figure 7.7 shows the C_{BZ}-t curve and S_{HE}-C_{BZ} curve of direct hydrogenation by Ru–Zn (10%) catalysts with the addition of different amounts of NaOH to the reaction system.

Table 7.9 shows the t_{40}, t_{50}, t_{60} and S_{40}, S_{50}, S_{60} of direct hydrogenation by Ru–Zn (10%) catalysts with the addition of different amounts of NaOH to the reaction system.

From Table 7.9, it can be seen that the effect of NaOH was the same as that of $Zn(OH)_2$. With the increase of NaOH, the times of the same benzene conversion of t_{40}, $t_{50,}$ and t_{60} were gradually prolonged and the selectivities of cyclohexene S_{40},

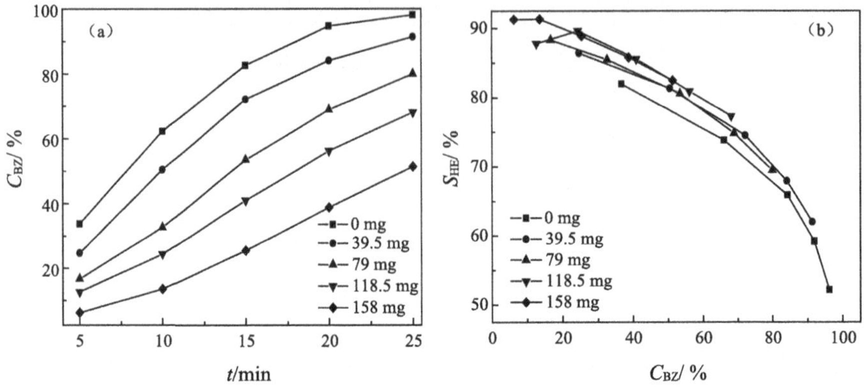

Fig. 7.7 The results of direct hydrogenation by Ru–Zn (10%) catalysts with the addition of different amounts of NaOH to the reaction system. **a** C_{BZ}-t curve; **b** S_{HE}-C_{BZ} curve

Table 7.9 The t_{40}, t_{50}, t_{60} and S_{40}, S_{50}, S_{60} of direct hydrogenation by Ru–Zn (10%) catalysts with the addition of different amounts of NaOH to the reaction system

NaOH/mg	t_{40}/min	t_{50}/min	t_{60}/min	S_{40}/%	S_{50}/%	S_{60}/%
0.0	5	7	9	81	78	75
39.5	8	10	12	83	81	78
79.0	12	14	17	84	81	78
118.5	15	18	22	86	83	80
158.0	20	24	29	86	83	80

$S_{50,}$ and S_{60} gradually increased when benzene conversion were 40, 50, and 60%. When the amount of NaOH reached a certain value, the selectivity of cyclohexene become the highest. Beyond this value, the catalyst activity was reduced while the selectivity of cyclohexene cannot be improved. For Ru–Zn (10%) catalyst, the best performance was obtained when NaOH was 158 mg.

According to the Stoichiometric equation,

$$2NaOH + ZnSO_4 = Zn(OH)_2 + Na_2SO_4$$

The 158 mg NaOH was corresponding to 196 mg $Zn(OH)_2$. The experimental results were in good agreement with the theory. The direct addition of NaOH not only in situ formed $Zn(OH)_2$ but also generated Na_2SO_4 in the reaction system, and Na_2SO_4 had no influence on the catalytic performance. As SO_4^{2-} originally existed in the slurry, the key factor should be the effect of Na^+. It has been reported [6] that the Na^+ can play as an additive for the Ru-based catalysts, and at least it is harmless on the catalytic system. Therefore, the use of NaOH on the catalyst is a practical method for modulation.

7.1.4 Modification of Alkaline Salt

Ru–Zn catalyst pretreatment can improve the selectivity and yield of cyclohexene. The essential reason is that the ZnO on the surface of the catalyst can interact with the zinc sulfate in the slurry to produce a poorly soluble base salt (BZSS), which was chemically adsorbed on the catalyst surface to form a stagnant layer. This layer can inhibit the reaction rate of the elementary steps of benzene hydrogenation so that it further inhibits the rate of hydrogenation of cyclohexene to cyclohexane, thereby increasing the cyclohexene selectivity [6, 7]. It is contemplated that if a layer of BZSS is previously covered on the catalyst surface, it will play the same role. Therefore, it can be expected to use BZSS directly on the catalyst for catalyst modulation to achieve the same results. BZSS can be directly added to the reaction system, or the prepared BZSS can be mixed with the catalyst first with stirring and adsorption for a certain time, and then they can be added to the reactor.

With the Ru–Zn (8%) and Ru–Zn (9%) catalyst for the controlled experiments, the catalyst and BZSS were added to the reaction with stirring for 1 h to make it fully adsorbed on the catalyst surface. Then the reaction slurry was added with heating to 140 °C and the addition of 140 mL of benzene for direct hydrogenation. The catalyst evaluation was performed using catalysts by adding different amounts of BZSS.

Figure 7.8 shows the C_{BZ}-t curve and S_{HE}-C_{BZ} curve of direct hydrogenation by Ru–Zn (8%) catalysts with the addition of different amounts of BZSS to the reaction system.

Table 7.10 shows the t_{40}, t_{50}, t_{60} and S_{40}, S_{50}, S_{60} of direct hydrogenation by Ru–Zn (8%) catalysts with the addition of different amounts of BZSS to the reaction system.

As can be seen from Table 7.10, with an increase in the amount of BZSS, the time of the same benzene conversion of t_{40}, $t_{50,}$ and t_{60} were gradually prolonged, which meant the catalytic activity decreased. While the selectivity of cyclohexene

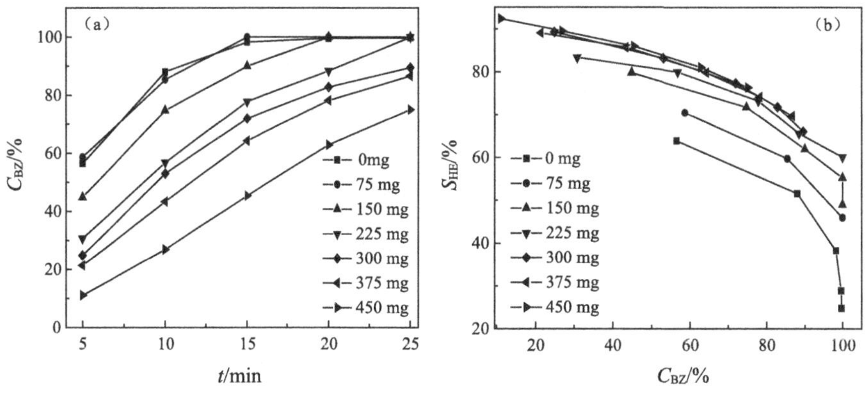

Fig. 7.8 The results of direct hydrogenation by Ru–Zn (8%) catalysts with the addition of different amounts of BZSS to the reaction system. **a** C_{BZ}-t curve; **b** S_{HE}-C_{BZ} curve

Table 7.10 The t_{40}, t_{50}, t_{60} and S_{40}, S_{50}, S_{60} of direct hydrogenation by Ru–Zn (8%) catalysts with the addition of different amounts of BZSS to the reaction system

$3Zn(OH)_2 \cdot ZnSO_4 \cdot xH_2O$(BZSS)/mg	t_{40}/min	t_{50}/min	t_{60}/min	S_{40}/%	S_{50}/%	S_{60}/%
0	3	4	6	68	66	62
75	3	4	6	77	73	70
150	4	6	8	81	78	76
225	7	9	11	82	81	79
300	8	10	12	86	84	81
375	9	12	14	86	84	81
450	14	16	20	87	85	82

S_{40}, S_{50}, and S_{60} gradually increased when benzene conversion was 40, 50, and 60% and this result was consistent with the result of the modification of $Zn(OH)_2$ and NaOH. When the amount of BZSS was 450 mg, the selectivity of cyclohexene was 87% at 40% benzene conversion and it was 85% at 50% benzene conversion. With further increase of BZSS dosage, the time to achieve the same benzene conversion of t_{40}, t_{50}, t_{60} gradually increased while the selectivity of cyclohexene cannot be improved. Thus, it can be concluded that the best modulation effect was obtained with the addition of 450 mg ZBSS for Ru–Zn (8%) catalyst.

Figure 7.9 shows the C_{BZ}-t curve and S_{HE}-C_{BZ} curve of direct hydrogenation by Ru–Zn (9%) catalysts with the addition of different amounts of BZSS to the reaction system.

Table 7.11 shows the t_{40}, t_{50}, t_{60} and S_{40}, S_{50}, S_{60} of direct hydrogenation by Ru–Zn (9%) catalysts with the addition of different amounts of BZSS to the reaction system.

As can be seen from Table 7.11, the influence of BZSS on Ru–Zn (9%) catalysts was similar to that on Ru–Zn (8%) catalysts. For both Ru–Zn (8%) and Ru–Zn (9%)

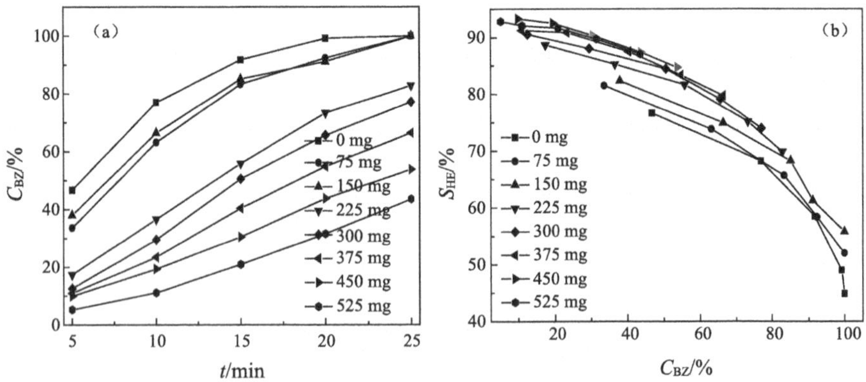

Fig. 7.9 The results of direct hydrogenation by Ru–Zn (9%) catalysts with the addition of different amounts of BZSS to the reaction system. **a** C_{BZ}-t curve; **b** S_{HE}-C_{BZ} curve

Table 7.11 The t_{40}, t_{50}, t_{60} and S_{40}, S_{50}, S_{60} of direct hydrogenation by Ru–Zn (9%) catalysts with the addition of different amounts of BZSS to the reaction system

$3Zn(OH)_2 \cdot ZnSO_4 \cdot 3H_2O$(BZSS)/mg	t_{40}/min	t_{50}/min	t_{60}/min	S_{40}/%	S_{50}/%	S_{60}/%
0	3	4	5	75	72	68
75	6	8	9	80	77	75
150	5	7	9	82	79	77
225	11	14	16	85	83	80
300	13	15	18	86	85	81
375	15	18	22	88	85	82
450	19	23	28	88	86	82
525	24	28	32	88	85	82

catalysts, 450 mg was the optimal amount of ZBSS. For Ru–Zn (8%) catalyst, the conversion of benzene was 40% and the selectivity of cyclohexene S_{40} was 87% at 14 min, and at 16 min, the conversion of benzene was 50% and the selectivity of cyclohexene S_{50} was 85%. For Ru–Zn (9%) catalyst, the conversion of benzene was 40% and the selectivity of cyclohexene S_{40} was 88% at 19 min, and at 23 min, the conversion of benzene was 50% and the selectivity of cyclohexene S_{50} was 86%.

7.1.5 *Comparison of the Effect of Catalyst Pretreatment and Basic Salt Modulation*

When the catalyst is used for direct hydrogenation, the catalytic activity is usually high and the cyclohexene selectivity is low. Generally, the pretreatment of catalysts for more than 22 h in the reaction system can reduce the catalyst activity and improve the selectivity of cyclohexene. The comparison of the results using catalysts with pretreatment and the addition of BZSS for direct hydrogenation is carried out in order to understand the nature of the pretreatment.

Figure 7.10 gives the C_{BZ}-t curve and S_{HE}-C_{BZ} curve of direct hydrogenation by Ru–Zn (9%) catalysts with pretreatment for 22 h and the addition of different amounts of basic salt.

Table 7.12 gives the t_{40}, t_{50}, t_{60} and S_{40}, S_{50}, S_{60} of direct hydrogenation by Ru–Zn (9%) catalysts with pretreatment for 22 h and the addition of different amounts of basic salt.

It can be seen from Table 7.12 that the result of direct hydrogenation using a catalyst with the addition of 225 mg of basic salt in the reaction slurry was very close to that of the catalyst with pretreatment for 22 h, indicating that reason for the improvement of the selectivity of cyclohexene by the catalyst pretreatment was closely related to the formation of base salt on the catalyst surface.

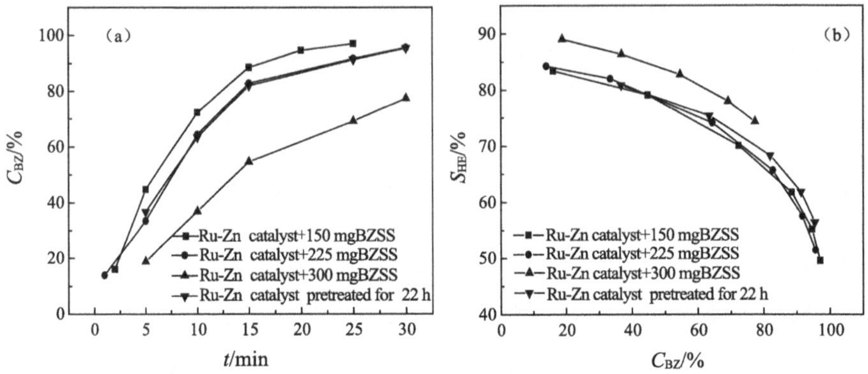

Fig. 7.10 The results of direct hydrogenation by Ru–Zn (9%) catalysts with pretreatment for 22 h and the addition of different amounts of basic salt. **a** C_{BZ}-t curve; **b** S_{HE}-C_{BZ} curve

Table 7.12 The t_{40}, t_{50}, t_{60} and S_{40}, S_{50}, S_{60} of direct hydrogenation by Ru–Zn (9%) catalysts with pretreatment for 22 h and the addition of different amounts of basic salt

Items	t_{40}/min	t_{50}/min	t_{60}/min	S_{40}/%	S_{50}/%	S_{60}/%
+150 mg BZSS	4	6	8	80	77	74
+225 mg BZSS	6	8	9	80	78	75
+300 mg BZSS	11	14	17	86	84	81
Pretreated for 22 h	6	8	9	80	78	76

When adding a catalyst in the industrial production process, the catalytic activity is often high in the initial state due to the lack of catalyst pretreatment. After a period of time, the catalyst selectivity can be restored to a normal level. In order to avoid such fluctuation, adding a layer of basic salt on the surface of the catalyst can have the same effect as a pretreatment process to ensure the stability of industrial operation.

7.1.6 Examples of Industrial Catalyst Modification

Table 7.13 shows the activity and selectivity of direct hydrogenation by industrial catalysts.

From Table 7.13, for the direct catalytic hydrogenation of industrial catalyst, 48.1% of benzene was converted and the selectivity of cyclohexene was 67.7% at 5 min. Obviously, the catalyst activity was relatively high and selectivity of cyclohexene was relatively low.

Table 7.14 shows the catalyst activity and selectivity of direct hydrogenation by adding 0.15 g $Zn(OH)_2$ to the reaction system.

It can be seen from Table 7.14 that benzene conversion was 43.9% and cyclohexene selectivity was 83.3% at 15 min, which indicated a good modulation effect.

Table 7.13 The activity and selectivity of direct hydrogenation by industrial catalysts

t/min	C_{BZ}/%	S_{HE}/%	Y_{HE}/%
5	48.09	67.73	32.57
10	76.56	57.24	43.82
15	91.78	45.50	41.76
20	95.85	36.94	35.41
25	98.95	29.42	29.11

Slurry composition 280 mL H_2O, 9.8 g ZrO_2, 45.7 g $ZnSO_4 \cdot 7H_2O$; Hydrogenation conditions: 150 °C, 5 MPa, 1400 r/min. The reaction started after adding 140 mL of benzene

Table 7.14 The results of direct hydrogenation after adding 0.15 g $Zn(OH)_2$ to the reaction system

t/min	C_{BZ}/%	S_{HE}/%	Y_{HE}/%
5	12.50	88.96	11.12
10	28.13	87.13	24.51
15	43.90	83.28	36.56
20	54.75	79.98	43.79
25	62.49	77.21	48.25

Slurry composition 280 mL H_2O, 9.8 g ZrO_2, 45.7 g $ZnSO_4 \cdot 7H_2O$; Hydrogenation conditions: 150 °C, 5 MPa, 1400 r/min, 140 mL benzene

In order to determine the optimal amount of $Zn(OH)_2$, Tables 7.15, 7.16, 7.17 show the catalytic performance with the addition of 0.20, 0.50, and 1.0 g $Zn(OH)_2$ to the reaction.

Figure 7.11 gives the C_{BZ}-t curve and S_{HE}-C_{BZ} curve of benzene hydrogenation by catalysts with the addition of different amounts of $Zn(OH)_2$ to the reaction.

It can be seen from Fig. 7.11 that the optimal amount of $Zn(OH)_2$ was 0.15 g. With further increase in the amount of $Zn(OH)_2$, the catalytic activity decreased and the selectivity of cyclohexene could not be improved. Thus, $Zn(OH)_2$ can significantly reduce the activity of the catalyst and improve the selectivity of cyclohexene. And

Table 7.15 The activity and selectivity of industrial catalyst for direct hydrogenation by adding 0.20 g $Zn(OH)_2$

t/min	C_{BZ}/%	S_{HE}/%	Y_{HE}/%
5	7.87	89.33	7.03
10	20.06	88.63	17.78
15	32.50	86.03	27.96
20	41.76	83.38	34.82
25	47.81	81.22	38.83

Slurry composition 280 mL H_2O, 9.8 g ZrO_2, 45.7 g $ZnSO_4 \cdot 7H_2O$; Hydrogenation conditions: 140 mL benzene, 150 °C, 5 MPa, 1400 r/min

Table 7.16 The activity and selectivity of the industrial catalyst for direct hydrogenation by adding 0.50 g $Zn(OH)_2$

t/min	C_{BZ}/%	S_{HE}/%	Y_{HE}/%
5	2.55	80.00	2.04
10	8.69	90.33	7.85
15	14.97	90.38	13.53
20	21.61	88.57	19.14
25	25.37	87.19	22.12

Slurry composition 280 mL H_2O, 9.8 g ZrO_2, 45.7 g $ZnSO_4{\cdot}7H_2O$; Hydrogenation conditions: 150 °C, 5 MPa, 1400 r/min, 140 mL benzene

Table 7.17 The activity and selectivity of the industrial catalyst for direct hydrogenation by adding 1.0 g $Zn(OH)_2$

t/min	C_{BZ}/%	S_{HE}/%	Y_{HE}/%
5	10.35	83.77	8.67
10	17.87	86.40	15.44
15	24.26	85.90	20.84
20	30.00	84.53	25.36
25	33.98	83.40	28.34

Slurry composition 280 mL H_2O, 9.8 g ZrO_2, 45.7 g $ZnSO_4{\cdot}7H_2O$; Hydrogenation conditions: 150 °C, 5 MPa, 1400 r/min, 140 mL benzene

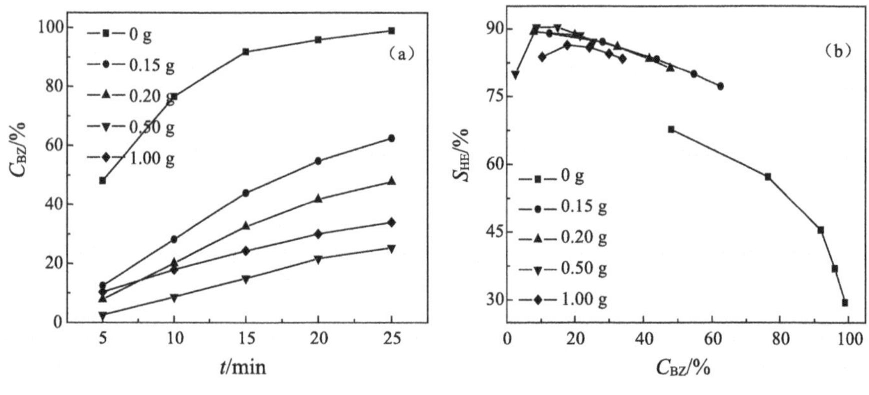

Fig. 7.11 The results of direct hydrogenation by catalysts with the addition of different amounts of $Zn(OH)_2$ to the reaction. **a** C_{BZ}-t curve; **b** S_{HE}-C_{BZ} curve

there was an optimal amount of $Zn(OH)_2$, above which the activity of the catalyst decreased while the selectivity of cyclohexene cannot be further improved.

Table 7.18 gives the result using a catalyst with the optimal amount of $Zn(OH)_2$, NaOH, and basic salt BZSS.

As can be seen from Table 7.18, the benzene conversion was more than 40% and the cyclohexene selectivity was lower than 80% with 1.96 g of Ru–Zn catalyst at

Table 7.18 The results using catalyst with the optimal amount of $Zn(OH)_2$, NaOH, and BZSS

	Catalyst	m^*/mg	t_{40}/min	t_{50}/min	t_{60}/min	S_{40}/%	S_{50}/%	S_{60}/%
$Zn(OH)_2$	Ru–Zn(8%)	196	11	14	16	86	84	82
	Ru–Zn(9%)	196	13	15	19	87	83	80
	Ru–Zn(10%)	196	16	19	23	87	83	80
NaOH	Ru–Zn(10%)	158	20	25	29	86	83	80
BZSS	Ru–Zn(8%)	450	14	16	20	87	85	82
	Ru–Zn(9%)	450	19	23	28	88	86	82
	Ru–Zn(10%)	450	9	12	14	85	83	81

*$Zn(OH)_2$, NaOH, and BZSS were added at the optimal amounts

10 min in the given reaction system. The addition of $Zn(OH)_2$, NaOH or basic salt can reduce the activity of the catalyst and improve the selectivity of cyclohexene. With the controlled conditions, the cyclohexene selectivity can be higher than 80% at 40% benzene conversion in 15 min. Using $Zn(OH)_2$ as the modulating substance, the optimal adding amount was 196 mg; using NaOH as the modulating substance, the optimal adding amount was 158 mg; using BZSS as the modulating substance, the optimal adding amount was 450 mg. In the process of industrial catalyst modification, the scale of the amount of catalyst can be amplified according to requirement.

It is $Zn(OH)_2$ which can adjust the pH of the slurry and influence the existence of Zn species and modulate the species and content of Zn on the surface of the catalyst. Thus, it directly affects the surface properties of the catalyst. NaOH, BZSS, and $Zn(OH)_2$ showed the same modulating mechanism and the effect of BZSS was the best which showed little influence on the activity of the catalyst but the most obvious influence on the selectivity of cyclohexene. The catalytic activity of Ru–Zn catalyst with the modulation of $Zn(OH)_2$, NaOH, and BZSS was studied. It was proved that the ZnO on the surface of the catalyst and the $ZnSO_4$ in the slurry formed the insoluble BZSS, which chemically adsorbed on the surface of the catalyst and played an important role on the activity and selectivity of Ru–Zn catalyst.

7.1.7 Modification Effect of H_2SO_4

According to the reversibility principle of chemical reaction, adding $Zn(OH)_2$ in the reaction system can reduce the activity of the catalyst and improve the selectivity of cyclohexene. However, adding sulfuric acid has the opposite effect, the activity of the catalyst can be improved and the selectivity of cyclohexene can be reduced. According to this principle, the measures for the deactivation of the catalyst in the industry are carried out by adding sulfuric acid, so that the selectivity of the catalyst is not reduced while the activity can be returned to normal levels.

The deactivation of industrial catalysts can be divided into reversible deactivation and irreversible deactivation. In general, deactivation due to aging of the catalyst, grain growth, and sulfur poisoning is often irreversible [8–10]. If it is due to improper operating conditions or the inducing *N*, *N*-Dimethylacetamide (DMAC) from benzene cycling, then the deactivation is reversible. The hydrolysis of DMAC produces a large amount of hydroxide, which interacts with the zinc sulfate in the slurry to produce a basic salt absorbed on the surface of the catalysts. A small amount of DMAC has the effect of improving the selectivity of cyclohexene, and excess DMAC can significantly reduce the catalyst activity. Therefore, judging the reason of catalyst deactivation is a prerequisite for the modulation.

Table 7.19 gives the catalytic activity and selectivity of direct hydrogenation by inactivated industrial catalysts.

As can be seen from Table 7.19, the catalyst activity was very low. Benzene was only converted to 6.33% at 25 min, but the cyclohexene selectivity was quite high. The catalyst was characterized by XRD and ICP. The results showed that a large amount of basic zinc sulfate salt was adsorbed on the surface of the catalyst, so that the catalyst activity could be returned to the normal level by adding H_2SO_4.

Table 7.20 shows the activity and selectivity of the direct hydrogenation after adding 1.5 g H_2SO_4 to the 1.96 g deactivated catalyst in the above catalyst slurry.

It can be seen from Table 7.20 that the catalyst activity was obviously restored, and the benzene conversion was increased from 1.13 to 6.00% at 5 min and from 6.33

Table 7.19 The catalytic activity and selectivity of direct hydrogenation by inactivated industrial catalysts

t/min	C_{BZ}/%	S_{HE}/%	Y_{HE}/%
5	1.13	82.30	0.93
10	2.01	88.06	1.77
15	3.09	88.03	2.72
20	5.04	90.08	4.54
25	6.33	90.05	5.70

Slurry conditions 8.8 g wet catalyst (equivalent to 1.96 g dry catalyst), 280 mL the original slurry. Hydrogenation condition: 150 °C, 5 MPa, 1400 r/min, 140 mL benzene

Table 7.20 The activity and selectivity of the deactivated industrial catalyst by adding 1.5 g H_2SO_4

t/min	C_{BZ}/%	S_{HE}/%	Y_{HE}/%
5	5.98	87.63	5.24
10	11.45	89.26	10.22
15	18.20	87.53	15.93
20	26.78	86.82	23.25
25	33.28	85.28	28.38

Slurry conditions 8.8 g wet catalyst (equivalent to 1.96 g dry catalyst), 280 mL the original slurry, 1.5 g H_2SO_4. Hydrogenation condition: 150 °C, 5 MPa, 1400 r/min, 140 mL benzene to 33.28% at 25 min, which reached the basic level of fresh catalyst. By continuing to add sulfuric acid to the reaction system, the amount of sulfuric acid increased by 1 time, that is, by adding 3.0 g H_2SO_4.

Table 7.21 shows the activity and selectivity of the direct hydrogenation after adding 3.0 g H_2SO_4 to the 1.96 g deactivated catalyst in the above catalyst slurry.

As can be seen from Table 7.21, by adding 3.0 g H_2SO_4, benzene was fully converted with less than 15 min, and the selectivity of cyclohexene was drastically decreased. The selectivity of cyclohexene was only 2.06% at 15 min which revealed that the sulfuric acid was excess.

In order to determine the optimum amount of sulfuric acid, Fig. 7.12 shows the catalytic activity and selectivity with the addition of 0.1, 1.2, and 1.4 g H_2SO_4.

Table 7.22 shows the t_{40}, t_{50}, t_{60} and S_{40}, S_{50}, S_{60} of direct hydrogenation addition by deactivated catalysts with different amounts of H_2SO_4 in the original slurry.

It can be seen from Table 7.22 that the activity of the catalyst was still very low by adding 0.1 g H_2SO_4, and the conversion of benzene was 9.00% at 25 min. After adding 1.4 g H_2SO_4, the activity of the catalyst was obviously restored and the selectivity of cyclohexene was slightly lower. The benzene conversion was 40%

Table 7.21 The activity and selectivity of the deactivated industrial catalyst by adding 3.0 g H_2SO_4

t/min	C_{BZ}/%	S_{HE}/%	Y_{HE}/%
5	78.59	18.58	14.60
10	98.73	8.68	8.57
15	100	2.06	2.06
20	100	0	0
25	100	0	0

Slurry conditions 8.8 g wet catalyst (equivalent to 1.96 g dry catalyst), 280 mL the original slurry, 3.0 g H_2SO_4. Hydrogenation condition: 150 °C, 5 MPa, 1400 r/min, 140 mL benzene

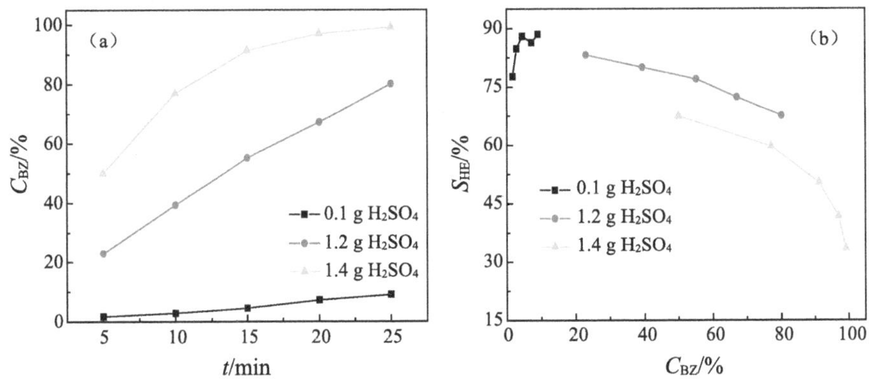

Fig. 7.12 The results of direct hydrogenation of the deactivated catalysts with the addition of different amounts of H_2SO_4. **a** C_{BZ}-t curve; **b** S_{HE}-C_{BZ} curve

Table 7.22 The t_{40}, t_{50}, t_{60} and S_{40}, S_{50}, S_{60} of direct hydrogenation addition by deactivated catalysts with different amounts of H_2SO_4

H_2SO_4/g	t_{40}/min	t_{50}/min	t_{60}/min	S_{40}/%	S_{50}/%	S_{60}/%
0.1	–	–	–	–	–	–
1.2	10	13	17	80	78	75
1.4	4	5	7	69	67	64

and the selectivity of cyclohexene S_{40} reached 69%. After adding 1.2 g H_2SO_4, the conversion of benzene was significantly improved. The conversion of benzene was 40% and the selectivity of cyclohexene S_{40} was 80%. The catalytic performance reached the original level.

The above examples show that sulfuric acid and zinc hydroxide have an opposite regulatory effect and they can significantly improve the catalytic activity and reduce the selectivity of cyclohexene. The key is to control the added amount appropriately.

In the preparation of the industrial catalyst, it is necessary to add the appropriate amount of BZSS to the Ru–Zn catalyst in order to prepare the Ru–Zn@BZSS-type catalyst. During the selective review of the active catalyst, it was found that the 29th batch catalysts had a very low activity after pretreatment while the selectivity of cyclohexene was high.

Table 7.23 gives the hydrogenation activity and selectivity for the 29th batch of catalyst after pretreatment.

It can be seen from Table 7.23 that benzene conversion was 7.4% and cyclohexene selectivity was 95% for the 29th batch of catalyst at 25 min. This can be determined to be caused by excess BZSS.

Figure 7.13 shows the XRD pattern of the 29th batch of catalyst. The typical diffraction peaks of the basic zinc sulfate on the 29th catalyst can be seen from Fig. 7.13, especially the diffraction peaks of $ZnSO_4{\cdot}6Zn(OH)_2{\cdot}4H_2O$ (JCPDS 00-0011-0280). It was found that the catalyst was not a single layer of dispersion, but the formation of a complete crystal, indicating a serious excess of basic salt.

Table 7.23 Hydrogenation activity and selectivity of the 29th batch of catalyst after pretreatment

t/min	T/°C	BZ/%	HA/%	HE/%	C_{BZ}/%	S_{HE}/%	Y_{HE}/%
5	138	99.15	0.05	0.79	0.80	94.18	0.75
10	140	97.15	0.24	2.24	2.36	90.53	2.13
15	142	96.00	0.26	3.73	3.80	93.63	3.56
20	143	94.42	0.35	5.23	5.31	93.87	4.98
25	144	92.22	0.40	7.37	7.41	94.97	7.04

Note The 29th batch of catalyst was 6.8 kg, basic salt BZSS was excess; BZ, HA and HE represented the mass percent of benzene, cyclohexane and cyclohexene in the product composition, respectively. C_{BZ} represented the molar percent conversion of benzene, S_{HE} represented the cyclohexene selectivity (mole fraction), Y_{HE} represented the yield of cyclohexene (mole fraction), and the following are the same

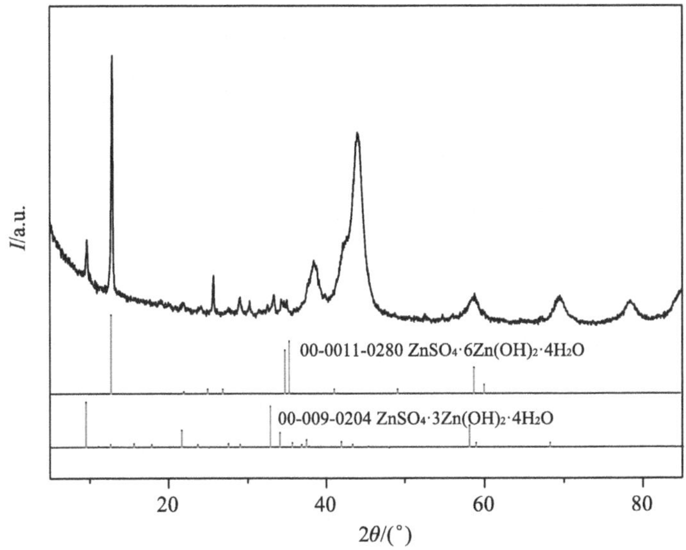

Fig. 7.13 The XRD pattern of the 29th catalyst

Considering that $Zn(OH)_2$ is soluble in acid and base, NaOH solution was added to the 29th batch of catalyst in order to reduce the BZSS content, and then resampling for direct hydrogenation.

Table 7.24 shows the activity and selectivity of the 29th batch of catalyst for direct hydrogenation after adding NaOH solution.

It can be seen from Table 7.24 that the addition of NaOH solution cannot wash away the basic zinc sulfate adsorbed on the surface of the catalyst and cannot reduce the Zn content in the catalyst. Although the catalyst activity was improved, benzene conversion was 25% and cyclohexene selectivity was 90% at 25 min. The catalyst activity was still low.

Table 7.25 shows the activity and selectivity of the 29th batch of catalyst for direct hydrogenation after washing with water.

Table 7.24 The activity and selectivity of the 29th batch of catalyst for direct hydrogenation after adding NaOH solution

t/min	T/°C	BZ/%	HA/%	HE/%	C_{BZ}/%	S_{HE}/%	Y_{HE}/%
5	144	97.05	0.35	2.48	2.69	87.89	2.37
10	155	90.81	0.72	8.34	8.65	92.23	7.98
15	158	85.39	1.25	13.24	13.87	91.56	12.70
20	155	78.22	2.21	19.45	20.81	90.02	18.73
25	150	73.97	2.7	23.21	24.95	89.80	22.40

Note The catalyst was washed by adding NaOH solution until the pH was close to 7

Table 7.25 The activity and selectivity of the 29th batch of catalyst for direct hydrogenation after washing with water

t/min	T/°C	BZ/%	HA/%	HE/%	C_{BZ}/%	S_{HE}/%	Y_{HE}/%
5	145	96.28	0.43	3.17	3.43	88.31	3.02
10	147	93.66	0.45	5.78	5.94	92.94	5.52
15	150	89.97	0.80	9.12	9.48	92.11	8.73
20	151	85.43	1.30	13.16	13.84	91.20	12.62
25	151	80.84	1.87	17.16	18.26	90.38	16.50

As can be seen from Table 7.25, benzene conversion was 18% and cyclohexene selectivity was 90% at 25 min. Compared with the washing effect of NaOH solution, the catalyst activity was reduced. This at least showed that the chemical force generated between the basic salt and the catalyst. It was not a simple physical adsorption or mixing, but the chemical adsorption on the catalyst surface.

Table 7.26 shows the activity and selectivity of the 29th batch of catalyst for direct hydrogenation after washing with H_2SO_4.

Table 7.26 The activity and selectivity of the 29th batch of catalyst for direct hydrogenation after washing with H_2SO_4

t/min	T/°C	BZ/%	HA/%	HE/%	C_{BZ}/%	S_{HE}/%	Y_{HE}/%
5	153	69.07	5.67	25.24	29.77	82.01	24.41
10	162	43.97	11.91	44.08	54.65	79.13	43.24
15	162	23.72	21.28	54.93	75.22	72.56	54.58
20	158	13.64	28.53	57.75	85.65	67.46	57.78
25	153	7.66	35.57	56.69	91.90	62.02	56.99

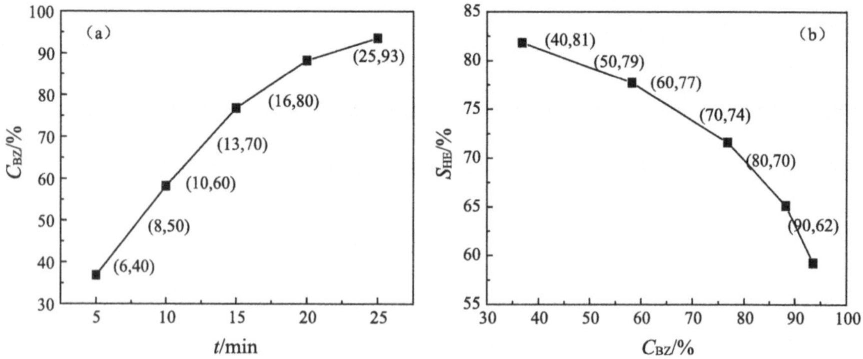

Fig. 7.14 The results of direct hydrogenation of the 29th batch of catalyst for direct hydrogenation after washing with H_2SO_4. **a** C_{BZ}-t curve; **b** S_{HE}-C_{BZ} curve

Figure 7.14 shows the C_{BZ}-t curve and S_{HE}-C_{BZ} curve of the 29th batch of catalyst for direct hydrogenation after washing with H_2SO_4.

From Fig. 7.14, it can be seen that benzene conversion was 40% and cyclohexene selectivity was 81% at 6 min; benzene conversion was 50% and cyclohexene selectivity was 79% at 8 min; benzene conversion was 60% and cyclohexene selectivity was 77% at 10 min. Apparently, this was due to excess H_2SO_4, indicating that the chemical adsorption of basic salt on the catalyst surface was completely reversible. With the role of acid, the basic salt can go into the slurry in the form of ions, thus restoring the catalytic activity.

7.2 Modulation Mechanism of Activity and Selectivity of the Catalyst for Benzene Selective Hydrogenation

7.2.1 Catalyst Structure and Texture Properties

Figure 7.15 shows the XRD patterns of Ru–Zn catalyst after adsorption of different amounts of BZSS.

In the benzene selective hydrogenation reaction system, in addition to Ru–Zn catalyst, there are zinc sulfate aqueous solution and dispersant micrometer ZrO_2 to prevent the agglomeration of nanometer Ru microcrystals and to prolong the service life of the catalyst [11]. Considering the influence of ZrO_2 diffraction peak after hydrogenation, no ZrO_2 was added to the reaction system.

It can be seen from Fig. 7.15a that all the catalysts were at $2\vartheta = 38.5°$, 44.0°, 58.3°, 69.2°, 78.4° (JCPDS 01-070-0274). The crystal phase of the catalyst did not change significantly when 75 mg BZSS was added. When the addition of BZSS was

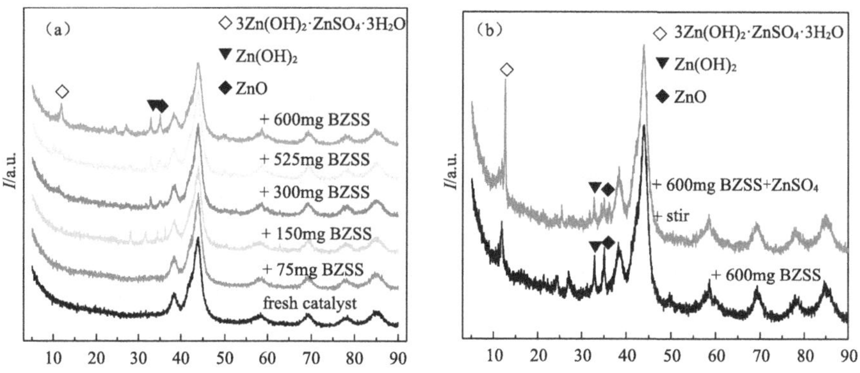

Fig. 7.15 X-ray diffraction pattern of Ru–Zn catalyst after adsorption of different amounts of basic zinc sulfate

increased to 150 mg, the strongest peak of $3Zn(OH)_2{\cdot}ZnSO_4{\cdot}3H_2O$ (JCPDS 00-044-0674) appeared at $2\vartheta = 12.6°$ on the sample, and the intensity increased with the increase of BZSS addition. However, the other diffraction peaks of BZSS were not obvious due to their relatively low intensity. In addition, the diffraction peaks of ZnO and $Zn(OH)_2$ were also present, which is related to Zn in the catalyst, indicating that Zn in the catalyst was mainly in the form of ZnO, and partially changed to $Zn(OH)_2$ under hydrogenation conditions.

Using the half peak wide of the strongest diffraction peak of Ru ($\vartheta = 44.0°$) and Scherrer's formula, the catalyst microcrystalline particle diameters are shown in Table 7.27. It was 3.4 nm before hydrogenation, and 3.4–3.7 nm after the addition of BZSS, indicating that BZSS did not significantly affect the crystallite size of the Ru–Zn catalyst.

The lower part of Fig. 7.15b is the XRD pattern of the catalyst after pre-adsorption BZSS. The above is the XRD pattern of the catalyst with pre-adsorption of BZSS after hydrogenation in the $ZnSO_4$ slurry. It can be seen the characteristic peak of BZSS at $2\vartheta = 12.6°$ becomes stronger after the hydrogenation in $ZnSO_4$ slurry, indicating that its microcrystalline particle size increased. The crystal phases of non-in situ prepared BZSS and in situ prepared BZSS were the same and BZSS content increased after hydrogenation.

Table 7.27 shows the texture parameters and microcrystalline particle sizes of the Ru–Zn catalyst with the pre-adsorption of BZSS catalyst before and after hydrogenation.

It can be seen from Table 7.27 that the BZSS was highly dispersed and uniformly distributed on the surface of the catalyst after the addition of BZSS, and the texture parameters of the catalyst do not change significantly before hydrogenation, which was consistent with the XRD results. With the increase of the amount of BZSS, the specific surface area, the average pore size, and the average pore volume decreased with the increase of the amount of BZSS. According to the simple pore structure geometry, the BZSS is formed on the inner and outer surfaces of the catalyst, thus reducing the size and pore volume and meanwhile reducing the physical adsorption

Table 7.27 The specific surface area S_{SET} and average pore volume V_p, pore size d_p, and microcrystalline particle size d of the Ru–Zn catalyst absorbed with BZSS

Samples	S_{BET}/(m²/g)	V_p/(cm³/g)	d_p/nm	d/nm
Ru–Zn(10%) catalyst before hydrogenation	59	0.16	6.4	3.4
Ru–Zn(10%) + 75 mg BZSS before hydrogenation	60	0.16	5.4	3.7
Ru–Zn(10%) + 300 mg BZSS before hydrogenation	58	0.13	4.5	3.6
Ru–Zn(10%) + 525 mg BZSS before hydrogenation	54	0.12	4.4	3.4
Ru–Zn(10%) catalyst after hydrogenation	54	0.14	5.3	–
Ru–Zn(10%) + 75 mg BZSS after hydrogenation	33	0.11	5.9	–
Ru–Zn(10%) + 300 mg BZSS after hydrogenation	32	0.10	5.5	–
Ru–Zn(10%) + 525 mg BZSS after hydrogenation	30	0.08	5.0	–

of N_2. It can also be seen from Table 7.27 that the specific surface area and the pore volume of the catalyst were significantly lower after hydrogenation, which can be explained by the fact that the BZSS generated during the hydrogenation process was further increased, blocking the channels of the catalyst.

Figure 7.16 shows the H_2-TPR spectra of the Ru–Zn catalysts after adding different amounts of BZSS. The sample was programmed in a hydrogen atmosphere with temperature reduction.

It can be seen from Fig. 7.16 that the Ru–Zn (10%) catalyst exhibited a reduction peak of Ru oxide with a shoulder at 50–122 °C, which was attributed to the stepwise reduction of $RuO_2 \rightarrow Ru_2O_3 \rightarrow RuO \rightarrow Ru$. The maximum reduction temperature was below 900 °C, where RuO was the predominant form. At 208–294 °C, the reduction peaks appeared at the highest reduction temperature around 230 °C. It was reported that the reduction of Zn oxide and Ru oxide have different degrees of synergistic effects in the reduction process [12]. When the easy reduction of Ru oxide was reduced to metal Ru, the overflow effect led to the reduction of the ZnO. Since the Ru reduction temperature was lower than the reaction temperature of 1500 °C and the Zn oxide reduction temperature was higher than 1500 °C, Ru was completely in the form of metal under hydrogenation conditions, and Zn existed mainly in the form of ZnO.

It can be seen from Fig. 7.16 that the reduction peaks of catalysts were similar to the adsorption of different amounts of BZSS. There were three shoulder peaks. With the increase of the adsorption capacity of BZSS, the reduction temperature gradually increased, indicating that the coalescence of BZSS on the catalyst surface delayed the reduction of RuO_x. From the shape of the reduction peak, it can be found that the reduction process of RuO_x has been changed due to the effect of BZSS.

Table 7.28 shows the hydrogen consumption calculated by the reduction peak area.

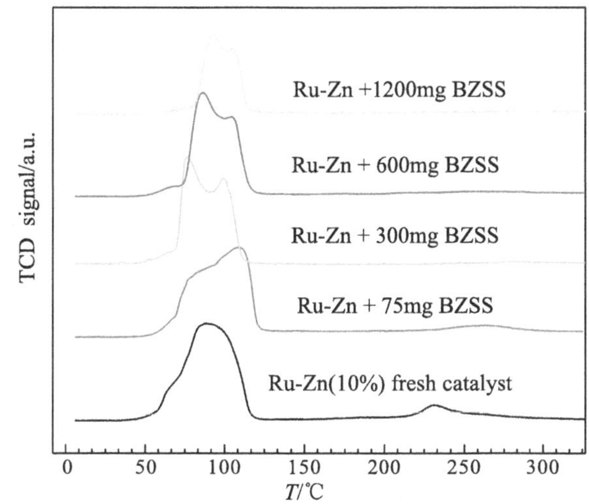

Fig. 7.16 The H_2-TPR spectra of Ru–Zn catalysts after adding different amounts of BZSS

Table 7.28 Hydrogen consumption of the catalyst absorbed with different amounts of BZSS

Catalyst	Hydrogen consumption/%
Ru–Zn(10%)	100
Ru–Zn(10%) + 75 mg BZSS	99
Ru–Zn(10%) + 300 mg BZSS	86
Ru–Zn(10%) + 600 mg BZSS	84
Ru–Zn(10%) + 1200 mg BZSS	51

In Table 7.28, it was assumed that the hydrogen consumption of the total reduction of RuO_x was 100%. With the increase of BZSS adsorption capacity, the amount of hydrogen consumption decreased gradually, which was related to the change of pore structure of the catalyst. In addition, it was related to the electronic nature of the active center.

7.2.2 SEM-EDX, XPS, and ICP-AES Analysis of the Catalyst

Figure 7.17 shows the photomicrograph (SEM) and the elemental energy-dispersive spectroscopy (CEDS) of the Ru–Zn catalyst, BZSS, BZSS adsorbed on the catalyst surface.

It can be seen from Fig. 7.17a that the Ru–Zn catalyst had a honeycomb-like structure with a large number of pore structures, which not only increased the specific surface area of the catalyst, but also increased the dispersion of the Ru active component, thus improving the activity of the catalyst [13].

As can be seen from Fig. 7.17b, BZSS exhibited a lamellar structure. From (c), (d), and (e), it can be seen intuitively that 600 mg BZSS was chemically adsorbed

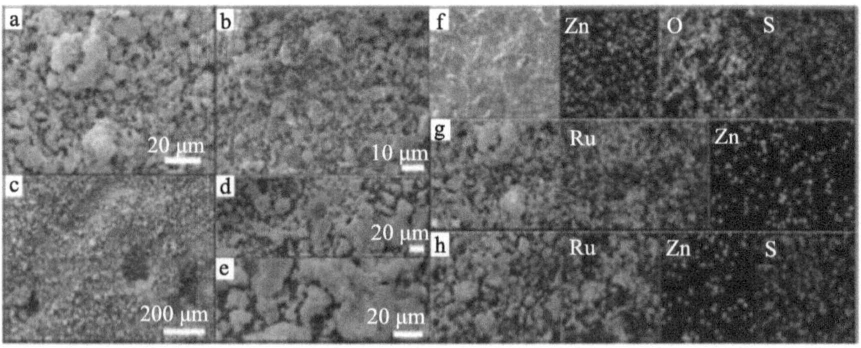

Fig. 7.17 **a** Ru–Zn catalyst; **b** BZSS; **c–e** the SEM spectra of Ru–Zn catalyst absorbed with 600 mg BZSS in the presence of 150 °C, 5 MPa, $ZnSO_4$, respectively; **f** EDS spectra of BZSS; **g** EDS spectra of Ru–Zn catalysts; **h** EDS spectra of Ru–Zn catalysts adsorbed with 600 mg BZSS

on the surface of the catalyst to form a uniform and smooth surface coating, which was confirmed by XRD results.

Figure 7.17f shows the energy spectrum of each element in BZSS of sample (b), where Zn, O, and S were marked in red, green, and blue, respectively. The atomic ratio of Zn and S and the atomic ratio of Zn and S was close to 4:1 in BZSS ($3Zn(OH)_2 \cdot ZnSO_4 \cdot 3H_2O$).

Figure 7.17g shows the energy spectrum of Ru and Zn in the sample (a) Ru–Zn catalyst. The atomic ratio of Zn and Ru calculated was 0.20. Figure 7.17h is the energy spectrum of the catalyst after adsorption of BZSS in sample (c). The atomic ratio of Zn and Ru was 0.19, indicating that BZSS did not improve the Zn content on the surface of Ru–Zn catalyst, and BZSS improved the selectivity of cyclohexene due to the change of the surface properties of the catalyst.

Figure 7.18 shows the X-ray photoelectron spectroscopy (XPS) of the Ru–Zn catalyst before hydrogenation and after the adsorption of different amounts of BZSS.

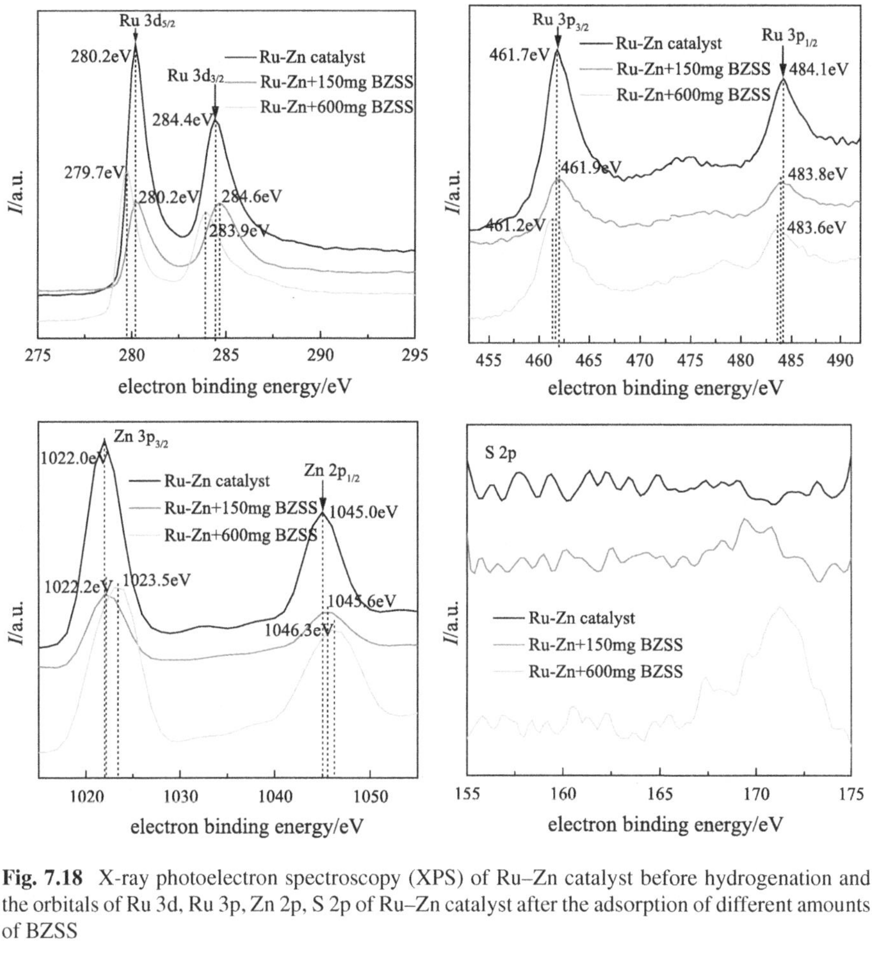

Fig. 7.18 X-ray photoelectron spectroscopy (XPS) of Ru–Zn catalyst before hydrogenation and the orbitals of Ru 3d, Ru 3p, Zn 2p, S 2p of Ru–Zn catalyst after the adsorption of different amounts of BZSS

From the Ru photoelectron spectroscopy in Fig. 7.18, the intensity of Ru spectroscopy decreased significantly with the adsorption of BZSS. This can be explained in two aspects. For one thing, BZSS was preferentially adsorbed in the Ru strongest active sites to play a shielding role on the Ru active centers. For another thing, BZSS was chemically adsorbed on the Ru–Zn catalyst surface, forming a layer of solid hysteresis water film under which these electrons on the 3p and 3d atomic orbitals of Ru were difficult to escape. From the Zn photoelectron spectroscopy, the intensity of Zn spectra was significantly reduced when adding 150 mg BZSS. The spectrum contained the Zn on the Ru–Zn catalyst surface and 2p photoelectron line of Zn in BZSS. On the one hand, the reduction of the intensity of 2p line of Zn on the catalyst surface was due to the shielding effect. The 2p photoelectron line of Zn in BZSS could not compensate for the reduction. After the addition of 600 mg BZSS, the intensity of 2p photoelectron line of Zn was enhanced compared with that after the addition of 150 mg BZSS, which indicated that the spectrum was mainly from the 2p photoelectron line of Zn in BZSS.

The change of intensity of Ru and Zn photoelectricity in the XPS spectrum reflected the chemical adsorption of BZSS on the catalyst surface. This adsorption affected the selectivity of cyclohexene from two aspects. The geometrical effect of the second metal changed the geometrical arrangement of the Ru atoms, or the BZSS shielded the partial Ru active centers. When the benzene molecule was adsorbed and activated by the Ru active site, the active center for the adsorption and dissociation of hydrogen molecules around the benzene molecule was relatively reduced. This was favorable for the hydrogenation of benzene to form cyclohexene and not conducive to direct hydrogenation to produce cyclohexane. On the other hand, BZSS was rich in crystal water, a layer of solid stagnant water film formed in the catalyst surface. Thus, the original hydrophobic Ru surface was changed to be hydrophilic to accelerate the desorption of cyclohexene. As a result, these two aspects improved the selectivity of cyclohexene.

It can also be seen from Fig. 7.18 that the Ru and Zn photoelectron lines were displaced after BZSS adsorption. Compared with Ru–Zn catalyst, the photoelectron spectrum of Ru was obviously shifted to the left with the increase of the amount of BZSS, which meant that the electron binding energy decreased. Then the 2p photoelectron line of Zn was obviously shifted right, which meant that the electron binding energy increased. This indicated that the electron moiety of Zn was transferred to Ru, leading to the electron-rich properties of the Ru active center. The electron-rich Ru active center weakened the adsorption of the π-electron cloud on the isolated double bonds of cyclohexene, which was beneficial to the desorption of in situ formed cyclohexene on the surface of the catalyst and not conducive to the reabsorption of olefins desorbed.

In addition, Fig. 7.18 also shows the 2p photoelectron spectrum of S. We could see that the fresh Ru–Zn catalyst had no 2p line of S and the obvious 2p line of S appeared after adding BZSS. However, this reliability was not high as the signal-to-noise ratio S/N was too low and it was not used for the analysis.

Table 7.29 gives the results of XPS qualitative analysis of Ru–Zn catalyst before and after adsorption of BZSS.

Table 7.29 The results of XPS qualitative analysis of Ru–Zn catalyst before and after adsorption of BZSS

XPS spectral line	Electron binding energy (BE)/eV			
	Standard data	Experimental data		
		Ru–Zn catalyst	Ru–Zn + 150 mg BZSS	Ru–Zn + 600 mg BZSS
Ru $3p_{1/2}$	485	484.1	483.8	483.6
Ru $3p_{3/2}$	463	461.8	461.9	461.2
Ru $3d_{3/2}$	286	284.4	284.6	283.9
Ru $3d_{5/2}$	282	280.2	280.2	279.7
Zn $2p_{1/2}$	1045	1045.0	1045.6	1046.3
Zn $2p_{3/2}$	1022	1022.0	1022.2	1023.5

The second column on the left in Table 7.29 gives the reference values for the electronic standard binding energy of Ru and Zn. It can be seen that the $3P_{3/2}$ electron binding energy of Ru was 461.8 eV for Ru–Zn catalyst, which was close to that of metal Ru and indicated that Ru existed in the metal state under the reaction condition. This was consistent with the TPR results. The weak shoulder peak appeared at around 464 eV of the Ru $3p_{3/2}$, which should be attributed to Ru oxide RuO_2 and a small part of the Ru on the surface was oxidized. The $2P_{3/2}$ electron binding energy of Zn was 1022.0 eV. The $2P_{3/2}$ electron binding energy of Zn (II) and metal Zn can be very close and the difference is only 0.1 eV [14]. Therefore, the valence of Zn was difficult to distinguish only by XPS. On the other hand, the kinetic energy of the auger electron of Zn LMM differed by more than 0.5 eV. The auger electron spectroscopy of Zn LMM can confirm that Zn was predominantly in the form of ZnO in the presence of 150 °C and 5 MPa H_2 [1].

It can also be seen from Table 7.29 that the electron binding energies of Ru $3P_{3/2}$ and Ru $3P_{5/2}$ were changed as 461. 8 eV → 461.9 eV → 461.2 eV and 280.2 eV → 280.2 eV → 279.7 eV, respectively, with the change of Ru–Zn → Ru–Zn + 150 mg BZSS → Ru–Zn + 600 mg BZSS. Meanwhile, the electron binding energy of Zn $2p_{3/2}$ was changed as 1022.0 eV → 1022.2 eV → 1023.5 eV. Particularly, the electron binding energy of Ru $3P_{3/2}$ decreased from 461.8 to 461.2 eV with the addition of 600 mg BZSS to Ru–Zn catalyst, and the binding energy decreased by 0.5 eV. Then the Zn $2P_{3/2}$ electron binding energy increased from 1022.0 to 1023.5 eV, and the binding energy increased by 1.5 eV, indicating that Ru obtained electrons in an electron-rich state and Zn was electronically deficient in electron state.

Inductively coupled plasma-atomic emission spectrum (ICP-AES) was used to determine the actual Zn content in the catalyst. The results were as follows:

The content of Zn in Ru–Zn catalyst was 5.98%;

Zn content of Ru–Zn catalyst with stirring adsorption of 300 mg BZSS was 13.69%. Assuming complete adsorption, the theoretical Zn content is 13.75%.

Zn content of Ru–Zn catalyst with 600 mg BZSS was 18.40%. Assuming complete adsorption, the theoretical Zn content is 20.23%.

It can be seen from the data above that the Zn content of the catalyst after adsorption of BZSS was significantly increased. When the BZSS addition was low, it was almost completely adsorbed and when the BZSS was added, the adsorption itself was a dynamic equilibrium and the adsorption reached a maximum.

ICP-AES results showed that BZSS could be chemically adsorbed on the surface of Ru–Zn catalyst. Modulation results showed that if BZSS was not the more the better, the catalyst activity reduced if it was beyond a certain value, and the selectivity of cyclohexene could not be improved.

Adding $Zn(OH)_2$, H_2SO_4, BZSS, etc., can modify the Zn content in catalyst and BZSS coverage on the surface of catalyst. The auxiliary agent Zn in the catalyst influences the geometrical arrangement and electronic properties of the Ru active center due to the electronic effect and the geometric effect. BZSS forms a layer of hydrophobic film on the surface of the catalyst, which affects the affinity of the catalyst with the reactants and the intermediate (cyclohexene). When the Zn content is suitable and BZSS is dispersed on the surface of the catalyst close to the monolayer, a higher selectivity and yield can be obtained.

References

1. Sun, H.J., Wang, H.X., Jiang, H.B., et al.: Effect of $(Zn(OH)_2)_3(ZnSO_4)(H_2O)_5$ on the performance of Ru–Zn catalyst for benzene selective hydrogenation to cyclohexene. Appl. Catal. A: Gen. **450**, 160–168 (2013)
2. Hu, S.C., Chen, Y.W.: Partial hydrogenation of benzene to cyclohexene on ruthenium catalysts supported on La_2O_3-ZnO binary oxides. Ind. Eng. Chem. Res. **36**, 5153–5159 (1997)
3. Sun, H.J., Guo, W., Xiao, X.L., et al.: Progress in Ru-based amorphous alloy catalysts for selective hydrogenation of benzene to cyclohexene. Chin. J. Catal. **32**, 1–9 (2011)
4. Struijk, J., Moene, T.V.D., Kamp J.J.F., et al.: Partial liquid-phase hydrogenation of benzene to cyclohexene over ruthenium catalysts in the presence of an aqueous salt solution: II. Influence of various salts on the performance of the catalyst. Appl. Catal. A: Gen. **89**, 77–102 (1992)
5. Wang, J.Q., Wang, Y.Z., Xie, S.H., et al.: Partial hydrogenation of benzene to cyclohexne on a Ru–Zn/m–ZrO_2 nanocomposite catalyst. Appl. Catal. A: Gen. **272**, 29–36 (2004)
6. Struijk, J., Angremond, M.R., Scholten, J.J.F.: Partial liquid-phase hydrogenation of benzene to cyclohexene over ruthenium catalysts in the presence of an aquesous salt solution. I. Preparation, characterization of the catalyst and study of a number of process variables. Appl. Catal. A: Gen. **83**, 263–395 (1992)
7. Struijk, J., Scholten, J.J.F.: Selectivity to cyclohexenes in the liquid-phase hydrogenation of benzene and toluene over ruthenium catalysts, as influenced by reaction modifiers. Appl. Catal. A: Gen. **82**, 277–287 (1992)
8. Wu, J.M., Yang, Y.F., Chen, J.L.: The research on inactivation of catalysts for benzene partial hydrogenation. Chem. Ind. Eng. Prog. **3**, 13–19 (2003)
9. Huang, X.P., Wang, J.P., Li, J.: Development of deactivation and generalization technologies of hydrogenation catalysts. Symp. 8th Natl. Ind. Catal. Technol. Appl. 69–71 (2011)
10. Xu, H., Li, C., Li, H.T.: Maintenance of catalyst property and improvement of hydrogenation efficiency. Chem. Prod. Technol. **11**(2), 46–48 (2004)
11. Zhou, G.B., Pei, Y., Jiang, Z., et al.: Doping effects of B in ZrO_2 on structural and catalytic properties of Ru/B–ZrO_2 catalysts for benzene partial hydrogenation. J. Catal. **311**, 393–403 (2014)

12. Hu, S.C., Chen, W.Y.: Partial hydrogenation of benzene on Ru–Zn/SiO_2 catalysts. Ind. Eng. Chem. Res. **40**, 6099–6104 (2001)
13. Liu, S.C., Luo, G., Xie, Y.L.: Study on Ru–Zn catalysts prepared by precipitation method for selective hydrogenation of benzene to cyclohexene. J. Mol. Catal. (Chin.) **16**(5), 349–354 (2002)
14. Peplinski, B., Unger, W.E.S., Grohmann, I.: Characterization of Cu/Zn/Al oxide catalysts in the precipitated, calcined and reduced state by means of XPS with the help of a finger-print data base. Appl. Surf. Sci. **62**, 115–129 (1992)

Chapter 8
Catalyst Deactivation and Regeneration in Benzene Selective Hydrogenation

For a set of 100,000–200,000 t/a industrial plant for benzene selective hydrogenation to produce cyclohexene and its downstream products, the initial loading of Ru catalyst is 200–400 kg for benzene selective hydrogenation, the life expectancy of the catalyst is about 2 years. During the operation of the device, the loss will be significant when the catalyst is inactivated due to improper operation or the introduction of impurities. If the deactivated catalyst cannot be regenerated in situ, it not only needs tens of millions of dollars to replace the catalyst, but also needs to discharge the mother liquor causing pollution. Therefore, it is very important to study the deactivation mechanism of catalyst for benzene selective hydrogenation and the regeneration method to ensure the normal production of industrial production.

The deactivation of Ru catalyst for benzene selective hydrogenation may result from carbon deposition, reactor wall corrosion, excessive salt adsorption, reductive deposition of metal ion, catalyst poisoning, and so on. Among them, sulfur, nitrogen compounds, and other trace impurities, especially *N*,*N*-dimethyl acetylene limbs (DMAC) as the extractant from the recycling of the benzene is one of the common causes of deactivation in industry. The catalyst can be regenerated by appropriate methods for different deactivation reasons. In 490–615 K with hydrogen reduction, the carbon is transferred into hydrocarbons to restore the pore volume, pore size distribution, and specific surface area of the catalyst. By controlling the pH and using corrosion-resistant material, one can avoid excessive adsorption of salts and reductive deposition of Fe, Cr, Ni, and other metal ions on the catalyst surface. Sulfur poisoning is irreversible which needs strict control of thiophene content in benzene and hydrogen. The Ru catalyst deactivation due to DMAC carried in recycling benzene is reversible and the catalyst can be restored to normal levels by the addition of an appropriate amount of H_2SO_4.

The deactivation of the catalyst is essentially due to the fact that the active sites on the surface are shielded and the abilities to adsorb and activate the reactants are lost. Or the internal pore is clogged and the reactant molecule loses the space for the reaction. The deactivation of Ru catalyst for benzene selective hydrogenation

Z. Liu et al., *Catalytic Technology for Selective Hydrogenation of Benzene to Cyclohexene*,
https://doi.org/10.1007/978-981-15-6411-6_8

results from carbon deposition, excessive adsorption of zinc sulfate and other salts, corrosion on reactor wall by Fe, Cr, Ni, and poisoning of sulfide and nitride.

The reactivity of Ru catalyst and the selectivity of cyclohexene decrease due to the ppm level thiophene in benzene feedstock and this is irreversible. Thus, it is necessary to strictly control the content of thiophene in benzene and hydrogen. The extractant induced deactivation of Ru catalyst belongs to the nitride poisoning, and H_2SO_4 can be used for in situ regeneration instead of using zinc sulfate solution to replace the nitride mother liquor for multiple times for regeneration. This is simple, resource saving, and environment friendly.

The main contents of this chapter are the catalyst deactivation in benzene selective hydrogenation, Ru catalyst deactivation and regeneration pilot study, deactivation, and regeneration of industrial catalyst.

8.1 Study on Deactivation of Benzene Selective Hydrogenation Catalyst

8.1.1 Deactivation of Ru Catalyst Caused by Carbon Deposition

Figure 8.1 shows the H_2-TPR spectra of the fresh Ru catalyst and the repeatedly used Ru catalyst [1].

It can be seen from Fig. 8.1a that the fresh catalyst was reduced at room temperature in a hydrogen atmosphere. It was completely reduced within 460 K, and the maximum reduction peak appeared at 365 K. The benzene hydrogenation reaction was carried out at 423 K, 5 MPa, so that Ru was present in the reduced state under the reaction conditions. Different TCD signals come from different Ru oxidation states on the surface. The lower TCD with FID signals of the catalyst at 440–570 K

Fig. 8.1 H_2-TPR spectra of fresh Ru catalyst. **a** Fresh catalyst (46 mg); **b** a reused catalyst (48 mg); the above is TCD signal, the below is FID signal, the heating rate is 10 K/min

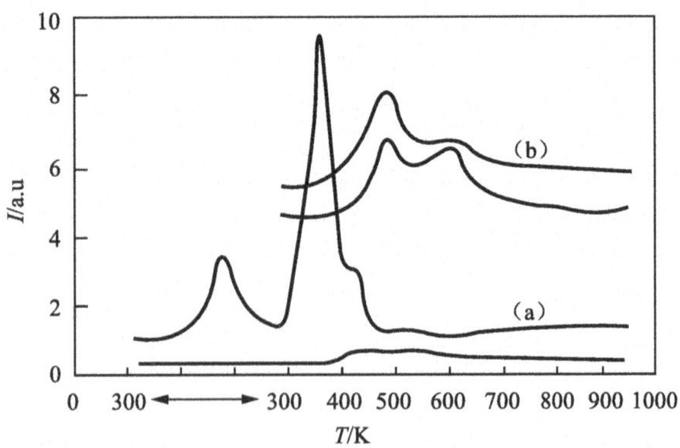

may be derived from the reduction of the organic matter adsorbed during the catalyst passivation process.

It can be seen from Fig. 8.1b that there was no Ru-reduced TCD signal due to Ru in the reduction state in the repeatedly used Ru catalyst, and the new hydrogen consumption peak appeared at 490 and 615 K with the generation of hydrocarbons. This resulted from the deposition of cracked straight-chains or aromatics in the catalyst channel during the reuse of the catalyst. Thus, when the Ru catalyst is used for a long time, the deposition of carbon species will occur so that the catalyst may lose the normal pore volume and pore size distribution, resulting in selectivity and activity decline in catalyst or even deactivation.

In order to avoid the deposition of carbon species, the catalyst pretreatment is very necessary. The catalyst is pretreated in the zinc sulfate slurry for more than 20 h. The zinc sulfate is pre-adsorbed on the surface of the catalyst, giving priority to take up the Ru active centers for cracking those of benzene, cyclohexene, and cyclohexane to avoid and prohibit the side reactions. In the case of deactivation of the catalyst caused by carbon deposition, the catalyst is regenerated by reduction of hydrogen at 490–615 K. The conversion of carbon to hydrocarbons can recover catalytic pore volume, pore size distribution, and specific surface area to regenerate the catalyst.

8.1.2 The Excessive Adsorption of Zinc Sulfate and Other Salts

In order to reduce the catalyst activity and improve the selectivity of cyclohexene, the zinc sulfate and other salts are often added to the hydrogenation slurry. The adsorption of zinc sulfate can increase the hydrophilicity and significantly improve the selectivity and yield of cyclohexene. When the concentration of zinc sulfate increased from 0 to 0.1 mol/L, the initial selectivity of cyclohexene increased from 2.5 to 60%. However, excessive adsorption of zinc sulfate will significantly reduce the catalyst activity. When the concentration of zinc sulfate increased from 0.1 to 0.2 mol/L, the selectivity of cyclohexene was almost unchanged. When zinc sulfate was above 0.16 mol/L, the activity of the catalyst decreased significantly due to the increase of the surface coverage of the catalyst.

The maximum yield of cyclohexene was 36.5% in the first hydrogenation by catalyst at the lower concentration of zinc sulfate (about 0.005 mol/L) with pretreatment for 15 h. The selectivity and yield of cyclohexene were 77.5 and 41.5%, while the reaction rate decreased gradually with the repeated use of the catalyst. The catalyst activity began to decrease with the increase of cyclohexene selectivity and maximum yield after 15 times use of catalyst. The analysis of BET surface area and salt coverage of the catalyst found that the BET surface area (75.3 m^2/g) of the catalyst was almost equal to that of the fresh catalyst (71.4 m^2/g), indicating that the deactivation of the catalyst was not due to sintering, but due to excessive adsorption of zinc sulfate. In addition, the Ru's Tamman temperature is 817 K and it is much higher than the actual reaction temperature of 423 K, at which the Ru migration rate is very low.

In addition, salts of Cd (I), Ga (III), and In (III) have a strong adsorption capacity on Ru catalysts. Even at very low concentrations (about 0.1%), a small amount of sulfate salt of Cd (II), Ga (III), especially In (III) can completely block the Ru active center, causing the catalyst deactivation [2].

For the catalyst deactivation due to excessive adsorption of salts, the solution is to strictly control the type and concentration of salt additives to maintain the appropriate pH of the reaction system.

8.1.3 Corrosion of Fe, Cr, Ni on the Reaction Wall

Atom absorption spectra showed that Fe coverage was 0.79 and Zn coverage was only 0.02 when the catalyst was reused for 15 times for benzene hydrogenation in a stainless steel reactor with a zinc sulfate slurry of 0.005 mol/L. The liquid pH was 4.9 after the first hydrogenation, and 5.1 after the 15th hydrogenation. Under acidic conditions, Fe on the wall of the stainless steel wall will slowly corrode and deposit on the catalyst surface. From the elemental analysis of slurry after hydrogenation, the Fe ion concentration was 10^{-3} mol/L and the Cr and Ni ions were only 4 and 2% of Fe, and the Zn concentration was 5×10^{-3} mol/L. The Fe and Zn ions completely absorbed on the surface of the catalyst and the adsorption of Fe ions was faster and stronger than that of the Zn ions, so that the Fe salt, like the zinc sulfate, can be a reaction modifier to improve the selectivity of cyclohexene whose maximum value was up to 30%. However, the continuous deposition of Fe on the surface of the catalyst will eventually shield the Ru active center, resulting in the deactivation of the catalyst.

For the reactor wall corrosion, the most fundamental way is to use corrosion-resistant material. At present, the Hastelloy (HC276) lining is used widely in industry. Meanwhile, controlling the slurry pH in the appropriate range, normally in the 5.6–5.8, and an appropriate high concentration of zinc sulfate is conducive to the inhibition of Fe adsorption.

The coverage of adsorption species induced by the metal M ions (e.g., Zn^{2+}, Fe^{3+}, C^{3+}, Ni^{2+}) on Ru active center can be expressed as follows, assuming that M is the divalent ion:

$$[\mathrm{M(H_2O)}_n]^{2+} + \mathrm{Ru}_s == [\mathrm{M_{ads}(H_2O)}_{n-x}]^{2+} + x\mathrm{H_2O} \tag{8.1.1}$$

$$[\mathrm{M(OH)(H_2O)}_m]^{+} + \mathrm{Ru}_s == [M_{\mathrm{ads}}\mathrm{(OH)(H_2O)}_{m-x}]^{+} + x\mathrm{H_2O} \tag{8.1.2}$$

$$\left[\mathrm{M(H_2O)}_n\right]^{2+} + \mathrm{H_2O} == [\mathrm{M(OH)(H_2O)}_{n-1}]^{+} + \mathrm{H_3O^+} \tag{8.1.3}$$

where n and m are the number of water molecules in the first hydration layer of M^{2+}; In order to achieve chemical adsorption, x is the number of water molecules leaving. The equilibrium movements of Eqs. (8.1.1) and (8.1.2), are obviously related to the nature of M^{2+}. On comparison of Fe ion with Zn ion, the Eqs. (8.1.1) and (8.1.2), will move obviously to the right due to the less activity of Fe hydroxide. A very small amount of Fe ions will cause the increase of coverage of $[Fe_{ads}(H_2O)_{n-x}]^{2+}$ or $[Fe_{ads}(OH)(H_2O)_{m-x}]^{+}$ on the Ru active center. Equation (8.1.3) indicates that the pH directly affects the ratio of $M^{2+}/[M(OH)]^{+}$, which affects the coverage of the metal ions on the catalyst surface and thus affects the activity of the catalyst and the selectivity of cyclohexene [3].

8.1.4 Deactivation of Catalyst Due to Other Factors

As in Eq. (8.1.3), the slurry pH affects the adsorbed species by a metal ion. $ZnSO_4$ is a strong acid weak alkali salt, and its hydrolysis reaction can be expressed as

$$Zn^{2+} + H_2O == [Zn(OH)]^{+} + H^{+} \tag{8.1.4}$$

In the 0.16 mol/L $ZnSO_4$ slurry, pH = 5.6–5.8, but small changes of pH occurred due to the residual amount of NaOH after the preparation of the catalyst. Before the start of the reaction, the pH = 2.4 with acidification with sulfuric acid, and the H^{+} concentration increased. The Eq. (8.1.4) equilibrated to the left, and most of the Zn was present in the form of Zn^{2+}, which had no significant effect on the catalyst performance. If a small amount of NaOH was added, the Eq. (8.1.4), equilibrated right and the catalyst surface was $[Zn(OH)]^{+}$, even $Zn(OH)_2$. The acid salt coverage of the sulfuric acid was increased, leaving only a small amount of Ru activity center, and the benzene hydrogenation reaction rate was significantly reduced. This part of the active center had the smallest adsorption heat for cyclohexene and the desorption activation energy of cyclohexene was very low. Thus, the desorption rate was very fast and the initial selectivity of cyclohexene was high, up to 80% or more. In contrast, if NaOH was not added, the selectivity of cyclohexene was only 55% at the same benzene conversion. Although the addition of NaOH can significantly improve the initial selectivity of cyclohexene, benzene conversion decreased significantly at the same cyclohexene selectivity, and high yield of cyclohexene could not be obtained.

The reduction behavior of metal ions can significantly affect the activity and selectivity of Ru catalyst. Under the condition of the benzene hydrogenation reaction, Zn^{2+} may react with the dissociated hydrogen atom H_{ads}

$$Zn^{2+}_{ads}(n-x)H_2O + 2H_{ads} == Zn^{0}_{ads} + 2H^{+} + (n-x)H_2O \tag{8.1.5}$$

$$Ru + Zn^{0}_{ads} == Zn/Rualloy \tag{8.1.6}$$

The equilibrium position of Eq. (8.1.5), is shifted right by the coupling reaction of Eq. (8.1.6). In the Eq. (8.1.5), if the H_{ads} is adsorption dissociated or overflow generated, the reaction is allowed from the thermodynamics. In fact, this reaction is difficult to occur due to the very low Zn^{2+}/Zn reduction potential (−0.76 V). If the Zn atom and the Ru are reduced in the Eq. (8.1.6), to form alloy under the benzene hydrogenation reaction conditions, and assuming that the enthalpy of formation is negative, this will promote the balance of the Eq. (8.1.5), to the right. As reported, Zn atom is commonly present in the Ru catalyst with the addition of Zn^{2+} either when the catalyst is prepared or when Ru catalyst is reduced in the zinc sulfate slurry [4, 5]. However, the Zn atom was about 1.1% of the total amount of Zn [6–8], which also explains the essential reason that Zn was considered to be the best additive agent.

And the reduction potentials of metal ions of $CuSO_4$, $SnCl_2$, $Pb(NO)_3$, $Fe(SO_4)_3$ are 0.34, −0.14, −0.13, and −0.04 V. Their reduction potential is positive or only slightly negative, and thus these ions are easily reduced to metal in the hydrogenation reaction as the Gibbs free energy of reduction reaction in the thermodynamics is negative. The metal will easily deposit on the surface of the Ru catalyst, clog the Ru active center, and deposit on the reactor wall to make the catalyst surface lose its hydrophilicity. Such salts should be avoided to use. Another type of salt such as Na_2SO_4, etc., even though the metal ions are difficult to be reduced, most of them exist in the slurry due to the low adsorption capacity of Ru as reaction modifiers. However, the improvement in the selectivity and yield of cyclohexene is not obvious [1, 2].

8.1.5 Catalyst Deactivation Caused by Sulfide

Thiophene is the main form of sulfide in benzene. With the increase of thiophene's concentration in benzene, the activity and selectivity of the catalyst decrease gradually. When the content of thiophene in benzene reaches 4 g/L, the decrease rate of the activity and selectivity of the catalyst will be obviously accelerated. However, thiophene was not detected in the reaction solution after hydrogenation, indicating that thiophene had a strong adsorption capacity on Ru catalyst. The reason is that Ru has an empty d orbit, while thiophene has lone pair electrons in its s orbit, which could form a stable complex between them. This phenomenon is similar to a chemical reaction, which could change the nature of the active center of Ru catalyst, causing a decrease of catalyst activity and cyclohexene selectivity, and this is irreversible. So the content of thiophene in benzene and hydrogen needs to be controlled strictly [3].

8.1.6 Catalyst Deactivation Caused by Nitride

The presence of hexene dinitrile, *N*,*N*-dimethyl acetamide (DMAC) and other nitrides in benzene can cause catalyst deactivation. Basically, the reason is that the solitary electrons on element N in the nitride have an affinity with the d orbit of Ru catalyst. However, this force is not as strong as it with the s orbit, hence the nitride-caused deactivation of Ru catalyst is reversible. It is possible to regenerate the catalyst by replacement of the mother liquor containing the nitride with the zinc phosphate slurry [3]. In industrial production, benzene or hydrogen containing sulfur or nitrogen compounds is one of the common causes for deactivation of Ru catalyst.

With the prolongation of the catalyst's service time, the growth of Ru microcrystal size, and the excessive hydrogen adsorption on the catalyst surface, activity, and selectivity of the catalyst reduce. Investigating the reasons for Ru catalyst deactivation and its regeneration methods could help to understand the deactivation mechanism and to prevent them. It is helpful to analyze the essential problem once deactivation occurs, and then adopt the corresponding regeneration methods to guarantee industrial production.

8.2 Pilot Investigation on Catalyst Deactivation and Regeneration

Figure 8.2 is the schematic illustration of the pilot plant. The catalytic performance, deactivation mechanism, and the regeneration methods of Ru catalyst were investigated on the continuous reactor cascade.

As shown in Fig. 8.2, the reaction was carried out simultaneously in two Hastelloy lined autoclaves (the volume of each reactor was 300 L). The optimal amounts of catalyst, slurry, zinc sulfate, water/benzene volume ratio, etc., were enlarged in proportion. 3.6 kg Ru catalyst, 17.9 kg ZrO_2, 500 L water, and 88.9 kg $ZnSO_4 \cdot 7H_2O$ were added into the reactor (7) via the catalyst inlet (5) and the storage tank (6) by overflowing. After replacing the air in the reactor with nitrogen three to five times, the nitrogen was replaced with hydrogen three to five times again, and then the hydrogen pressure was adjusted to 5 MPa with a stirring rate of 300 r/min. After the catalyst was pretreated for 22 h, benzene was added and the reaction was initialized. After benzene in the reactor (8) reached the preset residence time, it entered into the sedimentation tank (9, volume 770 L) via overflowing to achieve gas-liquid and oil-water separation, and then the slurry began to recycle. The water phase containing catalyst was recycled to the reactor (7) with a flow rate of 658 L/h powered by the solid-liquid pump (10). The oil phase containing benzene, cyclohexene, and cyclohexane entered into the subsequent extraction and distilling unit, where benzene was separated from cyclohexene and cyclohexane using extractant DMAC. Then the separated benzene was recycled to reactor (7), mixed with fresh benzene. Samples were taken every one hour to monitor the reaction conditions. Product composition was analyzed by gas chromatography to calculate benzene conversion and cyclohexene selectivity. The regeneration devices were turned on after the reaction was carried out for 24 h. During regeneration, about 1/10 of the slurry was decompressed via the screw valve (12 and 13) and entered to the flash tank (14), where the temperature was controlled

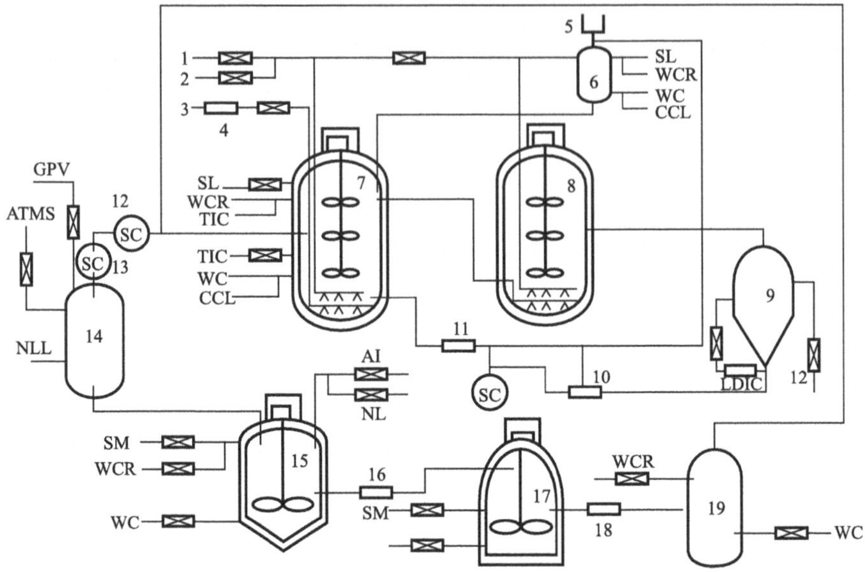

Fig. 8.2 Schematic illustration of the pilot plant. [controlling valve symbol] controlling valve; *SL* low pressure steam; *SM* medium pressure steam; *WCR* circulating water; *TIC* temperature indication controller; *WC* cooling water; *CCL* low pressure condensation cycle; *LDIC* liquid level indication controller; *SC* screw valve; *GPV* general valves; *ATMS* atmospheric pressure; *NLL* oxygen-poor nitrogen; *AI* air; *NL* low pressure nitrogen. 1. Hydrogen inlet; 2. Nitrogen inlet; 3. Benzene inlet; 4. Benzene feed pump; 5. Catalyst inlet; 6. Catalyst storage tank; 7. Reactor (1); 8. Reactor (2); 9. Sedimentation tank; 10. Slurry circulation pump; 11. Slurry flowmeter; 12, 13. Screw valve; 14. Flash tank; 15. Regenerator; 16. Delivery pump; 17. Boiling tank; 18. Regenerative pump; 19. Colling tank

at 100 °C for the flash to remove oil and hydrogen. Taking advantage of height difference, slurry would be sent to the regenerator (15), maintaining the liquid level of 90%, then aeration would proceed. The oxygen-lean air was provided to oxidize the hydrogen adsorbed on Ru surface (air flow rate: 50 L/h, N_2 flow rate: 400 L/h, temperature: 95 °C, agitation speed: 200 r/min, aeration time: 5 h). The slurry after aeration would be sent to the boiling tank (17) via the delivery pump (16) to be boiled for 1 h (150 °C, 200 r/min agitation). The slurry after boiling would be sent to the cooling tank (19) via the regeneration pump (18) to be cooled down to 50 °C, and finally, it would be sent to the hydrogenation reactor (7).

Measurements on activity and selectivity of catalyst showed that catalyst activity reduced significantly after 345 h of reaction. In order to investigate the reasons for catalyst deactivation and the regeneration methods, the catalysts were analyzed and characterized by XRD, ICP, and XPS, etc.

Figure 8.3 shows the TEM photographs of fresh catalyst(C), pretreated catalyst (PC), and deactivated catalyst(DC) and their microcrystalline dimensions.

In Fig. 8.3a, d Ru–Zn C represents the fresh catalyst without ZrO_2, while (b, e) Ru-Zn PC represents the pretreated catalyst, which has been pretreated for 22 h in the pilot plant, and (c, f) Ru-Zn DC represents the deactivated catalyst which has already

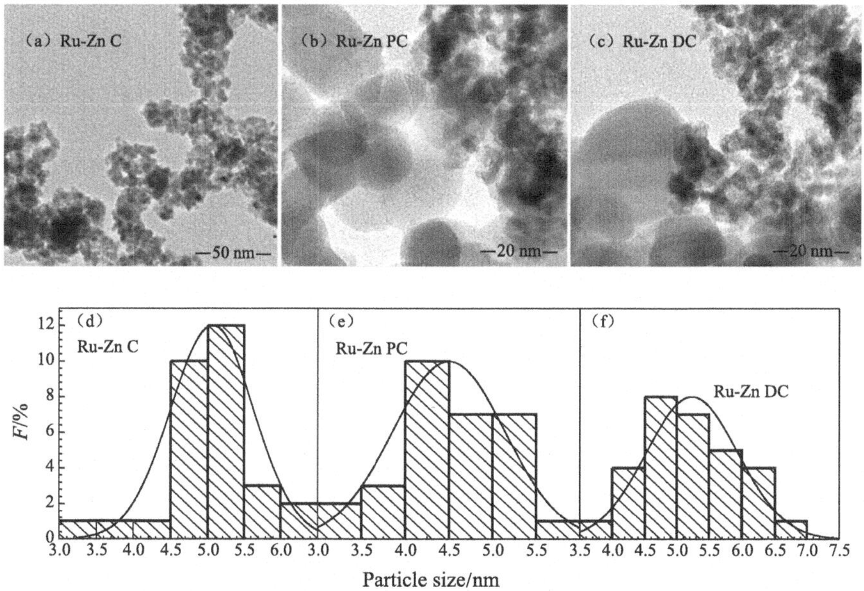

Fig. 8.3 TEM photographs of Ru-Zn catalyst and its microcrystalline dimensions. **a**, **d** Ru—Zn C; **b**, **e** Ru–Zn PC; **c**, **f** Ru–Zn DC

reacted for 345 h in the pilot plant. Both the pretreated catalyst and the deactivated catalyst contain ZrO_2.

As Fig. 8.3 shows, the crystallite size of the fresh catalyst was about 5.0 nm. However, after pretreatment using nano-ZrO_2 as dispersant, the crystallite size became 4.5 nm approximately. The crystallite size of the deactivated catalyst became 5.2 nm approximately after 345 h of reaction. The results showed that the increase of crystallite size was not the reason for catalyst activity loss, because the microcrystal size of Ru catalyst would obviously affect the catalyst activity and cyclohexene selectivity only after they grew to at least 20 nm. From Fig. 8.2b, c, we could see that Ru microcrystals distributed on micron-sized ZrO_2 particles, where ZrO_2 did not act as a carrier, but to disperse Ru microcrystals to prevent them from agglomeration.

In order to investigate the regeneration methods for the deactivated catalyst, 0.01–0.031 mol H_2SO_4 was added to 280 mL reaction slurry, and the activity and selectivity were evaluated in a 1 L autoclave. As a result, the activity of the catalyst recovered while the selectivity of cyclohexene decreased.

Figure 8.4 shows the XRD spectra of fresh catalyst, deactivated catalyst, and catalyst after hydrogenation with different amount of H_2SO_4.

As shown from Fig. 8.4a, for the XRD spectra of the fresh Ru–Zn (5.4%) catalyst, a strong and obvious Ru characteristic peak (JCPDS 01-070-0724) at $2\theta = 44.0°$ appeared. After hydrogenation, the ZrO_2 diffraction peak appeared, while the characteristic peak of Ru became wide. The microcrystalline size of Ru was determined to be 5.1 nm using the Scherrer formula, which was approximately consistent

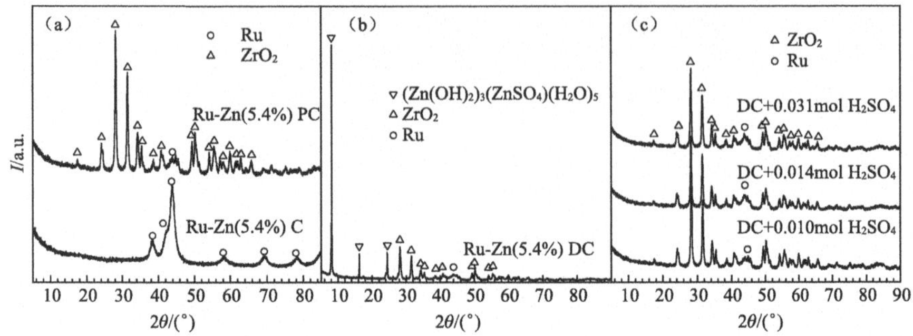

Fig. 8.4 XRD spectra of Ru-Zn catalyst. **a** Fresh catalyst; **b** alkaline complex salt; **c** catalyst after hydrogenation with different amount of H_2SO_4

with the TEM results. As shown from Fig. 8.4b, in addition to the characteristic peak of the dispersant monoclinic phase of ZrO_2 (JCPDS 00-024-1165) and the characteristic peak of metal Ru, the diffraction peak of $(Zn(OH)_2)_3(ZnSO_4)(H_2O)_5$ (CJCPDS 01-078-0246) appeared at $2\theta = 8°$ approximately. It was found that Zn existed in the form of ZnO in Ru–Zn catalyst [9], and after pretreatment, ZnO enriched on catalyst surface reacted with $ZnSO_4$ in the slurry, forming insoluble zinc sulfate basic salt, which could account for the appearance of the diffraction peak of $(Zn(OH)_2)_3(ZnSO_4)(H_2O)_5$ in Fig. 8.4b. As shown in Fig. 8.4c, for the XRD spectra of the deactivated Ru–Zn catalyst, the characteristic peak of the $(Zn(OH)_2)_3(ZnSO_4)(H_2O)_5$ disappeared after adding 0.010–0.031 mol H_2SO_4 into 280 mL of the intermediate slurry in the pilot plant, indicating the liberation of Ru active sites. In order to determine the reasons for catalyst deactivation furtherly, inductively coupled plasma (ICP) emission spectra analysis was applied to determine the elemental composition of the catalyst.

Table 8.1 shows the ICP results of Zn, Fe, and Ni content in Ru catalyst and Zn^{2+} concentration and the pH of the slurry.

Table 8.1 Metal element content in Ru–Zn catalyst and Zn^{2+} concentration and pH of the slurry

Sample	Elemental content/%[a]			Zn^{2+}/(mol/L)[a]	pH
	Zn	Fe	Ni		
Ru–Zn C	5.4	–	–	–	–
Ru–Zn PC[b]	2.6	0.09	0.008	0.54	5.4
Ru–Zn DC[b]	9.0	0.13	0.014	0.27	6.5
Ru–Zn DC + 0.010 mol H_2SO_4[c]	2.1			0.56	5.3
Ru–Zn DC + 0.014 mol H_2SO_4[c]	1.3			0.58	4.1
Ru–Zn DC + 0.031 mol H_2SO_4[c]	0.4			0.59	3.2

[a]Measured by ICP-AES
[b]Pretreated catalyst in the pilot plant
[c]Evaluated data of the deactivated catalyst in a 1 L autoclave

As shown in Table 8.1, the content of Ni and Fe in catalyst changed little during the period from pretreatment to deactivation, and the amount of Fe and Ni corroded from the vessel during hydrogenation was scarce, so it could be concluded that the deposition of Fe and Ni ions was not the cause of catalyst deactivation. While for Zn ions, the content of Zn in catalyst increased from 2.6 to 9.0%, and the concentration of Zn^{2+} in the slurry decreased from 0.54 to 0.27 mol/L. At the same time, the pH of the slurry increased from 5.4 to 6.5. This indicated that a large amount of Zn^{2+} in the slurry was chemically adsorbed onto catalyst surface, resulting catalyst deactivation. This phenomenon was consistent with the XRD results, indicating the formation of the basic sulfate salt and their excessive adsorption on Ru surface were the main reasons for catalyst deactivation during its long-time operation.

Meanwhile, in the pilot plant, it was found that the extractant DMAC could enter into the hydrogenation unit with benzene. In the slurry, which contains a large amount of sulfuric acid, DMAC could easily be decomposed into acetic acid and dimethylamine (DMA). DMA would be hydrolyzed to produce hydroxyl, and hydroxyl could react with $ZnSO_4$ furtherly in the slurry, forming $(Zn(OH)_2)_3(ZnSO_4)(H_2O)_5$. The reactions are as follows:

$$Zn^{2+} + HOH \rightarrow Zn(OH)^+ + H^+ \tag{8.2.1}$$

$$Zn^{2+} + 2HOH \rightarrow Zn(OH)_2 + 2H^+ \tag{8.2.2}$$

$$CH_3CON(CH_3)_2 + HOH \rightarrow CH_3COOH + HN(CH_3)_2 \tag{8.2.3}$$

$$HN(CH_3)_2 + HOH \rightarrow H_2N(CH_3)_2^+ + OH^- \tag{8.2.4}$$

$$4Zn^{2+} + 6OH^- + SO_4^{2-} + 5H_2O \rightarrow (Zn(OH)_2)_3(ZnSO_4)(H_2O)_5 \tag{8.2.5}$$

To prove the correctness of the above reactions, an appropriate amount of 98% concentrated sulfuric acid was added into 280 mL of deactivated catalyst slurry. With the increase of H_2SO_4 amount, both Zn content in catalyst and the pH of slurry gradually decreased, while the concentration of Zn^{2+} in the slurry gradually increased, indicating the amount of $(Zn(OH)_2)_3(ZnSO_4)(H_2O)_5$ dissolved from the catalyst surface increased. With the addition of 0.031 mol H_2SO_4, the content of Zn in the catalyst was reduced to 0.4% after hydrogenation, indicating that the $(Zn(OH)_2)_3(ZnSO_4)(H_2O)_5$ adsorbed on Ru surface was almost completely dissolved. The above experiments showed that the deactivation of Ru catalyst was mainly caused by the excessive adsorption of $(Zn(OH)_2)_3(ZnSO_4)(H_2O)_5$ due to the addition of extraction agent DMAC in hydrogenation unit. Thus, the application of concentrated sulfuric acid could be a simple and easy method for catalyst regeneration.

Figure 8.5 shows the benzene conversion, cyclohexene selectivity, and yield over deactivated catalyst when different amounts of H_2SO_4 were added into 280 mL of the test slurry.

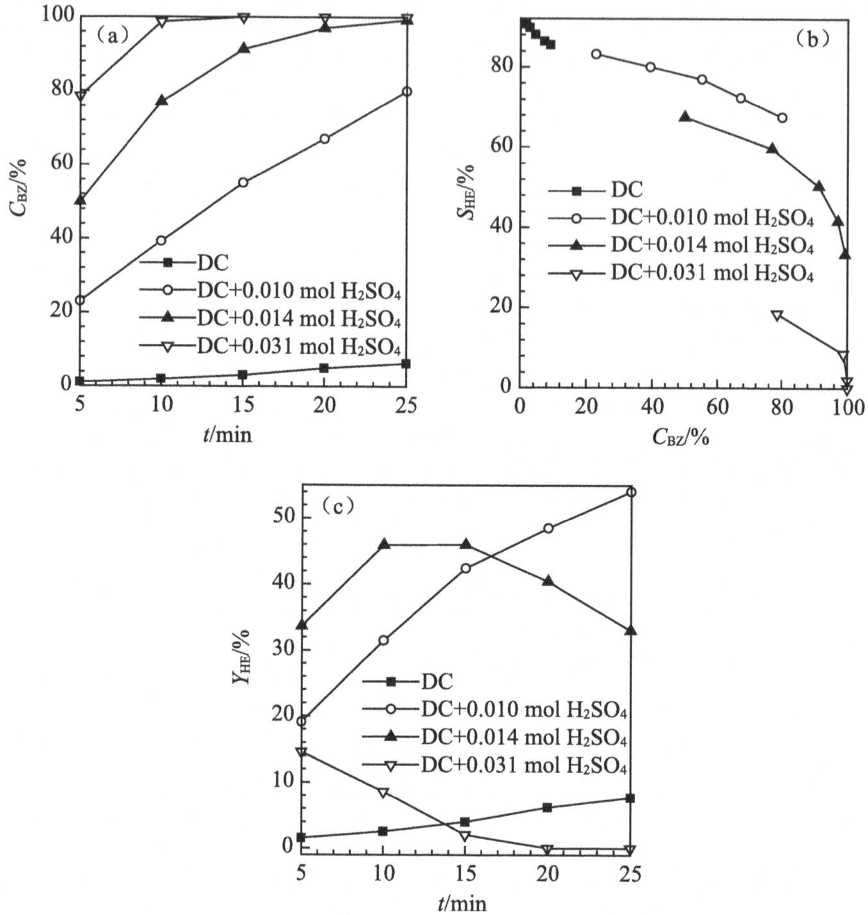

Fig. 8.5 Plots of the conversion of benzene, selectivity, and yield of cyclohexene over deactivated Ru catalyst versus the reaction time when different amount of H_2SO_4 was added. Reaction condition: 150 °C, 5 MPa, 1200 r/min, 280 mL of pilot slurry, 140 mL of benzene

As can be seen from Fig. 8.5, when the amount of H_2SO_4 added into 280 mL of the intermediate slurry increased from 0 to 0.031 mol, benzene conversion increased from 2.8 to 98.7% at 10 min, and the conversion of cyclohexene decreased from 89.8 to 8.7%. When 0.01 mol H_2SO_4 was added, benzene conversion and selectivity over deactivated catalyst was 39.3 and 80.0% at 10 min, respectively. The highest yield of cyclohexene reached 54.1% at 25 min, and the activity and selectivity of the catalyst were recovered. Scaling up the above results to the pilot plant, the datum of benzene conversion, cyclohexene selectivity, and yield over Ru catalyst were presented in Fig. 8.6.

As can be seen from Fig. 8.6, benzene conversion reduced to less than 10% at 345 h. Based on the above results, 1750 g of concentrated sulfuric acid (98%) was

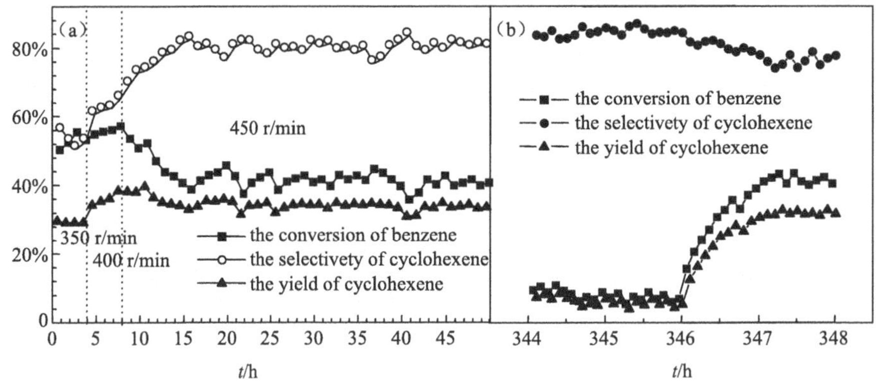

Fig. 8.6 Regeneration of Ru–Zn catalyst on pilot plant. Reaction condition: 145 °C, 4.5 MPa H_2, 3.6 kg Ru–Zn catalyst, 17.9 kg ZrO_2, 88.9 kg $ZnSO_4 \cdot 7H_2O$, 500 L H_2O, 1.75 kg H_2SO_4 was added at 345 h

added into the pilot plant after 346 h. At 347 h, Zn^{2+} concentration in slurry increased to 0.53 mol/L, and pH decreased to 5.3. At the same time, benzene conversion was around 40%, while cyclohexene selectivity and yield were restored to 80 and 32%, respectively, reaching the normal level of fresh catalyst. The results indicated that the catalyst deactivation induced by the nitride could be recovered by the addition of concentrated sulfuric acid into the reaction system. Compared with the previous reports where the mother liquor replacement method [3] was applied, this method is simple and easy to operate with lots of advantages such as resource saving, no waste discharge, and no environmental pollution.

8.3 Investigation of Deactivation and Regeneration of Industrial Catalyst

8.3.1 Unusual Deactivation of Industrial Catalysts

Catalyst unusual deactivation occurred in two sets of industrial device of benzene selective hydrogenation to produce adipic acid (production capacity: 100,000 t/a). To determine deactivation causes and regeneration methods, activity evaluation, selectivity evaluation, and characterization over the deactivated catalyst, and hydrogenation slurry were conducted using fresh catalyst after hydrogenation as a reference (deactivated catalyst and slurry were represented as $1^{\#}$ and $2^{\#}$, respectively).

Activity and selectivity evaluation: The amount of $1^{\#}$ and $2^{\#}$ were calibrated firstly, then the activity and selectivity when original slurry and freshly prepared slurry were used in GS-l Type Han Alloy Kettle (1 L) were determined, respectively. There were 1.96 g catalyst, 9.8 g ZrO_2, 49.2 g $ZnSO_4 \cdot 7H_2O$, and 280 mL H_2O in the slurry. 140 mL of benzene was introduced into the autoclave at 150 °C. The pressure was maintained at 5.0 MPa, and the stirring rate was kept at 1400 r/min. The catalyst activity and selectivity determined under the above conditions were comparable with that of the catalyst in the industrial scale.

Figure 8.7 shows the XRD patterns of the two kinds of inactivated catalysts 1#, 2#, and the reference samples (new catalyst after hydrogenation).

As shown in Fig. 8.7, the XRD patterns of the inactivated catalyst 1# and the fresh catalyst after hydrogenation were almost the same. Inactivated catalyst 2# displayed a strong diffraction peak of the alkali zinc sulfate phase of $(Zn(OH)_2)_3(ZnSO_4)(H_2O)_5$ (JCPDS 01-078-0247) at $2\theta = 8°$.

Table 8.2 shows the elemental composition of deactivated catalysts (1# and 2#) and reference catalyst determined by ICP.

As can be seen from Table 8.2, the content of Zn in catalyst 1# was lower than that of the fresh catalyst and was far lower than that of catalyst 2#. As for the content of S, catalyst 1# contained higher amount of S than that of fresh catalyst, while it was still much less than 2#. However, the contents of Fe and Ni were normal. Hence, it could be speculated that the deposition of metal ions corroded from the reactor wall was not the cause of catalyst deactivation. The catalyst 1# had a higher S content but a lower Zn content than the fresh catalyst, indicating sulfur poisoning may deactivate the catalyst. On the other hand, catalyst 2# had a much higher Zn content and S content

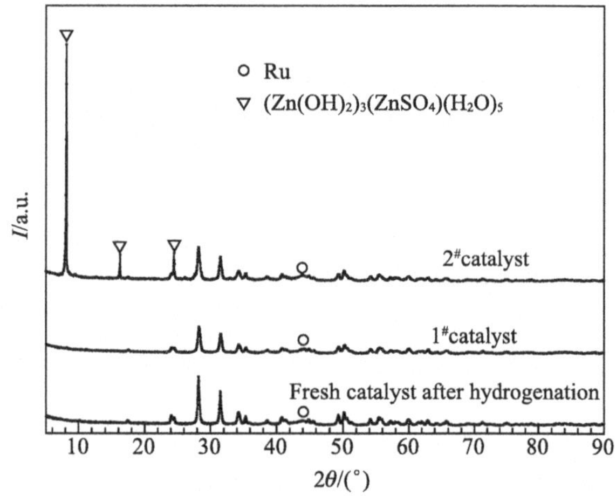

Fig. 8.7 XRD patterns of inactivated catalysts (1# and 2#) and reference fresh sample

Table 8.2 Composition of deactivated catalysts (1# and 2#) and reference catalyst after hydrogenation determined by ICP

Sample	Zn/%	S/%	Fe/%	Ni/%
Reference	2.43	0.23	0.07	0.008
1#	1.95	0.31	0.09	0.002
2#	9.04	1.30	0.08	0.014

compared with catalyst 1# and the fresh catalyst, demonstrating the adsorption of excessive alkali zinc sulfate on catalyst surface was the main reason for catalyst deactivation (with reference to the XRD pattern).

Table 8.3 shows the concentrations of the ions in the aqueous phase of catalysts 1# and 2# determined by ICP.

It can be seen from Table 8.3 that the amounts of Zn and S in the original slurry of catalyst 2# were much lower than that of 1#, and Fe and Ni were not detected, indicating some Zn^{2+} in slurry were adsorbed chemically on the catalyst, consistent with the above conclusions.

Table 8.4 shows the activity and selectivity of catalysts 1# and 2# in the original slurry.

As shown in Table 8.4, the activity and the selectivity of catalysts 1# and 2# were obviously different. The activity of catalyst 1# was better, which meant it had a high benzene conversion; however, the selectivity of cyclohexene was lower, which was a significant characteristic of sulfide poisoning. Meanwhile, catalyst 2# had a quite low activity, but high cyclohexene selectivity, which was a typical characteristic of the adsorption of a large amount of alkali zinc sulfate salt on the catalyst surface.

Taking the XRD, ICP data, activity, and selectivity results into consideration, it could be speculated that the abnormal deactivation of catalyst 1# was caused by sulfide poisoning, while the abnormal deactivation of catalysts 2# was caused by the extractant DMAC carried by the recycling benzene.

Different methods were applied to regenerate the catalysts based on different deactivation reasons for catalysts 1# and 2#. To avoid the introduction of impurity ions in hydrogenation slurry, activity, and selectivity of the catalyst were changed by acid-base modulation.

Table 8.3 Concentrations of the ions in the aqueous phase of catalysts 1# and 2# determined by ICP

Sample	Zn/(mol/L)	S/(mo l/L)	Fe/(mol/L)	Ni/(mol/L)
1# raw slurry	0.48	0.66	0	0
2# raw slurry	0.27	0.37	0	0

Table 8.4 Activity and selectivity of catalysts 1# and 2# in the original slurry

t/min	1#			2#		
	C_{BZ}/%	S_{HE}/%	Y_{HE}/%	C_{BZ}/%	S_{HE}/%	Y_{HE}/%
5	8.78	57.74	5.07	1.13	82.30	0.93
10	19.21	45.39	8.72	2.01	88.06	1.77
15	33.21	34.24	11.37	3.09	88.03	2.72
20	46.38	24.88	11.54	5.04	90.08	4.54
25	59.71	17.87	10.67	6.33	90.05	5.70

Note C_{BZ}: Benzene conversion; S_{HE}: cyclohexene selectivity; Y_{HE}: cyclohexene yield. Slurry composition: 1.96 g catalyst, 280 mL raw slurry. Hydrogenation condition: 150 °C, 5 MPa, 1400 r/min, 140 mL benzene

Table 8.5 Regeneration results over catalyst 1[#] which was sulfide poisoned

t/min	C_{BZ}/%	S_{HE}/%	Y_{HE}/%
5	11.61	72.27	8.39
10	27.97	63.21	17.68
15	43.12	53.73	23.17
20	55.88	45.81	25.60
25	64.87	39.94	25.91

Slurry composition: 1.96 g catalyst, 280 mL H_2O, 45.7 g $ZnSO_4$, 0.2 g NaOH, 5 g ZrO_2; Hydrogenation condition: 150 °C, 5 MPa, 1400 r/min, 140 mL benzene

8.3.2 Regeneration of Deactivated Catalysts Caused by Sulfide Poisoning

The purpose of acid (HCl) washing, salt (NaCl) washing, alkali (NaOH) washing, and hydrogen reduction in NaOH solution, etc., is to remove thiophene from Ru catalyst via the competitive adsorption in Ru active center (Cl^-, OH^- vs S) and hydrogenolysis. However, the results showed that single acid (HCl) washing, salt (NaCl) washing or alkali (NaOH) washing did not have a satisfactory effect. Even though Cl^- and OH^- have solitary electron pair, neither Cl^- nor OH^- could replace S from Ru via competitive adsorption because S has a much strong coordination function with Ru.

Table 8.5 shows the activity and selectivity of catalyst 1[#] after washed by 2% sulfuric acid twice, then washed by distilled water to neutral and finally reduced in 400 mL 5% NaOH solution for 3 h.

As can be seen from Table 8.5, for sulfide poisoning, catalyst activity could be recovered partially, however, the cyclohexene selectivity was still low. Cyclohexene selectivity could not be recovered to the level of fresh catalyst, demonstrating the sulfide poisoning was irreversible. Therefore, S content of benzene in raw materials should be strictly controlled, that is, the content of thiophene should be controlled below 0.05 ppm.

8.3.3 Regeneration of Deactivated Catalysts Caused by DMAC

Adding sulfuric acid to reduce the pH of slurry could restore the activity and selectivity of the deactivated catalysts caused by DMAC. However, the amount of sulfuric acid added into the slurry is crucial, which must be determined by experiments and then be scaled up to industrial production.

Table 8.6 shows the activity and selectivity of catalyst 2[#] after adding different amounts of sulfuric acid into the slurry.

Table 8.6 Activity and selectivity of catalyst 2# when different amounts of H_2SO_4 was added

t/min	1.5 g H_2SO_4			3.0 g H_2SO_4		
	C_{BZ}/%	S_{HE}/%	Y_{HE}/%	C_{BZ}/%	S_{HE}/%	Y_{HE}/%
5	5.98	87.63	5.24	78.59	18.58	14.60
10	11.45	89.26	10.22	98.73	8.68	8.57
15	18.20	87.53	15.93	100	2.06	2.06
20	26.78	86.82	23.25	100	0	0
25	33.28	85.28	28.38	100	0	0

Slurry composition: 1.96 g catalyst, 1.5 g and 3.0 g H_2SO_4 were added into 280 mL raw slurry; Hydrogenation condition: 150 °C, 5 MPa, 1400 r/min, 140 mL benzene

As can be seen from Table 8.6, catalyst activity was obviously restored and a high selectivity was obtained when 1.5 g sulfuric acid was added into the slurry. When 3.0 g sulfuric acid was added, catalyst activity increased sharply and cyclohexene selectivity decreased significantly, indicating excessive sulfuric acid not only dissolved the adsorbed alkali zinc sulfate, but also partially dissolved the zinc in the catalyst.

To determine the optimal amount of H_2SO_4 accurately, pure water was used as slurry and different amount of H_2SO_4 (from low to high amounts) was added into the slurry gradually. The activity and selectivity of the catalyst were evaluated.

The activity and selectivity of the catalyst after adding different amounts of sulfuric acid were given in Table 8.7.

As can be seen from Table 8.7, when the mother liquor was replaced with pure water and 0.1 g H_2SO_4 was added, the catalyst activity was very low, showing the amount of H_2SO_4 was insufficient. When the amount of H_2SO_4 increased to 1.4 g, the catalyst activity was very high, while the selectivity of cyclohexene was low, indicating that the amount of H_2SO_4 was excessive. Using a gradual reduction method, it was found that when 1.2 g H_2SO_4 was added, catalyst 2# showed optimal activity and cyclohexene selectivity.

Table 8.7 Activity and selectivity of catalyst 2# after adding different amounts of H_2SO_4

t/min	0.1 g H_2SO_4			1.4 g H_2SO_4		
	C_{BZ}/%	S_{HE}/%	Y_{HE}/%	C_{BZ}/%	S_{HE}/%	Y_{HE}/%
5	1.70	77.65	1.32	49.96	67.39	33.67
10	2.83	84.81	2.40	77.01	59.60	45.90
15	4.51	88.03	3.97	91.27	50.33	45.94
20	7.18	86.35	6.20	97.04	41.66	40.43
25	9.04	88.50	8.00	99.19	33.30	33.03

Slurry composition: 1.96 g catalyst, 280 mL pure water, 45.7 g $ZnSO_4 \cdot 7H_2O$, 0.1 g and 1.4 g H_2SO_4 were added respectively; Hydrogenation condition: 150 °C, 5 MPa, 1400 r/min, 140 mL benzene

Table 8.8 Activity and selectivity of catalyst 2# after regeneration when 1.2 g H_2SO_4 was added

t/min	C_{BZ}/%	S_{HE}/%	Y_{HE}/%
5	22.95	83.18	19.09
10	39.34	80.00	31.47
15	55.14	76.95	42.43
20	67.08	72.32	48.51
25	80.05	67.53	54.06

Slurry composition: 1.96 g catalyst, 280 mL pure water, 45.7 g $ZnSO_4{\cdot}7H_2O$, 1.2 g H_2SO_4; Hydrogenation condition: 150 °C, 5 MPa, 1400 r/min, 140 mL benzene

Table 8.8 shows the activity and selectivity of catalyst 2# after regeneration when 1.2 g H_2SO_4 was added.

As can be seen from Table 8.8, catalyst 2# showed good activity and selectivity after regeneration. After processing the data, it was found that when benzene conversion was 40% at 10 min, cyclohexene selectivity reached 80%, and the catalyst activity reached the level of fresh catalyst (activity index $\gamma_{40} = 151$).

Based on the above characterization and regeneration results, the final solution and regeneration methods for deactivated catalyst are determined. For catalyst 1#, whether thiophene content of benzene is excessive should be detected. The detection results showed that the thiophene content of benzene in industrial plant reached 5×10^{-6}–6×10^{-6}, which was much higher than the controlled amount, confirming the above results were correct. Relative enterprises took proper approaches to control the thiophene content of benzene in raw materials strictly and replaced the poisoning catalyst, achieving the successful operation of an industrial plant with production of 100,000 t/a.

For catalyst 2#, whether the circulating benzene contains extractant DMAC in the industrial plant should be detected. The results showed that the amount of DMAC was much higher than the required level due to improper extraction and separation conditions. Enterprises took proper actions (adjusted the extraction and separation conditions, reduced the amount of DMAC within the required level, and added the optimized amount of H_2SO_4 by scaling up the experiments results into the industrial device) and the catalyst activity and selectivity was restored to normal level successfully.

References

1. Struijk, J., D'Angremond, M., Lucas-de Regt, W.J.M. et al.: Partial liquid phase hydrogenation of benzene to cyclohexene over ruthenium catalysts in the presence of an aqueous salt solution i. preparation, characterization of the catalyst and study of a number of process variables. Appl. Catal. A Gen. **83**, 263–295 (1992)
2. Struijk, J., Moene, R., van der Kamp, T., et al.: Partial liquid-phase hydrogenation of benzene to cyclohexene over ruthenium catalysts in the presence of an aqueous salt solution II. Influence

of various salts on the performance of the catalyst. Appl. Catal. A Gen. **89**, 77–102 (1992)
3. Wu, J., Yang, Y., Chen, J.: Study on the causes of catalyst deactivation of benzene semi-hydrogenation. Chem. Ind. Eng. Prog. **22**, 295–297 (2003)
4. He, H.M., Yuan, P.Q., Ma, Y.M., et al.: Theoretical and experimental study on the partial hydrogenation of benzene over Ru-Zn/ZrO_2 catalyst. Chin. J. Catal. **30**, 312–318 (2009)
5. Yuan, P.Q., Ma, Y.M., Cheng, Z.M., et al.: Effect of Zn^{2+}/Zn layer on H_2 dissociation on ruthenium (0001) surface: a first principles density functional study. J. Mol. Struct. Theochem. **807**, 185–189 (2007)
6. Wang, J.Q., Wang, Y.Z., Xie, S.H., et al.: Partial hydrogenation of benzene to cyclohexene on a Ru–Zn/m-ZrO_2 nanocomposite catalyst. Appl. Catal. A Gen. **272**, 29–36 (2004)
7. Bu, J., Liu, J.L., Chen, X.Y., et al.: Ru/SBA-15 catalysts for partial hydrogenation of benzene to cyclohexene: tuning the Ru crystallite size by Ba. Catal. Commun. **9**, 2612–2615 (2008)
8. Liu, J.L., Zhu, L.J., Pei, Y., et al.: Ce-promoted Ru/ SBA-15 catalysts prepared by a "Two Solvents" impregnation method for selective hydrogenation of benzene to cyclohexene. Appl. Catal. A Gen. **353**, 282–287 (2009)
9. Sun, H.J., Jiang, H.B., Zhang, Y.X.: Effect of transition metals (Cr, Mn, Fe Co, Ni, Cu and Zn) on the hydrogenation properties of benzene over Ru-based catalyst. Appl. Catal. A Gen. **1**, 464–469 (2013)

Chapter 9
The Catalytic Technologies and Key Facilities for Benzene Selective Hydrogenation

In foreign countries, the agitated high-pressure reaction kettle is used in industrial devices for selective hydrogenation of benzene, however, only a few countries could produce the key components all over the world. Moreover, this large-scale autoclave manufacturing technology is monopolized by several multinational corporations, limiting China's industrialization process for selective hydrogenation of benzene.

The key facilities of a liquid phase selective hydrogenation of benzene catalytic processes include static mixer and external cyclic reactor and oil-water separator. The whole processes include the utilization of static mixing, external cyclic reaction, and Venturi effect to accelerate the interphase mass transfer between gas, oil, water, and solid to improve benzene conversion, cyclohexene selectivity, and gas-liquid or oil-water separation in the oil-water separator. Hydrogen was recycled. After the oil phase containing benzene, cyclohexene, and cyclohexane was separated by extractive separation, benzene was recycled. The water phase containing catalyst, additives, and dispersants returned to static mixers to form a continuous production process. This device can realize the strict control of lots of parameters such as temperature, pressure, flow rate, liquid level, etc., and could achieve the desired benzene conversion and cyclohexene selectivity. It could not only replace the agitated autoclave, but also reduce the costs of facilities' manufacturing, operation, and maintenance.

The key facilities of catalysts preparation devices include high-pressure reaction systems, feeding systems, dispensing, discharging, washing, packaging, and other complete systems. The main catalyst technical indexes were as follows: benzene conversion of 60%, cyclohexene selectivity of more than 80%, cyclohexene yield of more than 48%. After improvement, benzene conversion reached 70%, and cyclohexene selectivity and cyclohexene yield were higher than 80% and 56%, respectively. By decreasing the temperature of dispensing and sedimentation, and shortening the unit operation time, the production capacity of a single set could be increased by more than 3 times, which satisfies the catalyst's demand in the domestic market.

The main contents of this chapter are as follows: introduction of selective hydrogenation of benzene and the key facilities; introduction of the key devices and

Z. Liu et al., *Catalytic Technology for Selective Hydrogenation of Benzene to Cyclohexene*,
https://doi.org/10.1007/978-981-15-6411-6_9

technology of catalyst preparation; a brief introduction of the invention patent of benzene selective hydrogenation catalyst, which covers from the amorphous alloy to nanocrystal, from complex systems containing modifiers to simple core-shell structure catalyst.

9.1 Technologies and Key Facilities of Benzene Selective Hydrogenation

9.1.1 Key Facilities and Technological Processes for Liquid Phase Selective Hydrogenation of Benzene

Figure 9.1 shows the key facilities and technological processes for liquid phase selective hydrogenation of benzene.

As shown in Fig. 9.1, the compressed hydrogen, high-pressure benzene, and reaction slurry are premixed in the static mixer. The external cyclic reactor **R** includes

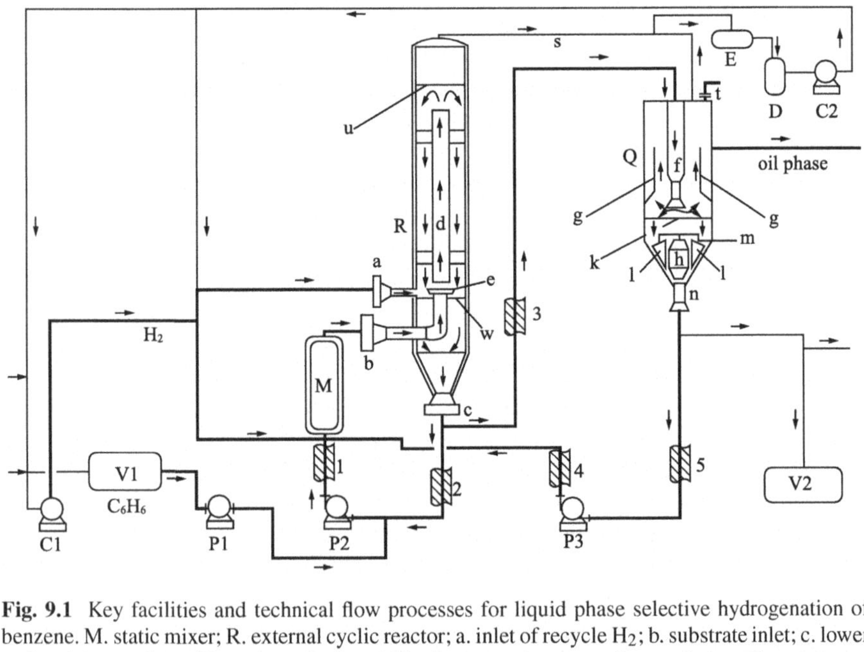

Fig. 9.1 Key facilities and technical flow processes for liquid phase selective hydrogenation of benzene. M. static mixer; R. external cyclic reactor; a. inlet of recycle H_2; b. substrate inlet; c. lower outlet; d. inner tube; e. Venturi nozzle; f. variable diameter tube; g. overflow weir; h. sedimental core; l. cone block; m. connecting pipe; n. water phase exit; s. balance tube; t. filling hole; u. outer tube's upper clapboard; w. outer tube' lower clapboard; Q. oil-water separator; C1, C2. gas compressor; P1. high-pressure benzene pump; P2, P3. solid-liquid pump; V1. benzene storage tank; V2. catalyst storage tank; E. heat exchanger; D. buffer tanks. 1, 2, 3, 4, 5. electric heating zone

the inner tube **d**, outer tube's upper clapboard **u,** and outer tube's lower clapboard **w**. Venturi nozzle **e** is located at the bottom of the inner tube, where it is the main site for selective hydrogenation of benzene to produce cyclohexene. Powered by the solid-liquid circulating pump **P2**, the substrates were premixed by pre-mixer **m** firstly. As a result of the Venturi injection effect, the premixed substrates would form turbulent flow in the inner tube to accelerate the mass transfer and eliminate the internal and external diffusion restriction, improving the benzene conversion. After the substrates flowed through the top of the inner tube and reached the benzene residence time, they entered the outer tube, where the turbulent flow became laminar flow due to the volume effect, and the reaction was terminated to avoid the continuous hydrogenation of cyclohexene to generate cyclohexane and enhance the cyclohexene selectivity. The function of the outer tube's upper clapboard **u** and its lower clapboard **w** is to prevent the substrates from flowing out with the gas flow and constraint the flow region of the substrates. The reaction flow left from the outer circulation reactor via the lower discharging port of the outer tube and then entered into the oil-water separator **Q**. The reaction materials entered into the variable diameter tube **f** from the top, then they were decompressed to achieve gas-liquid separation. Then hydrogen entered into the heat exchanger **E** and buffer tank **D**, then entered the compressor **C2** to realize recycling. On the other hand, the flow velocity would be increased because of the variable diameter effect of tube **f**, and then the oil phase would flow upward while the water phase containing catalyst, dispersant, and additives went down through the baffle plate hole to enter the cone part of **Q**. The special structure of the cone part could avoid catalyst deposition. The oil phase containing the unreacted benzene, the desired product cyclohexene, and the by-product cyclohexane would flow across the overflow weir **g** to enter the following rectification, separation, and purification units, while the aqueous phase containing the catalyst would be recirculated back to the reactor by the exit **n** and the solid-liquid pump **P3**.

Table 9.1 shows the cold-state experimental data in the plexiglass transparent device and the material flow in the outer loop reactor.

As can be seen from Table 9.1, the condition of the liquid flow in the outer loop reactor could be changed significantly by changing the flow rate of the circulated liquid and gas, and by changing the pressure. A satisfied flow condition could be achieved by choosing the proper gas flow rate, liquid flow rate, and proper pressure. An ideal performance of benzene hydrogenation could be achieved by improving the extent of substrates mixing.

The auxiliary facilities include compressors **Cl** and **C2**, pumps **Pl**, **P2**, and **P3**, and heat exchanger **E**, buffer tank **D**, containers **Vl** and **V2**, connecting piece **m**, and balance tube **s**. **Cl** provides 5–6 MPa of hydrogen, and some of them enter into the reactor directly to maintain the system pressure, while others go into the static mixer **M** to accelerate the mixing of materials by the bubbling effect. The emitted hydrogen is firstly subjected to heat exchanger **E** for heat exchange to remove the condensate, and then enters into the buffer tank **D**, followed by compressor **C2** to be compressed for recycling. Benzene in **V1** will be pressurized by **P1** to over 5 MPa. Then the pressurized benzene, the slurry (over 5 MPa) recycled from **P3**, and the hydrogen over 5 MPa from **Cl** will enter the mixer **M** together. After premixing, they will be

Table 9.1 The cold-state experimental data and the material flow in the outer loop reactor

Number of experiments	Circulation fluid		Gas		Fluid phenomenon in the reactor	Notes
	Flow rate/(m^3/h)	Pressure/MPa	Gas flow rate/(m^3/h)	Pressure/MPa		
1	6.0	0.04	6.0	0.52	Gas and liquid were mixed evenly and the liquid was over the upper clapboard around 400 mm	The pressure gauge of the circulating liquid is installed behind the outlet valve of the circulation pump and in front of the flowmeter; the gas pressure gauge is installed in front of the flowmeter; the gas is the purifying air; the water is softened water
2	6.0	0.04	5.0	0.61	Gas and liquid were mixed evenly and the liquid was over the upper clapboard around 200 mm	
3	12.0	0.14	6.5	0.44	Gas and liquid were mixed evenly and the liquid was over the upper clapboard around 300 mm	
4	9.98	0.10	4.5	0.66	Gas and liquid were mixed evenly and the liquid was over the upper clapboard around 300 mm	

(continued)

Table 9.1 (continued)

Number of experiments	Circulation fluid		Gas		Fluid phenomenon in the reactor	Notes
	Flow rate/(m^3/h)	Pressure/MPa	Gas flow rate/(m^3/h)	Pressure/MPa		
5	0.0	0.0	24.0	0.24	Mixed evenly; the space of the lower clapboard was around 150 mm; the liquid was over the upper clapboard around 400 mm	
6	0.0	0.0	18.0	0.10	Mixed evenly; no space on the lower clapboard; the liquid was over the upper clapboard around 200 mm	
7	0.0	0.0	10.0	0.40	Mixed evenly; no space on the lower clapboard; the liquid was over the upper clapboard around 150 mm	

(continued)

Table 9.1 (continued)

Number of experiments	Circulation fluid		Gas		Fluid phenomenon in the reactor	Notes
	Flow rate/(m^3/h)	Pressure/MPa	Gas flow rate/(m^3/h)	Pressure/MPa		
8	0.0	0.0	8.0	0.52	Mixed evenly; no gap in the lower clapboard; the liquid was over the upper clapboard around 50 mm	
9	0.0	0.0	0.0	0.70	The liquid is flushed with the inner guide tube	
10	7.86	0.06	0.0	0.70	Flow condition is good; and no liquid is above the clapboard. No gap in the lower clapboard	
11	10.9	0.10	0.0	0.70	Flow condition is good; and no liquid is above the clapboard. No gap in the lower clapboard	

(continued)

Table 9.1 (continued)

Number of experiments	Circulation fluid		Gas		Fluid phenomenon in the reactor	Notes
	Flow rate/(m^3/h)	Pressure/MPa	Gas flow rate/(m^3/h)	Pressure/MPa		
12	12.81	0.14	0.0	0.70	Flow condition is good; and no liquid is above the clapboard. No gap in the lower clapboard	
13	14.85	0.18	0.0	0.70	Gas-liquid backmixing mixes well and the liquid was over the upper clapboard around 200 mm	
14	14.50	0.18	0.0	0.70	Gas-liquid backmixing mix well and the liquid was over the upper clapboard around 200 mm. Lower clapboard begin to layer	
15	14–15	0.18	0.0	0.70	Gas-liquid backmixing mixes well and the liquid was over the upper clapboard around 200 mm. The gap in the lower clapboard is around 150 mm	

(continued)

Table 9.1 (continued)

Number of experiments	Circulation fluid		Gas		Fluid phenomenon in the reactor	Notes
	Flow rate/(m^3/h)	Pressure/MPa	Gas flow rate/(m^3/h)	Pressure/MPa		
16	10.0	0.09	5.0	0.70	Gas-liquid backmixing mixes well and it takes around 3–5 s for the stream flow through the upper inlet to the lower exit	

mixed with the recycled hydrogen and enter the reactor via inlet **b** subsequently. **Vl** is a benzene storage tank and **V2** is a catalyst tank to reserve the catalysts for device starting, terminating, and breakdown. 1, 2, 3, 4, and 5 on the pipe are electric heating bands.

The processes of liquid phase selective hydrogenation of benzene are as follows. The solid-liquid circulation pump provides energy and power, and the reaction materials including benzene, hydrogen, and the reaction slurry (the aqueous phase containing the Ru catalyst, the zinc sulfate additive, and the zirconium oxide dispersant) are fully premixed in a static mixer firstly and then enter into the inner tube of the outer circulation reactor. The flow rate of benzene and slurry are set according to the inner tube volume, water/benzene volume ratio, and benzene residence time. The appropriate high benzene conversion (over 40%) and preferably higher cyclohexene selectivity (more than 80%) would be controlled. The reaction flow and slurry containing cyclohexene, cyclohexane, unreacted benzene are separated by gas-liquid separation (at the upper part of the oil-water separator) and oil-water separation (at the lower part of the oil-water separator). The oil phase containing cyclohexene, cyclohexane, and benzene enters into the subsequent extraction distillation section. While the water phase containing catalysts, additives and dispersants would return back through the solid-liquid pump to the static mixer, and then be mixed with fresh benzene and hydrogen, entering into the external circulation reaction device to form a continuous production process.

Static mixers and external recirculation reactors are jacketed outside, accessible to steam or cooling water, and there are some heating devices distributed on pipelines to jointly control the reaction temperature. The device does not require mechanical agitation, and compared with the current foreign device, it not only is safe and efficient, but also greatly reduces the facilities' manufacturing, operation, and maintenance costs.

9.1.2 Operation Scheme and Performance

Operation scheme:

(1) Firstly, the system is blown by nitrogen, and then washed by the zinc sulfate aqueous solution (pH = 5–6). A certain amount of zirconia powder is added into the zinc sulfate aqueous solution to remove the oxide film and other residues on the wall.
(2) The reaction slurry containing water, catalyst, dispersant zirconia, and additive zinc sulfate is prepared strictly according to the required proportion. The prepared slurry is added into the reaction system via the filling hole **t** above **Q** as preparation for the whole device's first use.
(3) The pump **P3**, **P2**, and the electric heater are started.

(4) When the temperature of the outlet thermometer of **P2** and **Q** reaches 150 °C, **P1** and **C1** are started. Then hydrogen is firstly fed into **M** and **R** before benzene is fed.
(5) The electric heating power and the medium flowrate in the jacket of M and R are adjusted according to the temperature of **Q** to maintain the reaction temperature around 135–150 °C.
(6) Whether the amount of hydrogen is sufficient which is determined according to the emitted exhaust amount from **Q**. A certain gas flow is maintained to improve the flow conditions of the reaction materials.
(7) After the system is stabilized, the products and catalysts are sampled and analyzed on a regular interval.

The benzene stored in the benzene storage tank (**V1**, 10 m^3, atmospheric pressure) is pressurized by the pump **P1** (pumping head is 6.5 MPa; flow rate is 500 L/h, which could be adjusted to 50 L/h) firstly, then benzene together with the slurry from **P3** enter to the reactor **R** after they are fully mixed via the static mixer **M**. The reaction flow exported from **R** is pressurized by circulating pump **P2** (pumping head is 25 m, flow rate is 30 m^3/h, which can be adjusted to 10 m^3/h), and then is mixed with benzene from **P1** in the static mixer **M**. Via the static mixer **M** and external cyclic reactor **R**, the reactants are in full contact with the catalyst, and then the reaction of selective hydrogenation of benzene to produce cyclohexene begin. An electric heater (the length is around 8 m) is placed outside the outlet of **P1** and static mixer. **M** and **R** will be heated by steam or cooled down by circulating water in the outer jacket.

The reaction products are fed into the oil-water separator **Q**, where the hydrogen is separated. Then the hydrogen enters into the cyclic gas compressor (**C2**, pressure difference is 0.3 MPa, flow rate is 500 Nm^3/h) to be compressed by **C2**. Then the compressed hydrogen together with the fresh hydrogen from the hydrogen compressor (**C1**, flow rate is 80 Nm^3/h) enters into **M** and **R** after they are fully mixed. The oil phase and the aqueous phase containing the catalyst are separated in **Q**. After overflowing through the overflow weir, the oil phase is cooled down by the casing-type heat exchanger and enters into the subsequent process. The aqueous phase containing the catalyst is returned to the static mixer **M** and reactor **R** via **P3** (pumping head is 25 m, flow rate is 1000 L/h, which could be adjusted to 30 L/h). Hydrogen is introduced to the hydrogen compressor **C1** via the pipeline. The pump, valve, temperature, pressure, and liquid level as well as the flow rate could be controlled by controller, PLC, or the DCS automatic control system.

The operation results show that this device can control the operation parameters including temperature, pressure, flow rate, and fluid level strictly. The average reaction temperature was 135–155 °C, and the average reaction pressure was 4.5–5.5 MPa, both of which are controlled smoothly and the desired benzene conversion and cyclohexene selectivity could be achieved.

9.2 Key Facilities and Processes Flow for Catalyst Preparation

9.2.1 Key Facilities and Processes Flow

Figure 9.2 shows the schematic illustration of the key facilities and processes flow for catalyst preparation.

In Fig. 9.2, the key facility is the high-pressure reaction system which includes autoclave and accessory pipes and valves. The reaction kettle where the flange is installed is made of carbon steel, and is lined with Hastelloy HC276 inwardly. The reaction kettle has a volume of 400–800 L with a height/diameter ratio of (1.5–1.8):1, and there are 3 baffles built inside with an interval of 120°. The operation temperature is 130–160 °C, and the operation pressure is 4–6 MPa. Double-layer mixing is applied in the reaction kettle, where flat-type mixing is adopted for the

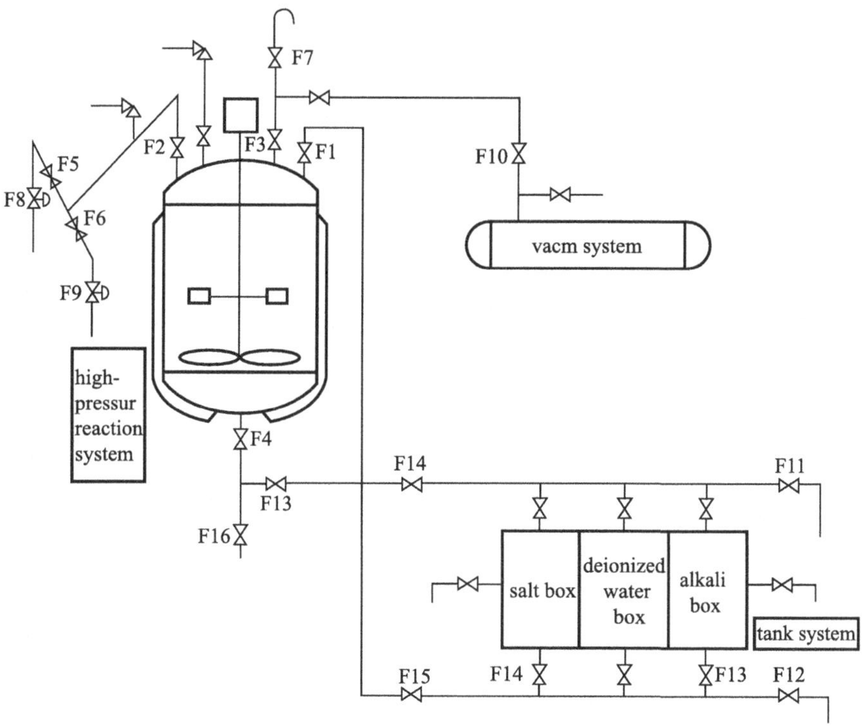

Fig. 9.2 Schematic illustration of the key facilities and processes flow for catalyst preparation F1. feed valve; F2. gas intake valve; F3. exhaust emission valve; F4. discharging valve; F5. hydrogen valve; F6. nitrogen valve; F7. three-way exchange valve; F8. hydrogen pressure regulating valve; F9. nitrogen pressure regulating valve; F10. vacuum valve; F11. blowdown valve; F12. blowdown valve; F13. lye control valve; F14. salt liquid control valve; F15. feed valve; F16. discharge valve

upper layer and propeller agitation is applied for the lower layer. The inclination angle of 45°, rotating speed of 500–800 r/min, and line speed bigger than 4 m/s are applied to ensure the full mixing of the reaction material from both radial and axial directions. Mechanical seal or magnetic seal is adopted for the high-pressure kettle, and the bearing lubrication adopts sulfur-free graphite. The autoclave is equipped with gas inlet, emission vent, liquid inlet, discharging port, etc. The gas inlet pipe runs from the top to the bottom of the reactor and a circular or hexagonal distributor with a downward opening is located at the bottom of the reactor. The upper part is equipped with liquid inlet, emission vent, pressure gauge port, and temperature measuring port, and material discharging vent is also designed in the lower part. Any parts contacted with the liquid, including the thermometer casing, inlet pipe, stirring shaft, stirring paddle, coupling, baffle, and others are made of Hastelloy HC276 alloy. Medium-pressure steam or heat-conducting oil is used for heating (controlling accuracy is ± 1 °C and heating rate is 1–2 °C/min). Natural cooling and forced cooling are adopted for cooling, and the cooling coil is placed in the jacket.

Vacuum system: The vacuum system is connected with the high-pressure reaction system through the exhaust port, and it is the key component of a multifunctional reactor/tank. Its main role is to feed the reaction materials into the reaction kettle automatically using the vacuum and atmospheric pressure, which is easy to operate. Besides, it could effectively improve the air replacement efficiency inside the kettle, saving high-purity nitrogen and high-purity hydrogen.

Control system: The control system includes temperature, pressure, agitation speed, heating rate, constant temperature, time setting, etc. Controlling is to ensure the stability of operating conditions and preserve the historical data.

Tank system: The batching tank could be used as washing tank, product treatment tank, etc. Tank system is connected with the high-pressure reaction system via the pipes and valves. The effective volume of two batching tanks is equal to the effective volume of the reaction system. The fabrication materials for the tank system could adopt 316 L stainless steel, polypropylene, or polycarbonate.

The technical processes include batching, reaction, aging, reduction, washing, and packaging.

Batching: The reaction materials are divided into salt liquid and lye, which are prepared in the salt box and alkali box with stirring, respectively. The salts used here refer to the salts containing active components and additives precursors, and the alkali refers to alkali metal hydroxide. To prepare the salt solution, the precursor of the active component ruthenium salts (commonly used ruthenium trichloride, $RuCl_3 \cdot 3H_2O$) and additives salts (soluble salts of Zn, Mn, La, Ce) are added into the salt box together according to the required proportion. A certain amount of 90 °C deionized water is added into the salt box to dissolve the ruthenium salts and the additive salts with agitation (the final temperature is about 80 °C). To prepare the lye, a certain amount of deionized water at room temperature is added into the alkali tank firstly, then the alkali metal hydroxide is added slowly with stirring. The dissolution of alkali metal hydroxide in water is a strong exothermic reaction, and the final temperature should be controlled around 80 °C. The temperature and volume of the two kinds of solutions are approximately the same.

Reaction: The precipitate reaction proceeds in the high-pressure autoclave. First, lye is imported into the autoclave using vacuum and atmospheric pressure. The temperature is controlled around 80 °C, and then salt solution is introduced into lye under agitation speed of 300–400 r/min using vacuum and atmospheric pressure. The reaction is conducted at atmospheric pressure and agitation for 3–8 h. After that, the reaction slurry is naturally cooled down to 50–60 °C. The pH of the slurry (ideally it should be around 11–13) is measured.

Aging: The reaction slurry is aged under static state at 50–60 °C for 8–12 h.

Reduction: Reduction is conducted for 6–12 h at the temperature of 130–150 °C, stirring speed of 300–400 r/min, and a hydrogen pressure of 4–5 MPa. The slurry is naturally cooled down to 80 °C approximately and the reduction performances are checked. The black solids in the slurry would be settled into layers quickly, and the pH of the supernatant is 11–13.

Washing: The catalysts after reduction are placed in two batching tanks, and washed many times until the pH of the detergent is 7 and there are no chlorine ions.

Packing: The catalysts after reduction are stored in a barrel lined with polypropylene. The catalysts in each barrel have a net weight of 100–150 kg. The solid content is 5–10%.

The united domestic and international industry standard method is adopted to evaluate the catalyst selectivity and process data.

9.2.2 Catalyst Preparation and Main Technical Specifications

The preparation of Ru nanometer microcrystalline catalyst is as follows:

Preparation of salt solution: first, 15 kg $RuCl_3 \cdot 3H_2O$ was measured and added into the salt box, and then 154 L deionized water (90 °C) was added into the salt box. The salt was dissolved under agitation (the final temperature was 82 °C). Preparation of alkaline solution: first, 80 L deionized water (room temperature) was added into the alkali box, and then 10 kg NaOH GR was added slowly with stirring. After that, 74 L deionized water (90 °C) was imported into it (the final temperature was 78 °C). Before the vacuum system was opened, the conditions of all the valves of the high-pressure kettle were checked to see whether they were in the right states. Then the vacuum system was opened and the lye solution followed by the salt solution under the agitation speed of 350 r/min was introduced. The reaction would be kept for 6 h with stirring under the atmospheric condition. After the reaction slurry was cooled down to 60 °C, whether the pH was higher than 12 was checked and then they were placed overnight for aging. After aging, all valves of the high-pressure kettle were closed. The vacuum system was opened again and the system was pumped to vacuum under 50 °C. Then high-purity nitrogen was introduced into the autoclave and the pressure was kept at 4 MPa for 10 min for stabilization to exclude air. After that, the nitrogen was discharged by opening the exhaust emission valve (the above procedures were repeated for at least 3 times). Subsequently, high-purity hydrogen was charged into the autoclave. When the pressure rose to 4 MPa, the condition was

Table 9.2 The evaluation results of the activity and selectivity of 1st batch of catalyst (5.6 kg catalyst)

Time/min	Benzene conversion/%	Cyclohexene selectivity/%	Cyclohexene yield/%
5	18.05	86.31	15.58
15	50.82	80.12	40.72
30	79.63	70.60	56.22
45	92.08	60.15	55.38
60	96.95	49.82	48.30

kept for 10 min for stabilization. After the hydrogen was discharged, high-purity hydrogen was fed into the system again and the pressure was elevated to 5 MPa. The stirring system was opened at the stirring speed of 350 r/min. The temperature was raised to 80 °C by the heating system. Pressure changes in the kettle were monitored and hydrogen was replenished in time to maintain the pressure of 5 MPa. The reaction of benzene hydrogenation was kept for 6 h after the temperature was raised to 140 °C and the stirring speed reached 400 r/min. After that, the system was cooled down naturally under a stirring speed of 150 r/min. When it was cooled down to 80 °C, the hydrogen in the system was discharged to keep the pressure in the autoclave of atmospheric pressure. The reaction slurry was discharged from the discharging port, where the black solids settled down quickly, and the pH of the supernatant should be over 12. The catalysts after reduction were put into two salt boxes and alkali boxes, and then was washed many times until there were no chlorine ions in the washing solution with a pH of 7. The final catalysts (5.6 kg) were preserved in the barrel lined with polypropylene with 100 kg water.

Tables 9.2, 9.3, 9.4, 9.5, 9.6, 9.7, 9.8, 9.9, 9.10 and 9.11 show the evaluation results of the activity and selectivity of 10 batches of catalysts (5.6 kg catalyst).

The main technical indexes of the 1st batch of catalyst can be obtained from Table 9.2:

$$t_{40} = 12\ \text{min};\ \gamma_{40} = 126;\ S_{40} = 82\%$$
$$t_{50} = 15\ \text{min};\ \gamma_{50} = 126;\ S_{50} = 80\%$$
$$t_{60} = 20\text{min};\ \gamma_{60} = 113;\ S_{60} = 77\%$$

Table 9.3 The evaluation results of the activity and selectivity of 2nd batch of catalysts (5.6 kg catalyst)

Time/min	Benzene conversion/%	Cyclohexene selectivity/%	Cyclohexene yield/%
5	20.16	85.67	17.27
15	53.94	78.79	42.50
30	82.40	67.53	55.64
45	93.63	55.58	52.04
60	97.27	44.36	43.15

Table 9.4 The evaluation results of the activity and selectivity of 3rd batch of catalyst (5.6 kg catalyst)

Time/min	Benzene conversion/%	Cyclohexene selectivity/%	Cyclohexene yield/%
5	18.85	86.07	16.23
15	54.70	78.95	43.31
30	84.22	63.16	44.92
45	94.85	49.26	41.49
60	98.34	38.07	36.11

Table 9.5 The evaluation results of the activity and selectivity of 4th batch of catalyst (5.6 kg catalyst)

Time/min	Benzene conversion/%	Cyclohexene selectivity/%	Cyclohexene yield/%
5	20.26	86.67	17.56
15	54.62	80.43	43.93
30	82.60	70.85	58.52
45	93.40	60.49	56.50
60	97.57	50.43	49.20

Table 9.6 The evaluation results of the activity and selectivity of 5th batch of catalyst (5.6 kg catalyst)

Time/min	Benzene conversion/%	Cyclohexene selectivity/%	Cyclohexene yield/%
5	15.21	88.66	13.49
15	48.16	82.96	33.95
30	77.22	74.78	57.75
45	90.66	65.16	59.07
60	96.26	55.64	53.56

Table 9.7 The evaluation results of the activity and selectivity of 6th batch of catalyst (5.6 kg catalyst)

Time/min	Benzene conversion/%	Cyclohexene selectivity/%	Cyclohexene yield/%
5	16.58	89.10	14.77
15	50.59	83.22	42.10
30	79.52	75.23	59.83
45	91.58	66.65	61.04
60	96.05	58.98	56.65

Table 9.8 The evaluation results of the activity and selectivity of 7th batch of catalyst (5.6 kg catalyst)

Time/min	Benzene conversion/%	Cyclohexene selectivity/%	Cyclohexene yield/%
5	12.48	89.77	11.21
15	39.43	85.81	33.84
30	66.17	80.41	53.20
45	79.99	74.58	59.66
60	89.95	68.22	61.37

Table 9.9 The evaluation results of the activity and selectivity of 8th batch of catalyst (5.6 kg catalyst)

Time/min	Benzene conversion/%	Cyclohexene selectivity/%	Cyclohexene yield/%
5	11.12	90.47	10.06
15	37.95	85.77	32.55
30	68.05	79.34	53.99
45	83.78	72.50	60.76
60	91.81	65.34	59.99

Table 9.10 The evaluation results of the activity and selectivity of 9th batch of catalyst (5.6 kg catalyst)

Time/min	Benzene conversion/%	Cyclohexene selectivity/%	Cyclohexene yield/%
5	13.34	89.94	12.00
15	41.94	85.43	35.83
30	71.81	79.45	57.05
45	86.21	71.00	61.21
60	93.42	63.28	59.12

Table 9.11 The evaluation results of the activity and selectivity of 10th batch of catalyst (5.6 kg catalyst)

Time/min	Benzene conversion/%	Cyclohexene selectivity/%	Cyclohexene yield/%
5	11.71	90.34	10.58
15	36.88	86.15	31.77
30	65.16	80.38	52.38
45	81.58	73.99	60.36
60	90.04	67.62	60.89

The main technical indexes of the 2nd batch of catalyst can be obtained from Table 9.3:

$$t_{40} = 11\ \text{min};\ \gamma_{40} = 137;\ S_{40} = 82\%$$
$$t_{50} = 14\ \text{min};\ \gamma_{50} = 135;\ S_{50} = 80\%$$
$$t_{60} = 18\ \text{min};\ \gamma_{60} = 126;\ S_{60} = 76\%$$

The main technical indexes of the 3rd batch of catalyst can be obtained from Table 9.4:

$$t_{40} = 11\ \text{min};\ \gamma_{40} = 137;\ S_{40} = 82\%$$
$$t_{50} = 14\ \text{min};\ \gamma_{50} = 135;\ S_{50} = 80\%$$
$$t_{60} = 18\ \text{min};\ \gamma_{60} = 126;\ S_{60} = 76\%$$

The main technical indexes of the 4th batch of catalyst can be obtained from Table 9.5:

$$t_{40} = 11 \text{ min}; \gamma_{40} = 137; S_{40} = 83\%$$
$$t_{50} = 14 \text{ min}; \gamma_{50} = 135; S_{50} = 81\%$$
$$t_{60} = 18 \text{ min}; \gamma_{60} = 126; S_{60} = 78\%$$

The main technical indexes of the 5th batch of catalyst can be obtained from Table 9.6:

$$t_{40} = 12 \text{ min}; \gamma_{40} = 126; S_{40} = 84\%$$
$$t_{50} = 16 \text{ min}; \gamma_{50} = 118; S_{50} = 82\%$$
$$t_{60} = 21 \text{ min}; \gamma_{60} = 108; S_{60} = 80\%$$

The main technical indexes of the 6th batch of catalyst can be obtained from Table 9.7:

$$t_{40} = 12 \text{ min}; \gamma_{40} = 126; S_{40} = 84\%$$
$$t_{50} = 16 \text{ min}; \gamma_{50} = 118; S_{50} = 82\%$$
$$t_{60} = 21 \text{ min}; \gamma_{60} = 108; S_{60} = 80\%$$

The main technical indexes of the 7th batch of catalyst can be obtained from Table 9.8:

$$t_{40} = 15 \text{ min}; \gamma_{40} = 101; S_{40} = 86\%$$
$$t_{50} = 21 \text{ min}; \gamma_{50} = 90; S_{50} = 84\%$$
$$t_{60} = 27 \text{ min}; \gamma_{60} = 84; S_{60} = 82\%$$

The main technical indexes of the 8th batch of catalyst can be obtained from Table 9.9:

$$t_{40} = 16 \text{ min}; \gamma_{40} = 94; S_{40} = 85\%$$
$$t_{50} = 21 \text{ min}; \gamma_{50} = 90; S_{50} = 83\%$$
$$t_{60} = 26 \text{ min}; \gamma_{60} = 87; S_{60} = 81\%$$

The main technical indexes of the 9th batch of catalyst can be obtained from Table 9.10:

$$t_{40} = 14 \text{ min}; \gamma_{40} = 108; S_{40} = 86\%$$
$$t_{50} = 19 \text{ min}; \gamma_{50} = 99; S_{50} = 84\%$$
$$t_{60} = 24 \text{ min}; \gamma_{60} = 94; S_{60} = 82\%$$

The main technical indexes of the 10th batch of catalyst can be obtained from Table 9.11:

$$t_{40} = 16\ \text{min};\ \gamma_{40} = 94;\ S_{40} = 86\%$$
$$t_{50} = 22\ \text{min};\ \gamma_{50} = 86;\ S_{50} = 83\%$$
$$t_{60} = 27\ \text{min};\ \gamma_{60} = 84;\ S_{60} = 81\%$$

As shown in Tables 9.2, 9.2, 9.3, 9.4, 9.5, 9.6, 9.7, 9.8, 9.9, 9.10 and 9.11, when benzene conversion of ten batches of catalysts was 40%, the activity γ_{40} was about 94–137, and selectivity of cyclohexene was 82–86%. When benzene conversion was 50%, activity γ_{50} was 86–135, and the selectivity of cyclohexene was 80–84%. When benzene conversion was 60%, activity γ_{60} was 84–126, and the selectivity of cyclohexene was 76–82%.

The above catalyst preparation technology has good repeatability and stability, with benzene conversion of 60%, cyclohexene selectivity of higher than 80%, and yield of higher than 48%.

The catalyst preparation technology for selective hydrogenation of benzene to cyclohexene and relative key facilities have been authorized with a national invention patent, and the authorization number was CN 103785379 A.

9.2.3 *Key Facilities and Processes After Improvement*

In Sect. 9.2.1, the key facilities and processes flow of catalyst preparation are introduced, where the high-pressure reactor is the most important facility. The high-pressure reactor provided with oil tanks and workstations is sealed by mechanical seal. The vacuum system is used to emit the exhaust, replace the air from the reactor, and add reaction materials. However, in this process, there still exist some problems. Firstly, during the evacuation process, the internal pressure of the mechanical seal is lower than the external pressure because the pressure in the kettle is lower than the outer atmosphere, resulting in lubricant leakage. Secondly, under high-speed stirring, the lubricant often adsorbs on the catalyst surface, changing the hydrophilicity of the catalyst surface, and making the catalyst poisoned if the lubricant contains sulfur. In terms of the preparation technology, the reaction temperature is relatively high, and the productivity of each batch is low. Besides, the reaction time is long. All of these factors lead to low overall productivity.

To solve the problem of lubricant leakage, the batching tank, the autoclave, and the washing tank are placed in order (from high to low position), utilizing the position difference and gravity effect to guild the materials from previous operation unit into the next unit, avoiding the use of vacuum system for feeding, exhausting, and excluding the air. The low-pressure chlorine gas is used to replace the air in reaction kettle, while low-pressure hydrogen is used to replace chlorine in kettle. At the same time, the technological processes are optimized. Firstly, the catalyst preparation formula is changed to improve the catalyst activity and cyclohexene selectivity. Then, the concentration of the reaction material is increased to enhance the catalyst productivity of a single batch. Thirdly, the temperature of batching and precipitation is lowered and the unit operation time is reduced. In detail, the operation time of sedimentation, aging, reduction reaction, maturing and cooling processes, and other unit operations is decreased. As a result, the catalyst production cycle for a single batch is reduced from 24 to 8 h.

When the improved key facilities and technological processes were used to produce catalysts, lots of breakthroughs of the main technical indexes including benzene conversion of 70%, cyclohexene selectivity of more than 80%, and cyclohexene yield of more than 56% were achieved. The catalyst production capacity was increased by 3 times because the temperature for catalyst proportioning and precipitation is reduced and the unit operation time is shortened.

Figure 9.3 shows the schematic illustration of the key facilities and processes after improvement on the catalyst preparation for selective hydrogenation of benzene.

In Fig. 9.3, the key facilities include: high-level tank, batching tank, high-pressure reactor, washing tank, etc. High-level tank provides reaction materials and washing water to high-pressure reactor. The reaction materials in the batching tank could be imported into the high-pressure reactor by gravity. The high-pressure reactor is used to prepare catalyst precursors and reduce catalyst, where the subsequent materials after reduction are imported into the washing tank by gravity.

The technological processes mainly include batching, precipitation, and reduction reaction. For batching and precipitation, NaOH solution is prepared in the batching tank and then is imported into the high-pressure reactor by gravity using the high-level tank. Then the mixed solution of ruthenium trichloride and zinc salts is prepared. Before it is introduced to the high-pressure reactor, the stirring system is opened, and then the reaction is kept for 2 h at 20–30 °C and aging reaction is kept for 1 h. After that, the pH of the reaction slurry is tested, which should be higher than 12.

For the reduction reaction, nitrogen is firstly used to exclude the air in the high-pressure reactor for at least 3 times and then hydrogen is used to replace the nitrogen in the kettle for at least 3 times again. The reduction reaction is kept for 3 h at 140 °C and 4.5 MPa of hydrogen pressure. The maturing reaction is kept for 1 h. After that, the pH of the reaction slurry is tested, which should be higher than 12. The temperature is lowered to around 50 °C after exhausting.

The reduced catalysts are introduced to the washing tank to be washed many times until no chlorine ions could be detected.

9.2.4 Main Technical Specifications of Improved Catalyst

Figure 9.4 shows the activity and selectivity of the representative catalysts after improvements.

According to Fig. 9.4, the main technical indexes of the catalysts are as follows:

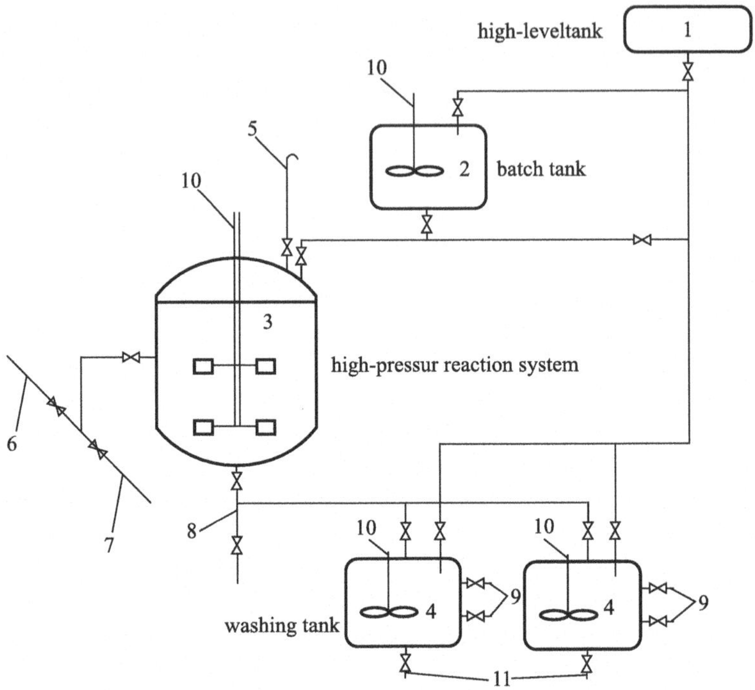

Fig. 9.3 Schematic illustration of the key facilities and processes after improvement on the catalyst preparation for selective hydrogenation of benzene 1. high-level tank (1000 L, made by 304 stainless steel); 2. batching tank (400 L, made by polypropylene plastic material); 3. high-pressure reactor (500 L, reactor body is made by carbon steel, inner lined with C276 Hastelloy Alloy); 4. two washing tanks (made by polypropylene plastic, 400 L); 5. oil tank and mechanical seal workstation; 6. nitrogen bottle; 7. hydrogen bottle; 8, 9. finished cans (galvanized bucket of 200 L, lined with polypropylene plastic bags); 10. heat conduction oil discharging facility; 11. Pipelines for air exhaust and replacement

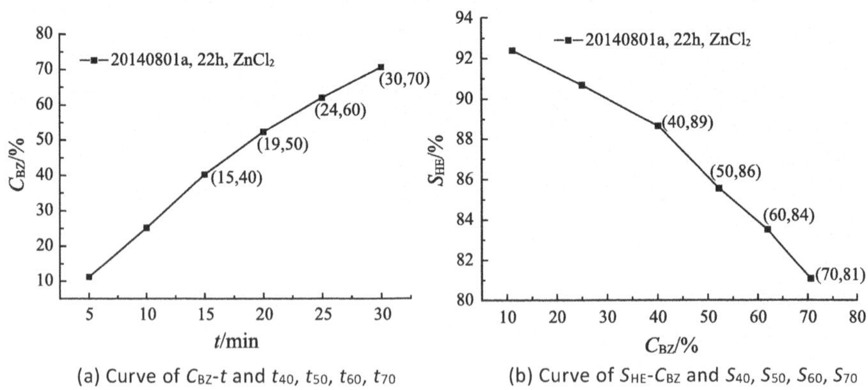

Fig. 9.4 Activity and selectivity of the catalysts after improvements

Table 9.12 Main technical indexes of the catalysts of the 7th, 8th, 11th, 27th, 31st batches after pretreatment and hydrogenation for 22 h

Batches	t_{40}/min	t_{50}/min	t_{60}/min	t_{70}/min	S_{40}/%	S_{50}/%	S_{60}/%	S_{70}/%
7	12	14	18	22	88	86	83	80
8	14	17	20	24	86	85	83	80
11	14	17	20	25	86	85	82	80
27	15	19	24	30	89	86	84	81
31	14	17	21	25	87	85	83	80

$$t_{40} = 15\text{ min};\ \gamma_{40} = 101;\ S_{40} = 89\%$$
$$t_{50} = 19\text{ min};\ \gamma_{50} = 99;\ S_{50} = 86\%$$
$$t_{60} = 24\text{ min};\ \gamma_{60} = 94;\ S_{60} = 84\%$$
$$t_{70} = 30\text{ min};\ \gamma_{70} = 88;\ S_{70} = 81\%$$

Table 9.12 shows t_{40}, S_{40}, t_{50}, S_{50}, t_{60}, S_{60}, and t_{70}, S_{70} of the 7th, 8th, 11th, 27th, 31st batches of catalysts after pretreatment and hydrogenation for 22 h.

As can be seen from Table 9.12, the time of benzene conversion of 40% was 12–15 min, with cyclohexene selectivity of 86–89%. The time of benzene conversion of 50% was 14–19 min, with cyclohexene selectivity of 85–86%. The time of benzene conversion of 60% was 18–24 min, with cyclohexene selectivity of 82–84%. The time of benzene conversion of 70% was 22–30 min, with cyclohexene selectivity of 80–81%. It has achieved the breakthrough of benzene conversion of 70% and cyclohexene selectivity higher than 80%.

9.3 Catalyst Preparation Technologies for Selective Hydrogenation of Benzene to Cyclohexene

9.3.1 Monolayer-Type Catalyst for Selective Hydrogenation of Benzene to Cyclohexene and Its Preparation Method [1]

The catalyst is mainly composed of active component Ru, alkali sulfate of additive M, and orthosilicic acid polymer. The alkali sulfate of M which is stabilized by the orthosilicic acid polymer disperses on Ru's outer surface. The preparation method outperforms the traditional co-precipitation method and impregnation method, and through the self-assembly of the silicate polymer, the alkali salt of additive M could be dispersed on the surface of Ru catalyst in monolayer type.

(1) Preparation of the nanosized Ru microcrystals. At 70–100 °C, the NaOH solution and $RuCl_3$ solution were simultaneously introduced into the reaction tank

by parallel flow with stirring (stirring was continued for 10–30 min after the completion of the precipitation). The pH of the mother liquor was about 12. The precipitates together with the mother liquor were then reduced for 1–12 h (100–200 °C, 3–5 MPa of hydrogen pressure). The resulting solids were washed until the filtrate did not contain Cl^-.

(2) Preparation of the nanosized Ru–$M_{2x/n}O_x$ (M = Mn, Fe, Ce, La, Zn etc.) oxide stabilized by the orthosilicic acid polymer. The mixture of ethyl silitate and acetate acid salts of M (M = Mn, Fe, Ce, La, Zn, etc.) was added to the mixture of nanosized Ru microcrystals and NaOH solution prepared from (1) under stirring for 5–30 min at 80 °C. After the reaction ended, it was filtered, and the resulting solids were dried at 50–100 °C for 1–12 h.

(3) Preparation of the alkali salt of the nanosized Ru–$M_{2x/n}(OH)_x \cdot M_{2y/n}(SO_4)_y \cdot zH_2O$ (M = Mn, Fe, Ce, La, Zn etc., and n represents the valence of the elements) stabilized by the orthosilicic acid polymer. The oxides of Ru–$M_{2x/n}O_x$ prepared from (2) were reduced at 50–150 °C for 1–12 h in the sulfate of M (1–4 mol/L) with hydrogen pressure of 1–5 MPa. The reduced solids were washed until no M ions were detected.

(4) Preparation of the monolayer-type nanosized Ru–M (M = Mn, Fe, Ce, La, Zn, etc.) catalyst stabilized by orthosilicic acid. By adjusting the amount of sulfate prepared from (3), the dispersed monolayer-type Ru–M catalyst stabilized by the original silicate polymer was obtained by self-assembly. The amount of the ethyl silicate was controlled because deficient ethyl silicate could neither protect the active compounds nor stabilize the additive, while excessive ethyl silicate would block the active site, reducing the catalyst activity. The appropriate mass ratio of Si/Ru was about 1–15%.

Structure and performance characteristics of dispersed monolayer-type catalyst for selective hydrogenation of benzene to cyclohexene are as follows:

(1) The active component Ru of the prepared catalyst and additive M disperses evenly. Ru–M catalyst with a core-shell structure is composed of the active component Ru (acts as the core) and alkali salt stabilized by original silicate (acts as the shell). The core-shell structure can prevent the collision and coalescence between the active components, prolonging the catalyst lifetime.

(2) The alkali salt of additive M disperses on the surface of the active component (Ru microcrystals) in the monolayer-type, and the catalyst shows the best cyclohexene selectivity and yield.

The alkali salts are difficult to dissolve, and compared with $ZnSO_4$, they are much easier to be chemically adsorbed on the surface of the catalyst and dispersant ZrO_2. The alkali salts containing crystalline water can increase the hydrophilicity of the surface of active component Ru and promote the desorption of the intermediate product cyclohexene to prevent its further adsorption. The alkali salts chemically adsorbed on Ru would preferentially occupy the strongest active site of Ru, beneficial to prevent the formation of cyclohexane by benzene one-step hydrogenation and improve cyclohexene selectivity. The majority of the active sites of Ru still have catalytic activity because of the monolayer dispersion, which could keep high benzene conversion and obtain a higher cyclohexene yield. The electronic and geometrical effects between M ions and active component Ru in alkali salts could change the electronic properties and geometrical arrangement of Ru, making the stepwise hydrogenation of benzene to cyclohexene become dominant.

(1) Example 1 of catalyst preparation

(1) Firstly, 20 g $RuCl_3$ was dissolved in 500 mL of water. 500 mL of 5% NaOH solution was added into it under stirring (stirring was kept for 3 h at 80 °C). And then the precipitates together with the mother liquor were reduced for 4 h at 150 °C and hydrogen pressure of 4 MPa. The resulting solids were washed many times until no Cl^- could be detected in the washing filtrate, and the nanometallic Ru microcrystals were obtained.
(2) The mixed solution containing 0.3 g ethyl acetate and metal M (M = Mn, Fe, Ce, La, Zn, etc., M/Ru atom ratio of 4%, 16%, 18%, 25%, and 28%, respectively) was added to the mixture of metal Ru and 500 mL 5% NaOH solution. Stirring was kept for 30 min under 80 °C, and then the solids were dried at 80 °C for 3 h.
(3) The solids were reduced in 500 mL of M sulfate (M = Mn, Fe, Ce, La, Zn, etc.) solution (0.5 mol/L) for 5 h at 150 °C and hydrogen pressure of 4 MPa. The reduced solids were washed many times until no M ions were detected, and finally the monolayer-dispersed nanosized Ru–M catalyst modified by the base salt which was stabilized by the orthosilicic acid was obtained.

Figure 9.5 shows the XRD pattern of the Ru–Zn catalyst.

As shown in Fig. 9.5, when the Zn/Ru atomic ratio was 18%, the basic salts dispersed on the Ru surface in the form of a monolayer. X-ray fluorescence spectroscopy spectra showed the ratio of Si/Ru was 3.4%, indicating the original silicic acid polymer was adsorbed on Ru surface chemically.

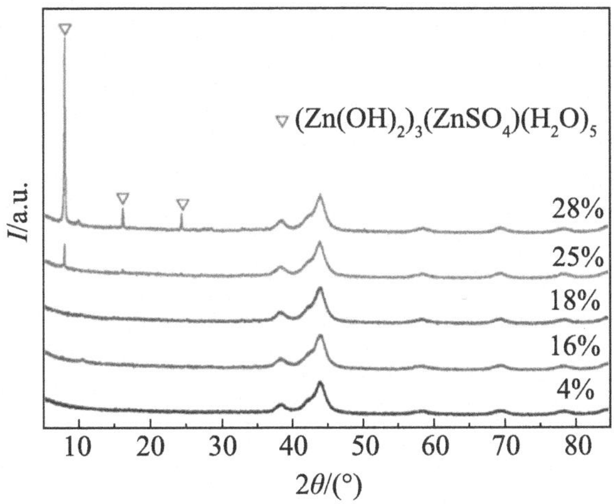

Fig. 9.5 XRD pattern of Ru–Zn catalyst with different Zn/Ru ratio

Table 9.13 shows the main performance indexes of Ru–Zn catalyst with different Zn/Ru atomic ratios.

As shown in Table 9.13, for the monolayer-dispersed nanosized Ru–Zn catalyst, when benzene conversion was 74.7% at $t = 15$ min, the selectivity and yield of cyclohexene were 79.7% and 59.5% respectively, and the highest cyclohexene yield was 64.7%.

Table 9.14 shows the activity and selectivity of the monolayer-dispersed Ru–Zn catalyst stabilized by the orthosilicic acid without $ZnSO_4$.

As shown in Table 9.14, with the absence of $ZnSO_4$, for the monolayer-dispersed Ru–Zn catalyst stabilized by the orthosilicic acid, when benzene conversion was 51.6% at $t = 20$ min, the selectivity and yield of cyclohexene were 81.1% and 41.8%, respectively.

(2) Example 2 of catalyst preparation

Another catalyst was prepared by replacing M with Fe in step (2) in example 1, while all the other procedures in example 1 were kept identical. The XRD pattern showed that when Fe/Ru atomic ratio was 72%, the basic salts were monolayer-dispersed on the Ru surface.

Table 9.15 shows the activity and selectivity of the monolayer-dispersed Ru–Fe catalyst stabilized by the orthosilicic acid.

Table 9.13 Main technical indexes of Zn–Ru catalyst with different Zn/Ru atomic ratios

Zn/Ru	$C_{BZ}/\%^{*}$	$S_{HE}/\%^{*}$	$Y_{HE}/\%^{*}$	$Y_{max}/\%^{**}$
4%	92.0	59.7	54.9	54.9
16%	81.6	67.7	55.2	56.3
18%	74.7	79.7	59.5	64.7
25%	54.0	82.3	44.4	54.0
28%	19.5	90.9	17.7	33.2

Experimental conditions: 150 °C, 5 MPa, 1200 r/min, catalysts of 1.96 g, ZrO_2 of 9.8 g, $ZnSO_4 \cdot 7H_2O$ of 49.2 g, benzene of 140 mL, H_2O of 280 mL

*benzene conversion C_{BZ}, selectivity of cyclohexene S_{HE} and yield Y_{HE} when $t = 15$ min; **the highest yield of cyclohexene Y_{HE} when $t = 25$ min, the following cases are the same

Table 9.14 Activity and selectivity of the monolayer-dispersed Ru–Zn catalyst stabilized by the orthosilicic acid without ZnSO4

t/min	$C_{BZ}/\%$	$S_{HE}/\%$	$Y_{HE}/\%$
5	18.5	86.0	15.9
10	31.5	86.6	27.3
15	43.1	83.8	36.1
20	51.6	81.1	41.8

Experimental conditions: 150 °C, 5 MPa, 1200 r/min, catalysts of 1.96 g, ZrO_2 of 9.8 g, benzene of 140 mL, H_2O of 280 mL. Time was recorded when benzene was added and samples were taken every 5 min

Table 9.15 Activity and selectivity of the monolayer-dispersed Ru–Fe catalyst stabilized by the orthosilicic acid

t/min	C_{BZ}/%	S_{HE}/%	Y_{HE}/%
5	17.7	87.0	15.4
10	31.6	84.4	26.7
15	50.3	81.6	41.0
20	65.2	79.4	51.8
25	75.0	76.8	57.6

Experimental conditions: 150 °C, 5 MPa, 1200 r/min, catalysts of 1.96 g, ZrO_2 of 9.8 g, $ZnSO_4 \cdot 7H_2O$ of 49.2 g, benzene of 140 mL, H_2O of 280 mL. Time was recorded when benzene was added and samples were taken every 5 min

As shown in Table 9.15, for the monolayer-dispersed Ru–Fe catalyst stabilized by the orthosilicic acid, when benzene conversion was 65.2% at $t = 20$ min, the selectivity and yield of cyclohexene were 79.4% and 51.8%, respectively. When benzene conversion was 75.0% at $t = 25$ min, the selectivity and yield of cyclohexene were 76.8% and 57.6%, respectively.

(3) Example 3 of catalyst preparation

A catalyst was prepared by replacing M with Mn in step (2) in example 1, while all the other procedures in example 1 were kept identical. The XRD pattern showed that when Mn/Ru atomic ratio was 15%, the basic salts were monolayer-dispersed on the Ru surface.

Table 9.16 shows the activity and selectivity of the monolayer-dispersed Ru–Mn catalyst stabilized by the orthosilicic acid.

As shown in Table 9.16, for the monolayer-dispersed Ru–Mn catalyst stabilized by the orthosilicic acid, when benzene conversion was 70.3% at $t = 20$ min, the selectivity and yield of cyclohexene were 76.2% and 53.6%, respectively. When benzene conversion was 78.9% at $t = 25$ min, the selectivity and yield of cyclohexene were 72.8% and 57.4%, respectively.

(4) Example 4 of catalyst preparation

Table 9.16 Activity and selectivity of the monolayer-dispersed Ru–Mn catalyst stabilized by the orthosilicic acid

t/min	C_{BZ}/%	S_{HE}/%	Y_{HE}/%
5	25.4	88.6	22.5
10	46.8	84.7	39.6
15	64.4	80.7	52.0
20	70.3	76.2	53.6
25	78.9	72.8	57.4

Experimental conditions: 150 °C, 5 MPa, 1200 r/min, catalysts of 1.96 g, ZrO_2 of 9.8 g, $ZnSO_4 \cdot 7H_2O$ of 49.2 g, benzene of 140 mL, H_2O of 280 mL. Time was recorded when benzene was added and samples were taken every 5 min

A catalyst was prepared by replacing M with Ce in step (2) in example 1, while all the other procedures in example 1 were kept identical. The XRD pattern showed that when Ce/Ru atomic ratio was 20%, the basic salts were monolayer-dispersed on Ru surface.

Table 9.17 shows the activity and selectivity of benzene hydrogenation over the monolayer-dispersed Ru–Ce catalyst stabilized by the orthosilicic acid.

As shown in Table 9.17, for the monolayer-dispersed Ru–Ce catalyst stabilized by the orthosilicic acid, when benzene conversion was 57.6% at t = 20 min, the selectivity and yield of cyclohexene were 82.1% and 47.3%, respectively. When benzene conversion was 61.8% at t = 25 min, the selectivity and yield of cyclohexene were 80.9% and 50.0%, respectively.

(5) Example 5 of catalyst preparation

A catalyst was prepared by replacing M with La in step (2) in example 1, while all the other procedures in example 1 were kept identical. The XRD pattern showed that when the atomic ratio of La/Ru was 15%, the basic sulfate salts were monolayer-dispersed on Ru surface.

Table 9.18 shows the activity and selectivity of benzene selective hydrogenation over the monolayer-dispersed Ru–La catalyst stabilized by the orthosilicic acid.

Table 9.17 Activity and selectivity of benzene hydrogenation over the monolayer-dispersed Ru–Ce catalyst stabilized by the orthosilicic acid

t/min	C_{BZ}/%	S_{HE}/%	Y_{HE}/%
5	25.0	82.8	20.7
10	40.0	83.7	33.5
15	51.3	82.0	42.1
20	57.6	82.1	47.3
25	61.8	80.9	50.0

Experimental conditions: 150 °C, 5 MPa, 1200 r/min, catalysts of 1.96 g, ZrO_2 of 9.8 g, $ZnSO_4 \cdot 7H_2O$ of 49.2 g, benzene of 140 mL, H_2O of 280 mL. Time was recorded when benzene was added and samples were taken every 5 min

Table 9.18 Activity and selectivity of benzene selective hydrogenation over the monolayer-dispersed Ru–La catalyst stabilized by the orthosilicic acid

t/min	C_{BZ}/%	S_{HE}/%	Y_{HE}/%
5	26.6	82.6	22.0
10	43.4	80.6	35.0
15	58.5	77.3	45.2
20	71.6	74.9	53.6
25	81.4	71.7	58.4

Experimental conditions: 150 °C, 5 MPa, 1200 r/min, catalysts of 1.96 g, ZrO_2 of 9.8 g, $ZnSO_4 \cdot 7H_2O$ of 49.2 g, benzene of 140 mL, H_2O of 280 mL. Time was recorded when benzene was added and samples were taken every 5 min

As shown in Table 9.18, for the monolayer-dispersed Ru–La catalyst stabilized by the orthosilicic acid, when benzene conversion was 71.6% at $t = 20$ min, the selectivity and yield of cyclohexene were 74.9% and 53.6%, respectively. When benzene conversion was 81.4% at $t = 25$ min, the selectivity and yield of cyclohexene were 71.7% and 58.4%, respectively.

9.3.2 *Catalytic System Containing Nanosized Ru Catalyst and Basic Zinc Sulfate and Its Application for Selective Hydrogenation of Benzene to Cyclohexene [2]*

The catalytic system contains nanosized Ru catalyst, basic Zinc sulfate, $ZnSO_4$, and H_2O. Among them, the mass ratio of the nanosized Ru catalyst:basic Zinc sulfate:$ZnSO_4$:H_2O is 1:(0.2–0.5):(13–17):140.

$ZnSO_4$ could be included in this catalytic system, or water only. For the system also containing $ZnSO_4$, the mass ratio of the nanosized Ru catalyst:basic Zinc sulfate:$ZnSO_4$:H_2O is 1:(0.2–0.5):(13–17):140. For the system containing water only, the mass ratio of the nano-Ru catalyst:basic Zinc sulfate:H_2O is 1:(0.2–5):(100–200).

Soluble $RuCl_3 \cdot 3H_2O$ (or $Ru(CH_3COO)_3$), one kind of alkaline among NaOH, Na_2CO_3, $NaHCO_3$, $NH_3 \cdot H_2O$ (or amines), and one kind or a mixture of surfactants polyethylene glycol, amines, acacia, polyacrylic acid, polyvinyl alcohol, and polyvinylpyrrolidone were used to produce a highly dispersed product of Ru hydroxide or hydrated oxide, and then the nanosized Ru microcrystals were obtained by hydrogen *in situ* reduction.

(1) Catalyst preparation using $RuCl_3 \cdot 3H_2O$, NaOH and polyvinyl alcohol-1750

(1) $RuCl_3$ solution (70–90 °C, 0.5–3 mol/L) was added into a container firstly. At the same time, NaOH solution (5–10 mol/L) and polyvinyl alcohol solution were mixed evenly in another container. Then the two kinds of solutions were controlled to flow into the stirred reactor at the same flow rate. After the completion of the reaction, stirring was continued for 30 min at 80 °C. Then it was cooled down to room temperature.
(2) The reaction mixture was transferred to a high-pressure reactor lined with polytetrafluoroethylene firstly, then it was reduced for 3 h (hydrogen pressure of 5 MPa, 150 °C, 800–1200 r/min). After the completion of the reaction, it was cooled down to room temperature, and black solids were obtained.
(3) The black solids were washed with distilled water to neutral. Then vacuum drying was applied to obtain Ru catalyst. XRD measurement showed the size of Ru microcrystals was 3–5 nm.

(2) Preparation of basic zinc sulfate by NaOH and excessive $ZnSO_4$

50.0 g $ZnSO_4 \cdot 7H_2O$ was dissolved into 200 mL water firstly, then under agitation, 15% NaOH solution (3.5–13.9 g NaOH was dissolved) was introduced into it quickly.

The mixture was heated, and the temperature of the mixture was kept at 80 °C for 1 h. Finally, it was cooled down to room temperature and rinsed with distilled water until no Zn^{2+}. XRD measurement showed the chemical formula of basic zinc sulfate was $(Zn(OH)_2)_3(ZnSO_4)(H_2O)_3$.

Table 9.19 shows the main performance indexes when the catalyst was prepared by different amounts of basic zinc sulfate and $ZnSO_4$.

For sample 1 in Table 9.19, in which only Ru catalyst was used, benzene was completely converted to cyclohexane at $t = 5$ min. As can be seen from samples 2, 3, 4, and 5, in the absence of $ZnSO_4$, the activity of the catalyst gradually decreased and the selectivity of cyclohexene gradually increased with the increase of the amount of basic salt. When the amount of basic salt was 0.75 g, the selectivity of cyclohexene was 87.1%, although benzene conversion was only 6.0%. As can be seen from the samples 6, 7, 8, and 9, Ru catalyst and the basic salt interacted with each other in $ZnSO_4$ slurry (0.6 mol/L). In these cases, the catalyst activity gradually decreased, while the selectivity of cyclohexene gradually increased. When the amount of basic salt was 0.75 g, benzene conversion was 68.2% at 20 min, and the selectivity and yield of cyclohexene were 77.2% and 52.7%, respectively. When benzene conversion was 76.2% at 25 min, the selectivity and yield of cyclohexene were 73.4% and 56.0%, respectively. When the amount of basic salt was 1 g, benzene conversion was 48.1%, and the selectivity and yield of cyclohexene were 83.6% and 40.3%, respectively. It can be seen from samples 10 and 11 that the catalyst activity increased and the cyclohexene selectivity decreased with the increase of $ZnSO_4$ concentration. While with the decrease of $ZnSO_4$ concentration, the catalyst activity decreased sharply and the selectivity of cyclohexene increased.

Table 9.19 Main catalyst performance indexes with different amounts of basic zinc sulfate and ZnSO4

Number	Basic zinc sulfate/g	$ZnSO_4$/(mol/L)	t/min	C_{BZ}/%	S_{HE}/%	Y_{HE}/%
1	0.0	0.0	5	100	0.0	0.0
2	0.5	0.0	25	5.1	78.8	4.0
3	0.75	0.0	25	6.0	87.1	5.2
4	1.0	0.0	25	4.9	92.1	4.5
5	6.0	0.0	25	6.6	87.6	5.8
6	0.0	0.6	5	70.7	46.7	33.0
7	0.5	0.6	15	91.4	57.8	52.8
8	0.75	0.6	20	68.2	77.2	52.7
			25	76.2	73.4	56.0
9	1.0	0.6	25	48.1	83.6	40.3
10	0.75	0.7	25	82.4	67.8	55.9
11	0.75	0.15	25	21.5	87.4	18.8

Reaction conditions: 150 °C, 5 MPa, 1200 r/min, catalyst of 1.96 g, ZrO_2 of 9.8 g, benzene of 140 mL, H_2O of 280 mL. Time was recorded when benzene was added and samples were taken every 5 min

9.3.3 Preparation, Modulation, and Regeneration Methods for the Catalyst for Selective Hydrogenation of Benzene to Cyclohexene [3]

The catalyst comprises active component Ru, rare earth element La, transition element Zn or Fe, and dispersant ZrO_2. The mass fractions of the components were as follows: Ru was 10–15% of ZrO_2, and La, Zn or Fe were 1–5% of Ru. The active component Ru, the rare earth element La, or the transition element Zn, Fe were in the form of metal oxide or hydroxide, and the size of dispersant ZrO_2 was at the micron level. The catalysts were prepared by chemical reduction method or precipitation method. Firstly, $RuCl_3{\cdot}3H_2O + LaCl_3{\cdot}3H_2O$ solution or $RuCl_3{\cdot}3H_2O + FeSO_4{\cdot}7H_2O$ solution or $RuCl_3{\cdot}3H_2O + ZnSO_4{\cdot}7H_2O$ solution was prepared, and then the reducing agent $NaBH_4$ was added into it. During the reduction, the colored solution changed to colorless. Then the mixture was set aside for stratification, where the supernatant was poured out and the lower layer containing solids was reserved. For selective hydrogenation of benzene, the catalyst showed good activity and cyclohexene selectivity, with adjustable modulation and regeneration. The activity and selectivity of the catalyst could be adjusted by changing the pH of the reaction system. The pH of the reaction slurry was kept at 4.0–5.0 by adding dilute H_2SO_4 when pH was high and adding $Zn(OH)_2$ when pH was low. When catalyst deactivation was caused by the adsorption of ions from the slurry, catalyst activity, and selectivity could be restored by washing away the adsorbed organic compounds by hydrogen peroxide aqueous solution (1–3%) firstly and then 0.3–1 mol/L hydrochloric acid or sulfuric acid solution.

9.3.4 Catalyst for Selective Hydrogenation of Benzene to Cyclohexene and Its Preparation Method [4]

(1) One part of the active ingredient precursor $RuCl_3{\cdot}H_2O$ was dissolved by pure water at room temperature firstly, then 0.01–0.1 part of the additives $FeCl_2$ or $FeSO_4{\cdot}7H_2O$ were added and dissolved under agitation to form solution 1.
(2) 5–10 parts of dispersants ZrO_2 were added in solution 1. Stirring was kept for 0.5–2 h to adsorb the active component and the auxiliary agent precursor onto the dispersant to form solution 2.
(3) 0.1–1 part of reducing agent $NaBH_4$ or $AlLiH_4$ or HCOH was added to solution 2 under agitation. Then filtration, washing, and drying were applied to obtain the catalysts whose active component had small particle size and high-level dispersion. After pretreatment of 22 h using 0.8 g catalyst, 14.4 g $ZnSO_4{\cdot}7H_2O$, 100 mL water under the condition of 140 °C, 5 MPa, 1000 r/min, the reaction was initiated when 58 mL benzene was added. In this case, benzene conversion was 40%, and the selectivity of cyclohexene was 85%.

9.3.5 *Ru–Y@Ni Catalyst for Selective Hydrogenation of Benzene to Cyclohexene and Its Application [5]*

Ru–Y@Ni catalyst has a shell-core structure, where the central core layer is composed of metal Ni, and the shell layer is composed of metal Ru and Y_2O_3. The catalyst showed high activity and cyclohexene selectivity at the presence of $ZnSO_4$, compensating for the drawbacks of the industrial no-load Ru–Zn catalyst with high cost due to the use of a large amount of noble metals. The catalyst uses active Ni as a carrier instead of inert oxide to achieve a combination of highly hydrogenated active metal Ni catalyst and high cyclohexene selective Ru catalyst. Ru–Y@Ni catalysts exhibit high activity and high cyclohexene selectivity. The catalyst preparation method is as follows:

400 mL $Ni(NO_3)_2$ solution (0.85 mol/L) was prepared firstly, then its pH was adjusted to 6–7 using ammonia. After that, 2 g of polyethylene glycol-10000 was added into it, and then 200 mL $NaBH_4$ solution (8.5 mol/L) was added slowly under agitation. Subsequently, stirring was kept for 30 min before the black precipitates were filtrated out. The black precipitates were washed many times until the supernatant was neutral, and the carrier nanosized metal Ni was obtained. A mixture of 400 mL $RuCl_3$ solution (0.05 mol/L) and $Y(NO_3)_3$ solution (0.003 mol/L) was prepared firstly, then 200 mL NaOH solution (0.2 mol/L) was added. After precipitation, it was stirred at 80 °C for 30 min. The resulting black precipitates were added into the reflow stirring device for heating, and then 0.8 g polyethylene glycol-10000 and 0.01 mol/L HCl were added until the mixture became homogeneous and transparent sol. 20 g support metal Ni was added into the sol solution. Water was steamed out using a rotary evaporation device. The black solids were washed with distilled water to neutral. The resulting solids and 400 mL distilled water were mixed into the reactor, and reduced at 140 °C for 3 h (5 MPa, 800 r/min). The resulting solids were Ru–Y(0.06)@Ni catalyst (0.06 represents the theoretical atomic ratio of Y and Ru), and the size of the catalyst microcrystals was about 5 nm.

Table 9.20 shows the activity and selectivity of Ru–Y(0.06)@Ni, Ru–Y(0.12)@Ni and Ru–Y(0.06)@ZrO_2 catalysts.

Table 9.20 Activity and selectivity of Ru–Y(0.06)@Ni, Ru–Y(0.12)@Ni and Ru–Y(0.06)@ZrO2 catalysts

Catalyst	t/min	C_{BZ}/%	S_{HE}/%	Y_{HE}/%
Ru–Y(0.06)@Ni	10	70.15	80.29	56.32
Ru–Y(0.12)@Ni	10	60.86	83.39	50.74
Ru–Y(0.06)@ZrO_2	10	60.73	78.34	47.58

Reaction conditions: 150 °C, 5 MPa, 1200 r/min, catalyst of 1.96 g, ZrO_2 of 9.8 g, $ZnSO_4·7H_2O$ of 49.2 g, benzene of 140 mL, H_2O of 280 mL. Time was recorded when benzene was added and samples were taken every 10 min

Table 9.21 Activity and selectivity of Ru–Y(0.06)@Ni catalyst when it was reused for 5 times

Reuse time	t/min	C_{BZ}/%	S_{HE}/%	Y_{HE}/%
1	10	70.15	80.29	56.32
2	10	71.43	79.29	56.64
3	10	72.52	78.12	56.65
4	10	69.81	82.61	57.86
5	10	71.56	79.09	56.60

Reaction conditions: 150 °C, 5 MPa, 1200 r/min, catalyst of 1.96 g, ZrO_2 of 9.8 g, $ZnSO_4 \cdot 7H_2O$ of 49.2 g, benzene of 140 mL, H_2O of 280 mL

As can be seen from Table 9.20, for Ru–Y(0.06)@Ni catalyst, when benzene conversion was 70.15% at 10 min, the selectivity and yield of cyclohexene were 80.29% and 56.32%, respectively, better than Ru–Y(0.12)@Ni.

Table 9.21 shows the activity and selectivity of Ru–Y(0.06)@Ni catalyst when it was reused for 5 times.

As can be seen from Table 9.21, the change of benzene conversion, cyclohexene selectivity, and cyclohexene yield was very slightly, indicating the catalyst had good repeatability and good thermal stability.

9.3.6 Supported Catalyst for Selective Hydrogenation of Benzene to Cyclohexene and Its Preparation Method [6]

The active components of the catalyst are composed of a noble metal Ru, a nonmetallic B, a metal or a metal oxide modifier M, with zirconia alumina mixed oxides support as the base. The active component consists of 0.5–10% (mass fraction) of Ru, 0.5–3% (mass fraction) of M, and 0.1–5% (mass fraction) of B. The Zr–Al–O complex oxide support with a large specific surface area, pore volume, and appropriate pore diameter is prepared by precipitation method and modified by surfactant, and then the precursor of the catalyst active component is reduced to obtain the catalyst with the active component of small particle size and high dispersion. The cyclohexene selectivity is 70–85% at the benzene conversion of 30–60%.

(1) Example 1 of catalyst preparation

Preparation of Zr–Al–O complex oxide support: 296 g of $Al(NO_3)_3 \cdot 9H_2O$ and 51 g $ZrOCl_2 \cdot 8H_2O$ are formulated as a deionized aqueous solution with a total molar concentration of 0.25 mol/L. 25% ammonia solution and carbonated hinge are used as raw materials to prepare the alkaline precipitant, in which the mass ratio of ammonia to carbonic acid is 2.5:1, and 4% (mass fraction) of PEG-600 modifier is added. Under stirring conditions, the prepared mixture solution of Al, Zr salts is added to the precipitant, or the precipitant is added to the mixture solution. With Stirring for

12 h, the reaction ended at pH = 9.1, then a solid was obtained by filter after aging for 10 h. The solid was washed with deionized water until the detergent was neutral, followed by washing with dehydrated ethanol for 2–3 times, and then the precursor of complex oxide support is obtained. Samples are dried at 110 °C for 12 h and then calcined in a muffle furnace at 550 °C for 6 h to obtain 74 g of Zr–Al–O complex oxide support.

0.18 g of $RuCl_3{\cdot}3H_2O$, 0.06 g of $Mn(NO_3)_2{\cdot}4H_2O$, and 0.09 g of $Fe(NO_3)_3{\cdot}9H_2O$ are dissolved in 20 mL of deionized water and formulated as a solution, 6.0 g of the Zr–Al–O complex oxide support is added and slowly stirred for 40 min. Afterward, under the rapid stirring conditions, 6 mL of sodium borohydride and sodium hydroxide solutions with a concentration of 1.0 mol/L and 0.5 mol/L, respectively, are added dropwise at room temperature. After completion of the dropwise addition, stirring is continued for 1 h to obtain the precursor mixture of the catalyst. The mixture is allowed to stand at room temperature for 4 h and filtered to obtain a solid. The solid is then washed with deionized water until neutral, followed by washing with anhydrous ethanol for 2–3 times. Samples are dried at 110 °C for 1 h, and calcined at 550 °C for 6 h in a muffle furnace to obtain the example 1.

Under the conditions of 120 °C, 2.0 MPa, 1000 r/min, 15.0 g catalyst, 2.5 g accelerator 18-crown-6, 80 mL benzene, and 50 mL 0.2 mol/L $ZnSO_4$, the conversion of benzene is 54% at 20 min, and the cyclohexene selectivity is 80%.

(2) Example 2 of catalyst preparation

Preparation of Zr–Al–O complex oxide support: 296 g of $Al(NO_3)_3{\cdot}9H_2O$ and 51 g $ZrOCl_2{\cdot}8H_2O$ are formulated as a deionized aqueous solution with a total molar concentration of 0.1 mol/L. 25% ammonia solution and carbonated hinge are used as raw materials to prepare the alkaline precipitant, in which the mass ratio of ammonia to carbonic acid is 2:1, and 1% (mass fraction) of PEG-600 modifier is added. Under stirring conditions, the prepared mixture solution of Al and Zr salts is added to the precipitant, or the precipitant is added to the mixture solution. A solid is obtained by stirring Stir for 7 h, reacting until pH = 9.1, filtering after aging for 20 h. The solid is washed with deionized water until neutral, then washed with anhydrous ethanol for 2–3 times, and then the precursor of complex oxide support is obtained. Samples are dried at 100 °C for 20 h and then calcined in a muffle furnace at 400 °C for 10 h to obtain 74 g of Zr–Al–O complex oxide support.

0.18 g of $RuCl_3{\cdot}3H_2O$, 0.06 g of $Mn(NO_3)_2{\cdot}4H_2O$, and 0.09 g of $Fe(NO_3)_3{\cdot}9H_2O$ are dissolved in 20 mL of deionized water and formulated as a solution, 6.0 g of the Zr–Al–O complex oxide support is added and slowly stirred for 5 h. Afterward, under the rapid stirring conditions, 6 mL of sodium borohydride and sodium hydroxide solutions with a concentration of 1.0 mol/L and 0.5 mol/L, respectively, are added dropwise at room temperature. After completion of the dropwise addition, stirring is continued for 5 h to obtain the precursor mixture of the catalyst. The mixture is allowed to stand at room temperature for 20 h and filtered to obtain a solid. The solid is washed with deionized water until neutral, followed by washing with anhydrous ethanol for 2–3 times. Samples are dried at 100 °C for 20 h, and calcined at 400 °C for 10 h in a muffle furnace to obtain the Catalyst 2.

Under the conditions of 80 °C, 1.0 MPa, 800 r/min, 15.0 g catalyst, 2.5 g accelerator 18-crown-6, 57 mL benzene, and 50 mL 0.2 mol/L $ZnSO_4$, the conversion of benzene is 60% at 40 min, and the cyclohexene selectivity is 85%.

(3) Example 3 of Catalyst Preparation

Preparation of Zr–Al–O complex oxide support: 296 g of $Al(NO_3)_3 \cdot 9H_2O$ and 51 g $ZrOCl_2 \cdot 8H_2O$ are formulated as a deionized aqueous solution with a total molar concentration of 0.3 mol/L. 25% ammonia solution and carbonated hinge are used as raw materials to prepare the alkaline precipitant, in which the mass ratio of ammonia to carbonic acid is 2:1, and 5% (mass fraction) of PEG-600 modifier is added. Under stirring conditions, the prepared mixture solution of Al and Zr salts is added to the precipitant, or the precipitant is added to the mixture solution. A solid is obtained by stirring for 12 h, reacting until pH = 9.1, and filtering after aging for 15 h. The solid is washed with deionized water until neutral, then washed with anhydrous ethanol for 2–3 times, and then the precursor of complex oxide support is obtained. Samples are dried at 110 °C for 10 h and then calcined in a muffle furnace at 600 °C for 4 h to obtain 74 g of Zr–Al–O complex oxide support.

0.18 g of $RuCl_3 \cdot 3H_2O$, 0.06 g of $Mn(NO_3)_2 \cdot 4H_2O$, and 0.09 g of $Fe(NO_3)_3 \cdot 9H_2O$ are dissolved in 20 mL of deionized water and formulated as a solution, 6.0 g of the Zr–Al–O complex oxide support is added and slowly stirred for 3 h. Afterward, under the rapid stirring conditions, 6 mL of sodium borohydride and sodium hydroxide solutions with a concentration of 1.0 mol/L and 0.5 mol/L, respectively, are added dropwise at room temperature. After completion of the dropwise addition, stirring is continued for 1 h to obtain the precursor mixture of the catalyst. The mixture is allowed to stand at room temperature for 10 h and filtered to obtain a solid. The solid is washed with deionized water until neutral, followed by washing with anhydrous ethanol for 2–3 times. Samples are dried at 110 °C for 10 h, and calcined at 600 °C for 4 h in a muffle furnace to obtain the Catalyst 3.

Under the conditions of 180 °C, 10 MPa, 1500 r/min, 15.0 g catalyst, 2.5 g accelerator 18-crown-6, 54 mL benzene, and 80 mL 0.1 mol/L $ZnSO_4$, the conversion of benzene is 63% at 35 min, and the cyclohexene selectivity is 75%.

9.3.7 *Adsorbent for Finely Removing Sulfides in Benzene as Well as Its Preparation Method and Application [7]*

The sulfur poisoning of Ru catalyst for benzene-selective hydrogenation is irreversible, control of S content in the raw material benzene to ppb level, is the fundamental way to avoid catalyst poisoning. The acidic ionic liquid [BMIM]Cl–$AlCl_3$ is supported on the ZSM-5 zeolite with the pore diameter of 5–30 nm, so that the content of acidic ionic liquid [BMIM]Cl–$AlCl_3$ is 5–30%. Finally, the adsorbent loaded with ionic liquid is obtained, which can remove the sulfides in the petrochemical benzene from the level of ppm to ppb.

(1) Preparation of three kinds of ionic liquids

Anhydrous *N*-methylimidazole and n-n-butane are mixed at a molar ratio of 1:1 and then added to a toluene solvent, reacting for 48 h under the reflux conditions of 80 °C. The product is recrystallized with acetonitrile and ethanol until a white crystal is obtained. After vacuum drying at 95 °C for 12 h, the intermediate 1-Butyl-3-methylimidazolium chloride ([BMIM]Cl) is obtained. In hydrogen protection and mechanical stirring, anhydrous $AlCl_3$ is slowly added to a certain amount of [BMIM]Cl, to synthesize the ionic liquid [BMIM]Cl–$AlCl_3$ with a molar ratio of $AlCl_3$ and [BMIM]Cl being 0.67, 1.0, 2.0, respectively. The ionic liquids are then reserved in a dryer.

(2) Preparation of ZSM-5 carriers with four different Si/Al ratios

Using tetrapropyl ammonium bromide (TPABr) as the templating agent, and NaOH to adjust the alkalinity of solution, $Al_2O_3 \cdot 18H_2O$, and 25% silica sol as the aluminum source and silicon source, respectively, carriers of zeolite with the following four different Si/A1 ratios are prepared with the molar ratio of Na_2O:SiO_2:$A1_2O_3$:TPABr:H_2O of (4–8):(30–80):1:(2–10):(2000–4000).

Carrier 1: 2.5 g NaOH, 7.2 g TPABr, and 3.0 g $Al_2O_3 \cdot 18H_2O$ are dissolved in 200 mL deionized water, then 43.0 g of 25% silica sol is added and stirred well for about 1 h. The resulting white gel is then transferred to stainless steel hydrothermal reaction still, and crystallized for 36 h at 80 °C. The reaction kettle is water cooled to room temperature, the crystallized product is filtered and washed until the filtrate is neutral. Finally, the product is dried at 115 °C for 20 h, calcined at 450 °C for 4 h in muffle furnace, and reserved in a drier.

Carrier 2: 3.0 g NaOH, 10.8 g TPABr, and 2.7 g $Al_2O_3 \cdot 18H_2O$ are dissolved in 150 mL deionized water, then 58.0 g of 25% silica sol is added and stirred well for about 1 h. The resulting white gel is then transferred to stainless steel hydrothermal reaction still, and crystallized for 10 h at 180 °C. The reaction kettle is water cooled to room temperature, the crystallized product is filtered and washed until the filtrate is neutral. Finally, the product is dried at 115 °C for 20 h, calcined at 300 °C for 6 h in muffle furnace, and reserved in a drier.

Carrier 3: 2.0 g NaOH, 5.8 g TPABr, and 2.4 g $Al_2O_3 \cdot 18H_2O$ are dissolved in 200 mL deionized water, then 52.0 g of 25% silica sol is added and stirred well for about 1 h. The resulting white gel is then transferred to stainless steel hydrothermal reaction still, and crystallized for 40 h at 100 °C. The reaction kettle is water cooled to room temperature, the crystallized product is filtered and washed until the filtrate is neutral. Finally, the product is dried at 115 °C for 20 h, calcined at 600 °C for 2 h in muffle furnace, and reserved in a drier.

Carrier 4: 3.5 g NaOH, 16.0 g TPABr, and 5.2 g $Al_2O_3 \cdot 18H_2O$ are dissolved in 400 mL deionized water, then 73.5 g of 25% silica sol is added and stirred well for about 1 h. The resulting white gel is then transferred to stainless steel hydrothermal reaction still, and crystallized for 28 h at 140 °C. The reaction kettle is water cooled to room temperature, the crystallized product is filtered and washed until the filtrate is neutral. Finally, the product is dried at 115 °C for 20 h, calcined at 450 °C for 4 h in muffle furnace, and reserved in a drier.

Three above-mentioned acidic ionic liquids [BMIM]Cl–$AlCl_3$ are loaded on the above four carriers, to obtain 12 kinds of absorbents loaded with ionic liquid.

(3) Preparation of absorbent

The absorbent 1–3 loaded with carrier 1:

0.5 g acidic ionic liquids [BMIM]Cl–$AlCl_3$ with $AlCl_3$/[BMIM]Cl molar ratios of 0.67, 1.0, and 2.0, respectively, are dissolved in 15 mL benzene, and 3.0 g carrier 1 is immersed in the solution. Under the conditions of mechanical stirring, a constant temperature of 50 °C and reduced pressure, the ionic liquid is uniformly adsorbed on carrier 1. After benzene is totally volatilized, the system is heated up to 80 °C, and the vacuum of the system is simultaneously increased, lasting for 4–6 h. The adsorbents 1–3 are finally obtained after drying for 12 h at 120 °C under the protection of hydrogen.

The absorbent 4–6 loaded with carrier 2:

0.5 g acidic ionic liquids [BMIM]Cl–$AlCl_3$ with $AlCl_3$/[BMIM]Cl molar ratios of 0.67, 1.0, and 2.0, respectively, are dissolved in 15 mL benzene, and 3.0 g carrier 2 is immersed in the solution. Under the conditions of mechanical stirring, a constant temperature of 50 °C and reduced pressure, the ionic liquid is uniformly adsorbed on the carrier 2. After benzene is totally volatilized, the system is heated up to 80 °C, and the vacuum of the system is simultaneously increased, lasting for 4–6 h. The adsorbents 4–6 are finally obtained after drying for 12 h at 120 °C under the protection of hydrogen.

The absorbent 7–9 loaded with carrier 3:

0.5 g acidic ionic liquids [BMIM]Cl–$AlCl_3$ with $AlCl_3$/[BMIM]Cl molar ratios of 0.67, 1.0, and 2.0, respectively, are dissolved in 15 mL benzene, and 3.0 g carrier 3 is immersed in the solution. Under the conditions of mechanical stirring, a constant temperature of 50 °C and reduced pressure, the ionic liquid is uniformly adsorbed on the carrier 3. After benzene is totally volatilized, the system is heated up to 80 °C and the vacuum of the system is increased simultaneously, lasting for 4–6 h. The adsorbents 7–9 are finally obtained after drying for 12 h at 120 °C under the protection of hydrogen.

The absorbent 10–12 loaded with carrier 4:

0.5 g acidic ionic liquids [BMIM]Cl–$AlCl_3$ with $AlCl_3$/[BMIM]Cl molar ratios of 0.67, 1.0, and 2.0, respectively, are dissolved in 15 mL benzene, and 3.0 g carrier 4 is immersed in the solution. Under the conditions of mechanical stirring, a constant temperature of 50 °C and reduced pressure, the ionic liquid is uniformly adsorbed on the carrier 4. After benzene is totally volatilized, the system is heated up to 80 °C and the vacuum of the system is simultaneously increased, lasting for 4–6 h. The adsorbents 10–12 are finally obtained after drying for 12 h at 120 °C under the protection of hydrogen.

(4) Removing sulfides from benzene by using absorbents 1–12

Under the conditions of room temperature, atmospheric pressure, liquid space velocity 5.0 h^{-1}, and fixed bed, adsorbents 1–12 are used, respectively, to finely remove sulfides in benzene. The total sulfur content is determined by microcoulometric method using petroleum benzene as raw material. Results show that when the total sulfide content in the benzene feedstock is 14 ppm, the sulfide is finely removed by the adsorbent 1 to 12, without sulfides being detected, indicating that the content is reduced to ppb level. The desulfurization by adsorbent can be carried out at a temperature of about 50 °C, and can also be carried out in the pressurized state. Both can obtain a good desulfurization effect at a liquid space velocity of 2–30 h^{-1}, and sulfide is not detectable.

9.3.8 A Catalyst for Selective Hydrogenation of Benzene to Cyclohexene as Well as Its Preparation Method and Application [8]

The surface modification plays an important role in the properties of the catalyst. In the preparation process of the catalyst, various measures are taken to modify the surface of the catalyst, and the procedure is complex. The combination of catalyst preparation with industrial applications, and the *in situ* surface modification of the catalyst by the interaction of the catalyst with the hydrogenation slurry simplifies the catalyst preparation procedures, and improves the cyclohexene selectivity and yield.

$RuCl_3 \cdot 3H_2O$ is used as the precursor of active component Ru, and zinc salt [$ZnSO_4 \cdot 7H_2O$, $ZnCl_2$, $5ZnO \cdot 2CO_2 \cdot 4H_2O$, Zn powder, ZnO or $Zn(Ac)_2$] is used as the precursor of Zn additive, with alkali metal NaOH or KOH acting as a precipitant.

(1) Material preparation: Salt and lye are prepared, respectively, at room temperature. The salt solution is an aqueous solution containing the active component Ru and the precursor of Zn promoter, and alkaline solution is an aqueous solution of NaOH or KOH. The mass fraction (w_B) of Zn and Ru in salt solution is 1–5%, while the mass fraction (w_B) of NaOH or KOH in the lye is 20–30%, and the volume ratio of salt and lye is 1:1.
(2) Precipitation reaction: The precipitation reaction between salt solution and alkaline solution occurs by using parallel flow at room temperature, after complete mixing, continue to stir for 0.5–1 h, detect the pH value of material to ensure it is over 12. Then, transfer the material to the autoclave.
(3) Reduction reaction: At 130–150 °C, with hydrogen pressure of 4–5 MPa, and stirring speed greater than 4 m/s, Ru, Zn hydroxides in the material are reduced for 3–6 h.
(4) Precipitation washing: The material quickly shows layers after the reduction, with solid being settled. The supernatant is removed and the sediment is washed with high purity water or deionized water to neutral, and no chloride ions are detected.
(5) Preparation: Prepare 0.05–0.1 mol/L aqueous solution of NaOH or KOH, the catalyst after washing is kept in the alkaline solution, in which the mass fraction (w_B) of solid is 4–10%.

(6) Usage: The catalyst together with the alkaline solution is added to the hydrogenation reactor, which contains 0.4–0.6 mol/L aqueous solution of zinc sulfate. The mixture is stirred for 4–6 h at room temperature, and catalyst runs for 10–16 h at the temperature of 140–150 °C, hydrogen pressure of 4–5 MPa. Afterward, the hydrogenation reaction starts after adding benzene, regular sampling, testing the catalyst activity and cyclohexene selectivity, calculating the cyclohexene yield.

9.3.9 Modulation Methods for the Activity and Selectivity of Ru–Zn Catalyst for Selective Hydrogenation of Benzene to Cyclohexene [9]

In view of the commonly used Ru–Zn catalysts in industry, two simple and effective methods are proposed.

When benzene conversion is higher than 40%, while the cyclohexene selectivity is less than 80%, after the addition of NaOH or basic zinc sulfate salt (BZSS), the benzene conversion is reduced to 40%, cyclohexene selectivity increased to over 80%, and the main technical indicators of the catalyst back to normal levels. The basic zinc sulfate salt formed by the reaction of NaOH and $ZnSO_4$ in the slurry is chemisorbed on the catalyst surface, while the resulting sodium sulfate is present in the slurry, showing no negative impact on the system. The salt can selectively cover on Ru's strongest active sites to enhance the hydrophilicity of catalyst, inhibit the catalyst activity, and improve the cyclohexene selectivity. The key to using NaOH or BZSS modulation is to control the pH of the slurry at 5–7, with an optimum range of 5.5–6.5. The premise of using NaOH modulation is the presence of a large amount of zinc sulfate in the hydrogenation reaction slurry. By contrast, BZSS can be used with the presence of a large amount of zinc sulfate in the slurry, and it also can be used in the case of the slurry without zinc sulfate, which is more universal. The optimum amount of NaOH used for the modification is with the mass ratio of Ru–Zn catalyst to NaOH being 100:(6–8), while the optimum amount of BZSS is with the mass ratio of Ru–Zn catalyst to BZSS being 100:(19–22.5).

9.3.10 Production System and Preparation Method of the Catalyst for Selective Hydrogenation of Benzene to Cyclohexene [10]

The catalyst production system includes high troughs, dosing tanks, autoclaves, sinks, pipes, valves, etc. The high trough provides ingredients, and high purity water or deionized water used in autoclaves and washing process. It is made of 304 stainless steel, with a heated coil, and connected to the dosing tank, autoclave, and wash tank through piping and valves. Pipes and valves are all made with polytetrafluoroethylene or polypropylene plastics, to avoid the poisoning of catalyst caused by the introduction of Fe, Cr, Ni, and other impurity ions. The dosing tanks are used for the preparation of alkaline and salt solution, located below the high tank, water that is used for material preparation automatically inflows from the high tank under the gravity. Above the dosing, the tank is installed with stirring, and the bottom with a discharge port. The autoclave is used for the precipitation and reduction reaction, located below the ingredients tank. Materials can automatically flow in the autoclave, moreover, by using new self-priming autoclave and magnetic seal, the hydrogen pressure is automatically maintained constant. The autoclave is lined with polytetrafluoroethylene or HC276 Hastelloy, two leaves are fixed on the mixing shaft, to ensure the full contact of the gas, solid, and liquid. The working pressure and temperature are designed to be 5 MPa and 150 °C, respectively. There are exhaust pipe, liquid feed pipe, spare port on the above, and the temperature tube, exhaust pipe in the autoclave. The interior of the autoclave in contact with materials is fully lined with Hastelloy. The washing tank is located below the autoclave, with the reduced material automatically inflowing, to wash the reduced catalyst. There are a stir, water inlet on the above, and discharge mouth on the beneath, two drains on the side, so that the washing liquid can be automatically discharged. The catalyst preparation process includes proportioning, precipitation reaction, reduction reaction, washing, etc.

9.3.11 *An* in Situ *Regeneration Method of the Catalyst for Selective Hydrogenation of Benzene to Cyclohexene [11]*

For the nitride poisoning of catalyst for benzene-selective hydrogenation caused by extraction agent DMAC, an *in situ* regeneration method is proposed. The method only needs to add a chemical substance during the normal operation of the device, the inherent activity, and selectivity of the catalyst is completely restored through a chemical reaction. The regeneration method does not need to stop the device, it is time and labor saving, as well as resource saving, moreover, it shows good regeneration effect.

Acidic substances are used to neutralize the nitrides on the catalyst surface and re-release the active sites, thus the catalyst activity and cyclohexene selectivity are recovered. Among them, sulfuric acid is the best choice, because the neutralization product is zinc sulfate and water which are necessary for the mother liquor, without the introduction of any new ions and compounds, showing a little adverse effect on the catalytic system. The key is to determine the amount of sulfuric acid, the excessive amount will cause the increase of catalyst activity, as well as the decrease of cyclohexene selectivity, and vice versa the catalyst activity and selectivity cannot be fully recovered.

References

1. Liu, Z.Y., Liu, S.C., Sun, H.J., et al.: A monolayer-dispersed catalyst for selective hydrogenation of benzene to cyclohexene and its preparation method. Chinese patent, CN 10060765.5 (2012)
2. Liu, Z.Y., Liu, S.C., Li, Z.J., et al.: A catalytic system containing nano Ru catalyst and basic zinc sulfate salt and its mechanism for selective hydrogenation of benzene to cyclohexene. Chinese patent, CN 102553024 (2012)
3. Liu, S.C., Li, L.M., Wang, X.Y., et al.: A catalyst for selective hydrogenation of benzene to cyclohexene and its preparation, adjusting as well as regeneration method. Chinese patent, CN 100604510 (2004)
4. Liu, S.C., Li, L.M., Wang, X.Y., et al.: A catalyst for selective hydrogenation of benzene to cyclohexene and its preparation method. Chinese patent, CN 01122085 (2011)
5. Sun, H.J., Chen, L.X., Yuan, P., et al.: Ru-Y@Ni catalyst for selective hydrogenation of benzene to cyclohexene and its preparation method as well as usage. Chinese patent, CN 102046724 (2014)
6. Wang, Y.H., Zhang, Y.X., Tang, J.S., et al.: A supported catalyst for selective hydrogenation of benzene to cyclohexene and its preparation method. Chinese patent, CN 103941040 (2013)
7. Wang, Y.H., Zhang, Y.X., Wang, J.W., et al.: An absorbent of sulfides in refined benzene and its preparation method as well as usage. Chinese patent, CN 105257826 (2013)
8. Liu, Z.Y., Li, Z.J., Liu, S.C., et al.: A catalyst for selective hydrogenation of benzene to cyclohexene and its preparation method as well as usage. Chinese patent, CN 104816954 (2015)
9. Liu, Z.Y., Li, Z.J., Liu, S.C., et al.: An adjusting method of activity and selectivity of Ru-Zn catalyst for selective hydrogenation of benzene to cyclohexene. Chinese patent, CN 104799094 (2015)
10. Liu, Z.Y., Li, Z.J., Liu, S.C., et al.: Production system and preparation method of catalyst for selective hydrogenation of benzene to cyclohexene. Chinese patent, CN 104799840 (2015)
11. Liu, Z.Y., Li, Z.J., Liu, S.C., et al.: An in-situ regeneration method of Ru-Zn catalyst for selective hydrogenation of benzene to cyclohexene. Chinese patent, CN 10479707X (2015)

Chapter 10
Selective Hydrogenation of Benzene to Cyclohexene and Incorporate Device of Its Downstream Products

In 2010, the selective hydrogenation of benzene to cyclohexene and the catalytic technology of its downstream products were industrialized in China. In the patented invention patent, "A device of benzene selective hydrogenation" is conducive to control the reaction temperature and improve the reaction rate, and thus the catalyst has high selectivity and cyclohexene yield. "A reaction device and process of benzene selective hydrogenation to cyclohexene", including practical new-type patent of the reaction device, is smooth and easy to control, and shows higher benzene conversion and cyclohexene selectivity compared to the prior art. "An optimized benzene partial hydrogenation process with recyclable catalyst" reduces the loss of hydrogenation catalyst and additive zinc sulfate, and thus effectively reduces the production cost. "A gas-liquid-liquid-solid reaction device" achieves the separation and circulation of the catalyst and reaction solution in the reactor, enhances the catalytic effect, and improves the production capacity. "A production method of cyclohexene with high-purity benzene as raw material" changes the traditional idea that the separation of benzene, cyclohexane, and cyclohexene requires two-step extraction, instead, a three-component separation is achieved through a single extraction, and the energy consumption is thus greatly reduced. "A continuous production method of cyclohexene" ensures that the catalyst is always in a state with high reactivity and selectivity, and the long-period and continuous production of cyclohexene is achieved. "A production method of caprolactam with high-purity benzene as raw material" is resource saving and environment friendly. "A production method of cyclohexanone with a high yield" radically puts an end to the security risks, and the utilization rate of carbon is almost 100%. Moreover, neither wastewater nor exhaust gas is discharged/emitted. Compared with the traditional method with cyclohexane as raw material, the production cost per ton of cyclohexanone can be reduced by 1000–1500 RMB. In the field of this technology, the novel production method of cyclohexanone breaks the monopoly of foreign countries, so China has become the second country in the world achieving the industrialization of selective hydrogenation of benzene to cyclohexene and the catalytic technology of its downstream products, and has completely independent intellectual property.

Z. Liu et al., *Catalytic Technology for Selective Hydrogenation of Benzene to Cyclohexene*,
https://doi.org/10.1007/978-981-15-6411-6_10

10.1 The Production Technology of Cyclohexanone Through Benzene Selective Hydrogenation to Cyclohexene

In the years 1995 and 2005, the foreign catalytic technology of selective hydrogenation of benzene to cyclohexene was transferred twice to China. In 2009, the enterprise declared an invention patent of cyclohexane production method with coking benzene as a raw material [1].

Figure 10.1 shows a process flow chart for the production of cyclohexanone with coking benzene as a raw material.

In Fig. 10.1, the production method of cyclohexanone with coking benzene as a raw material includes **A** hydrogenation of coking benzene to cyclohexene, **B** extractive distillation, **C** hydration of cyclohexene to cyclohexanol, **D** separation and purification of cyclohexanol, **E** dehydrogenation of cyclohexanol to cyclohexanone, and **F** alcohol-ketone refining. The process is described as follows:

A. Hydrogenation of coking benzene to cyclohexene

(1) Pretreatment of coking benzene: In the preprocessor of benzene, the benzene flow rate is 13 t/h. Under the conditions of 150 °C and 0.8 MPa, by passing two sets of tandem fixed beds with the upper part being filled with 7.5 t activated bauxite, and the lower part with 3.5 t Pd catalyst, the thiophene and other sulfides, rust as well as dust in the raw materials of benzene are adsorbed and removed, to prevent the poisoning of the catalyst.

(2) Compression of H_2: Compress the hydrogen from outside to 5–7.5 MPa, and then send it into the hydrogenation reactor.

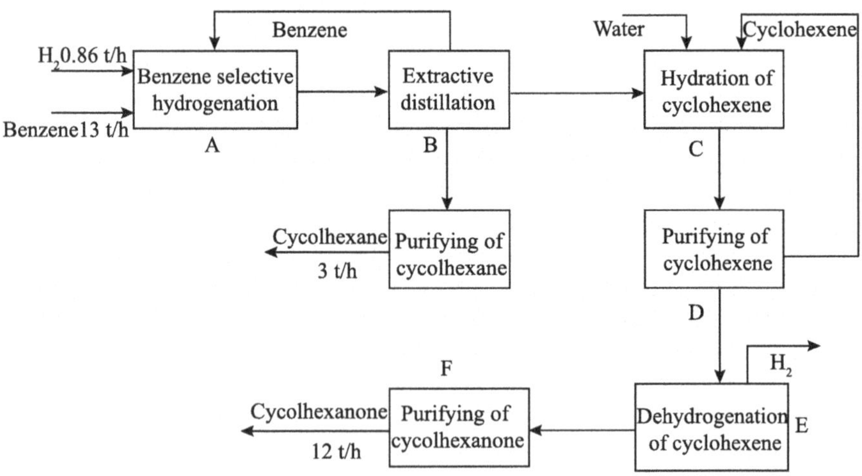

Fig. 10.1 Flow chart for the production of cyclohexanone with coking benzene as a raw material

(3) Hydrogenation reaction: Benzene reacts with compressed hydrogen according to the mass ratio of 100:6.6 by partial hydrogenation reaction to produce cyclohexene, meanwhile, generating by-product cyclohexane. The benzene partial hydrogenation takes place in the liquid phase containing Ru–Zn catalyst. In order to prevent the catalyst from poisoning by metal ions, HC-276 is used as the equipment substrate.
(4) Process flow: The pretreated benzene is pressurized to about 8 MPa in the feed pump of the hydrogenation reactor, and then fed to the bottom of the first hydrogenation reactor. Hydrogen from the compressor is fed to the bottom of the first and second hydrogenation reactors, respectively. The hydrogenation catalyst is a commercially available Ru–Zn catalyst from Asahi Kasei Chemicals Co., Ltd, with a usage of 3.5 g/t cyclohexanone, and the mass ratio of benzene to hydrogen being 100:6.6. The Ru–Zn catalyst slurry from the hydrogenation sedimentation tank is fed to the bottom of the first hydrogenation reactor by a slurry circulation pump. The benzene that has been sent to the first hydrogenation reactor is mixed with hydrogen and Ru–Zn catalyst slurry with a stirrer to react.

The reaction temperature is controlled between 120 and 160 °C by the cooling system of the reactor. The reaction pressure is then controlled in the range of 4.0–6.0 MPa by adjusting the reaction temperature and catalyst activity.

The resulting mixture of cyclohexene, cyclohexane, and catalyst slurry flows up to the top of the first hydrogenation reactor. After passing over the overflow weir, it is fed to the bottom of the second hydrogenation reactor under the force of gravity. In the second hydrogenation reactor, the same reaction occurs to produce a mixture of cyclohexene and cyclohexane, and the mixture is fed into the settling tank together with the catalyst slurry. The resulting cyclohexene, cyclohexane, and catalyst slurries are separated by settling. The separated cyclohexene and cyclohexane overflow to the posttreatment system. The catalyst slurry is withdrawn and returned to the first hydrogenation reactor. Finally, the cyclohexene selectivity of 77.7% is obtained at the benzene conversion of 51%. The increase in hydrogen adsorption results in the catalyst degradation during hydrogenation, and hydrogenation catalyst can be regenerated.

B. Extractive distillation of mixtures

(1) Separation/recovery of benzene: The hydrogenation product is a mixture of cyclohexene, cyclohexane, and unreacted benzene. With *N*, *N*-dimethylacetamide (DMAC) as extractant, feed ratio of the extractant and mixture being 86:26, the benzene is separated from the mixture by extractive distillation and it is purified and recycled after distillation.
(2) Separation/recovery of cyclohexene: The separated mixture of cyclohexene and cyclohexane is extracted with simultaneous distillation, with *N*, *N*-dimethylacetamide (DMAC) as extractant, feed ratio of the extractant and mixture being 111:15, thus cyclohexene and cyclohexane are separated.

C. Preparation of cyclohexanol by hydration of cyclohexene

(1) Cyclohexene washing system: Nitrides in the cyclohexene are removed, such as the extractant DMAC, by the use of high-purity water with a flow rate of 1 t/h, which can effectively prevent the poisoning of hydration catalyst and the formation of unqualified cyclohexanol.
(2) Cyclohexene hydration reaction: Catalyst for cyclohexene hydration is the crystalline aluminum silicate slurry, the initial capacity is 80 t, and the mixture is stirred and recycled in a hydration reactor followed by the addition of cyclohexene. Cyclohexene is hydrated to cyclohexanol, and meanwhile, some by-products are generated, such as methyl cyclopentene, methylcyclopentanol, and cyclohexyl.

Cyclohexene is heated to about 122 °C using cyclohexanol vapor in the feed preheater. The first hydration reactor is installed with a deflector so that the hydration catalyst slurry is sufficiently stirred and cycled by the agitator of the first hydration reactor. Cyclohexene is uniformly added to the circulating catalyst slurry in the form of oil droplets. Catalyst slurry flows down in the diversion bucket, and with an upward flow out of the barrel. By recirculating and stirring, the hydration catalyst slurry reacts with cyclohexene. The produced cyclohexanol and cyclohexene climb to the separation section of catalyst slurry (settling section) under the difference in the specific gravity of the catalyst slurry. The catalyst slurry descends and begins to recycle. The cyclohexanol and cyclohexene separated from the catalyst slurry are discharged through an overflow weir by gravity and then fed into the second hydration reactor.

The operation in the second hydration reactor is the same as that of the first hydration reactor, and the reaction is further carried out. The reaction temperature is maintained between 115 and 125 °C with the pressure of 0.5 MPa, to prevent cyclohexanol and cyclohexene from boiling in the reaction device.

Under the action of crystalline aluminum silicate catalyst, the hydration of cyclohexene occurs in the aqueous phase, with the reaction time of 30–40 min, and the main reaction generating cyclohexanol is a reversible exothermic reaction. The cyclohexanol conversion from cyclohexene increases with the increasing temperature, and meanwhile, the reverse reaction of cyclohexene from cyclohexanol is also accelerated.

When the reaction temperature is excessively increased, the reverse reaction will exceed the demand reaction, the cyclohexene conversion is reduced, and will also accelerate the side reactions, resulting in a decrease in the selectivity. The addition of organic compounds to the hydration catalyst will cause the catalyst deactivation, thus regeneration is needed. The conversion of cyclohexene is 9.0% and the selectivity of cyclohexanol is 99.0%.

D. Separation and purification of cyclohexanol

The cyclohexanol formed after the hydration reaction is separated; the concentration of cyclohexanol in the isolate is 12% (mass fraction), and it is 81% (mass fraction) for the concentration of cyclohexene.

With high-pressure steam of 3.3 MPa as a heat source, a mixture containing cyclohexanol and cyclohexene is gasified in a reboiler and then fed into the separation tower to separate cyclohexanol. After separation, the concentration of cyclohexanol is 72% (mass fraction), with a small amount of hydration catalyst solid remaining, thus, the further separation and purification are required. The medium pressure steam of 1.2 MPa is used as a heat source to evaporate the isolate. The cyclohexyl alcohol is collected at the top of the purifying column and cooled to below 40 °C at a concentration of 99% (mass fraction).

E. Preparation of cyclohexanone by dehydrogenation of cyclohexanol

The cyclohexanol steam in the dehydrogenation reactor is partially dehydrogenated to cyclohexanone and hydrogen under the effect of copper-silicon catalyst purchased from Germany BASF company. The reaction temperature is controlled in the range of 220–270 °C, with a space velocity of 0.7 h^{-1}.

The conversion of cyclohexanol in the dehydrogenation reaction is about 45–55%. In the initial stage of the dehydrogenation catalyst, the reaction temperature must be strictly controlled at 220 °C, and maintained for no less than 3 months. The activity of the dehydrogenation catalyst will gradually decline over time. In practice, the decrease of catalyst activity can be compromised by progressively increasing the reaction temperature.

The dehydrogenation catalyst is designed to operate for 2 years. In the later stage, since a high operating temperature must be maintained, by-products of dehydrogenation reaction will increase, so the catalyst needs to be replaced regularly.

F. Refining unit of alcohol-ketone

The cyclohexane and cyclohexanol in the dehydrogenation reaction product are separated, and the conversion of cyclohexanol is 50%, and the concentrations of cyclohexanone and cyclohexanol are both 49.5% (mass fraction).

The mixture of cyclohexanol and cyclohexanone is sent to the ketone column for rectification. The pressure on the tower top is maintained to be 5 kPa, and temperature of 70 °C; the gas phase product cyclohexanone maintains a reflux of 15% after condensation, with the rest as the cyclohexanone products. The liquid of crude cyclohexanol in the ketone tower is sent to an alcohol tower for distillation, to maintain the tower top pressure of 6 kPa, temperature of 89 °C. The gas phase product cyclohexanol maintains a reflux of 10% after condensation, with the rest being refined cyclohexanol, which returns to the last step for dehydrogenation to produce cyclohexanone. The liquid in the alcohol tower is about 1% (mass fraction) of the alcohol-ketone mixture, which can be used as fuel for export. Catalysts are commercially available and the catalyst regeneration process is also available.

From the abovementioned cyclohexanone technology of benzene selective hydrogenation of cyclohexene production, it can be seen that all the catalysts involved depend on imports. In particular, for Ru–Zn catalyst for benzene selective hydrogenation, only the transferee of the overall foreign manufacturers has the right to buy, and there is only limited supply. Therefore, it is imperative to develop the catalytic technology, key equipment, and incorporate devices of benzene selective hydrogenation to cyclohexene as well as its downstream products, with China's completely independent intellectual property.

10.2 Selective Hydrogenation of Benzene to Cyclohexene and Its Downstream Product Sets

In 2010, the catalytic technology of selective hydrogenation of benzene to cyclohexene and relative downstream products had achieved industrialization in China. A number of invention patents for selective hydrogenation of benzene to cyclohexene and relative downstream product sets are authorized by China.

10.2.1 A Unit for Selective Hydrogenation of Benzene [2]

A unit for selective hydrogenation of benzene comprises a liquid phase mixer, a gas phase mixer, a tubular reactor, a separation tank, and membrane filters. The tube feeding port of the tubular reactor is connected with the discharging port of the gas phase mixer through the first connection pipeline, and a jacket is provided outside the tubular reactor. The upper portion of the separation tank is connected with the outlet of the tubular reactor via the second connection pipeline. The top of the separation tank is connected with the hydrogen discharging pipe, and its bottom is connected to the inlet of the circulating fluid of the liquid phase mixer via the circulating pipeline. The bottom of the membrane filter is connected with the oil phase outlet above the separation tank via the reaction discharging pump. And the top of the membrane filter is connected with the liquid inlet port above the separation tank via the liquid pipeline. Product discharging port is also provided on the upper part of the membrane filter. The device is conducive to control the reaction temperature, improve reaction rate, enable the catalyst to achieve high selectivity, and improve the yield of cyclohexene.

Figure 10.2 shows a schematic diagram of a benzene selective hydrogenation device.

In Fig. 10.2, the liquid phase mixer **1** is connected to the feeding pipeline, which is equipped with many flowmeters and flow control valves. The feeding port of the gas phase mixer is connected to the feeding port of the liquid phase mixer through the feeding pump, and it is also connected to the hydrogen inlet pipe which has a hydrogen flowmeter and a hydrogen flow control valve. The tubular reactor inlet is connected with the discharging port of the gas phase mixer through the first connecting pipeline. A jacket with a temperature sensor is provided outside the tubular reactor, and inside the jacket, it is filled with a heat transfer medium. The upper part of the separation tank is connected to the discharging pipe of the tubular reactor through the second connecting pipeline. The top of the separation tank is connected with the hydrogen

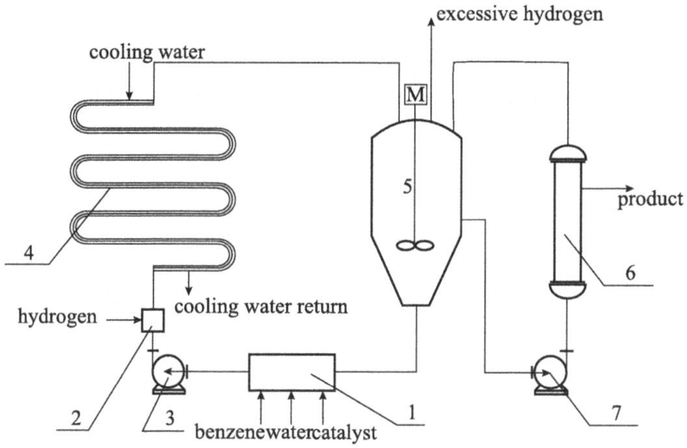

Fig. 10.2 Schematic diagram of a benzene selective hydrogenation device. 1. Liquid phase mixer; 2. Gas phase mixer; 3. Feed pump; 4. Tubular reactor; 5. Separation tank; 6. Membrane filters; 7. Solid-liquid circulation pump

discharging pipeline which is equipped with a pressure regulating valve and pressure meter. The bottom of the separation tank is connected to the circulating fluid inlet of the liquid phase mixer through the circulating line on which the flowmeter is provided. A separation tank is equipped with a level gauge and an interface meter. The bottom feeding port of the membrane filter is connected with the oil phase outlet on the separation tank through the reaction discharging pump, and the top of the membrane filter is connected to the upper inlet of the separation tank via the fluid pipeline. There are products discharging ports on the top of the membrane filter. All the valves, pumps, sensors, and interface meters in the device are controlled automatically using DCS, achieving the continuous production of selective hydrogenation of benzene, where benzene and hydrogen are continuously fed into the device, and the products flow out continuously.

The device has lots of advantages. Firstly, the feeding benzene and catalyst are premixed before they are mixed with hydrogen. Secondly, a jacketed tubular reactor is applied in this system, beneficial to control the reaction temperature, improve the reaction rate, and improve the selectivity and yield of cyclohexene. Thirdly, the separation system is a combination of sedimentation and filtration units, ensuring the complete separation of oil and water, avoiding the loss of noble metal catalyst, and achieving continuous production by the feeding pump. In this system, a variety of parameters could be controlled in the optimized range, realizing full contact between benzene, catalyst, and hydrogen. Besides, high selectivity and high yield of cyclohexene could be achieved by controlling the ratio of benzene and catalyst. The physical and chemical properties of the reaction system are fully considered when choosing the fabrication materials of the tubular reactor, where stainless steel which is equipped with Teflon or Hastelloy alloy in its inner wall to avoid catalyst poisoning and extends the service life of the catalyst is used in this system.

10.2.2 The Reaction Devices and Technology of Selective Hydrogenation of Benzene to Cyclohexene [3, 4]

The reaction devices of selective hydrogenation of benzene to cyclohexene include a benzene refiner used for purification of raw benzene, a gas-liquid mixer for mixing hydrogen and catalyst slurry, a static mixing reactor for selective hydrogenation reaction, and a separation tank for the separation of the reaction products and catalysts from static mixing reactor and so on. Raw benzene will be fed into benzene refiner for purification before it is heated to 100–130 °C. The catalyst slurry and hydrogen in a certain proportion are fed into the gas-liquid mixer with a certain pressure for fully mixing. Then the purified benzene and the catalyst slurry after fully mixing are introduced into a static mixing reactor, where the reaction temperature, reaction pressure, and liquid flow rate are controlled precisely so that all materials in the static mixing reactor could react to obtain a mixture containing cyclohexene. Finally, the mixture is fed into a separation tank. Compared with the previous technology, the facilities and processes are easy to control, and benzene conversion as well as the selectivity of cyclohexene become better.

Figure 10.3 is a schematic diagram of the technological devices and processes of selective hydrogenation of benzene to cyclohexene.

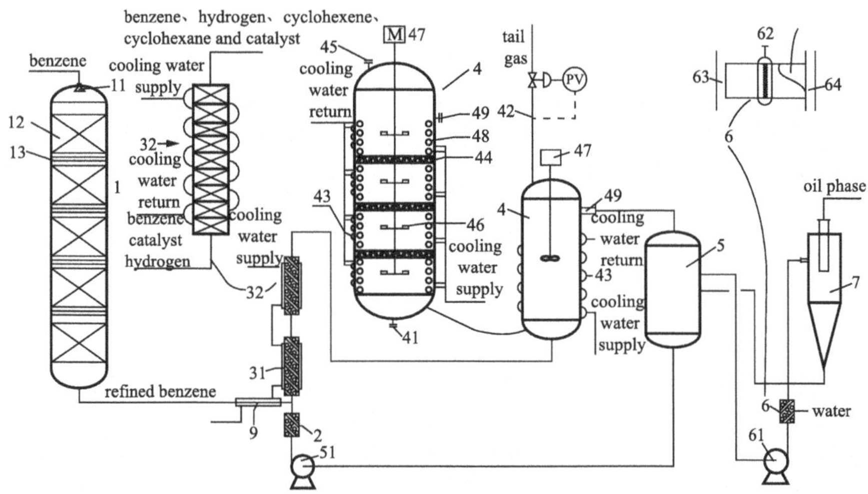

Fig. 10.3 Schematic diagram of the technological devices and processes of selective hydrogenation of benzene to cyclohexene. 1. Schematic diagram of the structure of the benzene refiner; 2. Gas-liquid mixer; 4. Auxiliary reactor; 5. Separation tank; 6. Oil-water mixer; 7. Rotary liquid separator; 9. Jacket heat exchanger; 11. High-efficiency nozzle; 12. Desulfurization fixed bed; 13. Liquid redistributor; 31. First static mixing reactor; 32. Second static mixing reactor; 41. Feed port; 42. Hydrogen discharge pipe; 43. Heat exchange coil; 44. Baffle; 45. Exhaust air vent; 46. Four stirring heads; 47. Stirrer; 48. Cooling water coil; 49. Discharging port; 51. Circulating pump; 61. Discharge pump; 62. Water inlet; 63. Oil phase inlet; 64. Outlet

In Fig. 10.3, the benzene refiner **1** is a fixed bed refiner equipped with an adsorbent filler layer **12**, which is filled with a desulfurization adsorbent. The internal inlet of the benzene refiner is equipped with a high-efficiency nozzle **11** and a liquid redistributor **13**. The liquid redistributor has four layers, which are distributed among the adsorbent packing layers. The liquid redistributor is a disk sieve structure with a pore size of 2–8 mm, and the total surface area of the sieve is 5–20% of the cross-sectional area of the benzene refiner. A jacket heat exchanger **9** is provided outside of the outlet of the benzene refiner, which is connected to the inlet of the first static mixing reactor **31**.

The gas-liquid mixer **2** is equipped with a hydrogen inlet and a catalyst slurry inlet for mixing hydrogen and catalyst slurry, and its outlet is connected to the static mixing reactor.

Static mixing reactors **31**, **32**, which are the main reaction zones for selective hydrogenation of benzene, consist of two static mixing reactors serially. The static mixing reactor is the "SV" type, which is filled with a "V"-type filler. Outside of the static mixing reactor is the coil heat exchanger for heat exchange which could be replaced by a jacketed heat exchanger. The gas-liquid mixture enters the first static mixing reactor **31** from its bottom, then enters the second static mixing reactor **32**. The outlet of the second static mixing reactor **32** is connected to the material inlet of the auxiliary reactor. The temperature is controlled efficiently by filling cooling water into the coil or jacketed heat exchanger.

In order to inhibit the generation of the cyclohexane, it is required to control the ratio of benzene/slurry to less than 2, or the residence time to less than 20 min, both of which could be realized by adapting 2–4 static mixing reactors in series. The physical and chemical properties of the materials were taken into consideration when designing the inner structure of the static mixing reactors in order to achieve sufficient mixing of the reaction materials.

In order to make full use of the temperature difference between the cooling water in different devices and to reduce the energy consumption, cooling water of 90–115 °C will be fed into the coil of the second static mixing reactor firstly, and after heat exchange, the cooling water of 100–125 °C will enter the coil of the first static mixing reactor. Then after heat exchange, the cooling water of 115–145 °C will enter the inlet of the jacketed heat exchanger from the outlet of the coil heat exchanger of the first static mixing reactor, heating the refined benzene to 120 °C.

The height/diameter ratio of the auxiliary reactor **4** equipped with a stirrer **47** is 2:1. There are four stirring heads **46** on the stirrer, which are spaced by the baffle layers. The feeding port **41** is connected to the discharging port of the second static mixing reactor **32**, while the discharging port **49** is connected to the inlet of the separation tank. At the top of the auxiliary reactor, there is an exhaust gas vent **45**, which is connected to the hydrogen discharging pipe **42** through the pipeline (the discharging pipe **42** is provided with a pressure regulating valve and a pressure gauge). Outside and inside of the auxiliary reactor **4**, there are heat transfer coil **43**. For cooling water coil **48**, it is placed along the inner peripheral wall of the auxiliary reactor. The heat transfer coil and cooling water coil could not only ensure the desired conversion of the reaction, but also control the reaction temperature accurately to avoid the deep hydrogenation. The heat exchange coil **43** could be replaced by a jacketed heat exchanger. Three baffles **44** for preventing back-mixing of the reaction liquid are provided in the auxiliary reactor. The diameter of the hole on baffle **44** is 20–40 mm and the total area of the holes is 10–30% of the surface area of the baffle. The baffle **44** includes a main plate and the cover plate, where there are through-holes in the middle of the main plate. The size of the through-hole matches the size of the stirring head on the stirrer, and the cover plate covers the through-holes. The shaft holes in the middle of the cover plate are designed for the stirring shaft passing through the cover plate, which are beneficial for the installation of the stirrer **47**.

Separation tank **5** is used for separating the reaction products and catalysts sent by the static mixer. The separation tank is provided with an oil phase outlet at the upper part and a catalyst outlet at the bottom, which is connected to the inlet of the gas-liquid mixer via the circulating pump **51**. The inlet of the separation tank is connected to the outlet of the auxiliary reactor **4**. The middle of the sidewall is also equipped with a catalyst inlet. The inner part of the separation tank is provided with a liquid guide tube (height/diameter ratio is 2:1–3:1), of which the bottom is of a cone shape. The flow rate of the circulating slurry from the bottom could be controlled by the circulation pump.

The oil-water mixer **6** contains two inlets, one of which is connected to the water pipe inlet **62** and the other one is connected to the oil phase inlet of the separation tank **63** via the discharging pump **61**. The outlet of the oil-water mixer **64** is connected to the inlet of the rotary separator **7**. In addition to supplying water for the reaction system, the function of the oil-water mixer is to separate the catalysts entrained in the oil phase during sedimentation and recover $ZnSO_4$ in the oil phase.

The rotary separator **7** with the height/diameter ratio of 5:1 is used to separate the catalysts from the oil phase. The catalyst outlet at the bottom of the rotary separator is connected to the middle catalyst inlet of the separation tank **5**, and the oil outlet at the top of **7** is connected to the product separation system.

All equipment in contact with the catalyst, such as auxiliary reactor, separation tank, and rotary separator, are fabricated with Hastelloy alloy to avoid metal ions entering the reaction slurry and prevent catalyst deactivation, as well as avoid catalyst poisoning to prolong catalyst time scale.

10.2.3 Technology for Partial Hydrogenation of Benzene Which Could Recover the Catalyst [5]

A technology for partial hydrogenation of benzene which could recover the catalyst is as follows. Raw materials benzene, hydrogen, catalyst, and additives enter the reactor first. After the reaction products enter the oil-water separator, the reaction liquid of the oil phase (which contains the catalyst and water phase) enters the flash tank (an oil and water separation tank is provided behind the flash tank). Another line (the first line) on the pipeline which is discharged from the flash tank to the wastewater treatment system is connected to the oil and water separation tank. Then the reaction fluid was pumped to the catalyst regeneration tank via the centrifugal pump. After that, together with the regenerated catalyst, it is returned to the reactor through the metering pump. The second pipeline on the outlet of the centrifugal pump (which is placed behind the oil and water separation tank) is connected to the inlet of the flash tank. The flash tank, oil and water separation tank, and pipes, as well as the pumps, are made of stainless steel lined with polytetrafluoroethylene (PTFE). This process could reduce the losses of the catalyst and additives (in other words, save a lot of catalysts and additives) and decrease the production cost effectively.

The current selective hydrogenation of benzene to cyclohexene processes can be summarized as follows. The fresh benzene together with the recirculated benzene, hydrogen, and the catalyst enters the reactor first. With stirring, four phases of gas-oil-water-solid (ruthenium-based noble metal nanosized catalysts) are mixed sufficiently. Besides, it is required to add a zinc sulfate solution into the reactor to adjust the pH to maintain the best working condition of the catalyst. After the reaction, the products enter the separation tank for gas-liquid separation and oil-water separation, and the aqueous phase containing the catalyst withdraws from the bottom of the separator (most of which is recycled to the reactor for recycling, and a small portion is fed to the catalyst regeneration system for continuous regeneration). The oil phase containing cyclohexene, cyclohexane, and benzene overflows from the upper part of the separator and enters the flash tank to achieve the full separation of the gas (trapped in the oil phase) and the reaction fluid. The small amount of catalyst and water carried by the oil phase is further separated here, where water is treated as wastewater and discharged to the wastewater treatment system directly. Some problems such as the huge consumption of the catalyst exist. Normally, especially during the system starting or stopping or when the hydrogenation process is unstable, catalyst loss is around 5%. If the amount of catalyst is 500 kg/a and the catalyst price is 150,000 RMB/kg, the economic loss caused by the catalyst loss will be around 3.75 million RMB each year. At the same time, the wastewater containing lots of zinc sulfate not only results in resource waste, but also increases the difficulty for wastewater treatment as well as the operation costs. The present invention is characterized by adding an oil-water separation tank and two pipelines behind the flash tank, which could effectively solve the problem above.

Figure 10.4 is a schematic illustration of the technology for partial hydrogenation of benzene which could recover the catalyst.

As shown in Fig. 10.4, after the raw materials benzene, hydrogen, catalyst, and the additive zinc sulfate solution entered reactor **1** and reached the residence time of benzene (15–20 min), the reaction product entered the oil-water separator **2**, achieving gas-liquid separation and oil-water separation. Considering that the aqueous solution containing the catalyst and the additive zinc sulfate would enter the flash tank **3** along with the oil phase, an oil and water separation tank **4** was placed after the flash tank **3**. On the pipeline which was discharged from the flash tank **3** to the wastewater treatment system, the first pipeline **5** was connected to the oil and water separation tank **4**. Then the reaction fluid was pumped to the catalyst regeneration tank **7** via the centrifugal pump **6**. After that, together with the regenerated catalyst, the catalyst was returned to the reactor **1** via the metering pump **8**. The

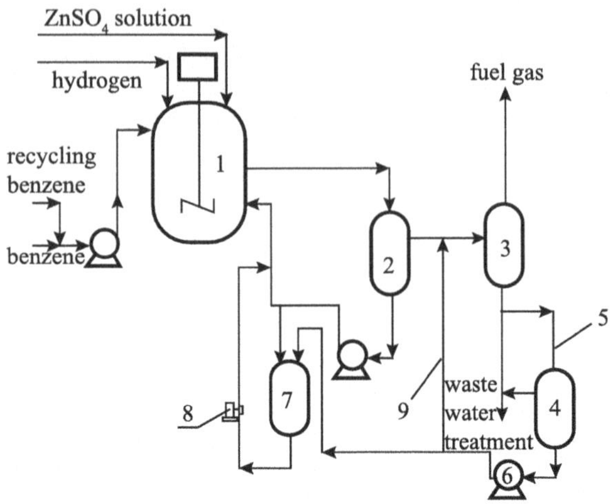

Fig. 10.4 Schematic illustration of the technology for partial hydrogenation of benzene which could recover the catalyst. 1. Reactor; 2. Separator; 3. Flash tank; 4. Oil-water separation tank; 5 First pipeline; 6. Centrifugal pump; 7. Catalyst regeneration tank; 8. Metering pump; 9. Second pipeline

second pipeline **9** which was connected to the inlet of the flash tank was placed at the outlet of the centrifugal pump **6** (which was behind the oil-water separation tank **4**) to effectively prevent the oil phase (from the flash tank **3**) entering the subsequent catalyst regeneration tank **7**. When a large amount of the reaction liquid entered the oil-water separation tank **4**, the reaction materials from the centrifugal pump **6** were fed to the flash tank **3** instead of to the catalyst regeneration tank **7**. When a small amount of the reaction liquid entered the oil-water separation tank **4**, the separated water could be discharged into the wastewater treatment system via the overflow port. Hastelloy alloy was used to fabricate the reactor **1**, the catalyst regenerator **7**, and other parts which have contact with the catalyst directly. For flash tank **3**, the newly added pipelines, the oil-water separation tank **4**, and the pumps are made of stainless steel lined with polytetrafluoroethylene (PTFE).

10.2.4 A Gas-Liquid-Liquid-Solid Reaction Device [6]

A gas-liquid-liquid-solid reaction device includes a cylindrical cylinder, a draft tube, a stirrer, a heat transfer jacket, heat exchange coil, gas-liquid distributor, baffle, oil phase outlet, overflow weir, etc. Both top and bottom of the cylindrical cylinder are equipped with heads. Inside of the cylinder, there is a guide tube, which is equipped with a stirrer inside the guild tube (the stirrer has three blades). A heat exchange jacket is placed outside of the cylinder, and a multistage heat exchanger coil is equipped inside of the cylinder. The gas phase distributor connected to the gas feeding pipe and the liquid phase distributor connected to the liquid feeding pipe are installed below the stirrer in the cylinder. A baffle is mounted in the vertical direction between the cylinder and the draft tube. An oil phase outlet is provided in the upper area which is side-closed by the cylinder, the draft tube, and the baffle. An overflow weir is located near the oil outlet in the cylinder, and on the other side of the baffle, at least one opening gaps are provided in the upper area of the draft tube. External circulation of the catalyst is unnecessary when this device is adapted, where separation and circulation of the reaction solution are achieved inside the reactor. Using this device, the catalyst performance and productivity are enhanced.

Figure 10.5 is the schematic illustration of the gas-liquid-liquid-solid reaction device for selective hydrogenation of benzene to cyclohexene.

As shown in Fig. 10.5, both top and bottom areas of cylinder **4** are provided with heads. A motor **7** is provided near the upper head, and a guide tube **5** which is provided with drafts is mounted on the cylinder. The stirring shaft equipped with three blades is connected to the motor axis. Among the three blades, the lower blade is an axial flow blade (a propeller blade is preferred), while the middle blade and upper blade are runoff blades where multiple stages of blade agitation are applied (preferably disk turbine blades). In addition to stirring, the blades could also control the flow of the reaction materials. The lower blades form upward flow, whereas the middle layer and the upper blade reinforce gas-liquid-liquid-solid mixing. A heat transfer jacket **11** and a multistage heat exchange coil **12** are provided outside of the cylinder to achieve partial heating or partial cooling as well as more uniform temperature control. The gas distributor **1** (connected to the gas inlet) and liquid distributor **2** (connected to the liquid inlet) are placed below the agitator in the cylinder. A baffle plate **13** is mounted vertically between the cylinder and the guide tube. An oil phase outlet **10** is provided in the upper part of the area which is enclosed by the cylinder, guild tube, and baffle. At least one opening gaps **6** is located on the upper section of the guild tube (in the opposite direction of the baffle). There is an exhaust outlet **8** on the upper head. A level gauge and interface meter are mounted on the cylinder. A catalyst addition port **14** and another port **15** for catalyst discharging and sampling are provided in the lower part of the cylinder. The position of the guild tube is higher

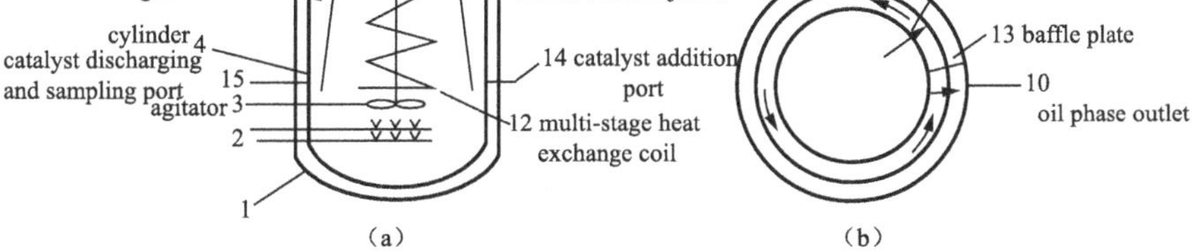

Fig. 10.5 Schematic illustration of the gas-liquid-liquid-solid reaction device for selective hydrogenation of benzene to cyclohexene. **a** Schematic illustration of the gas-liquid-liquid-solid reaction device; **b** the cross-sectional view of the reaction device; **c** the upper section of the guide tube

than the overflow weir **9**, while the bottom of the guide tube gap is below the overflow weir. The baffle is placed at the same height as the guide tube. The above structure can realize the separation of the water phase and the oil phase in the reactor, achieving catalyst recycle.

The overflow weir, guide tube, baffle, and other components in this device could control material flow, making the materials rise from the reactor bottom first, then pass through the guide tube gap to enter the area between the guild tube and the reactor wall, and finally move to the baffle along the circumferential direction. The upper oil phase overflows from the oil phase outlet **10** through the overflow weir, while the aqueous phase containing the catalyst is stratified during the process of circumferentially moving, and then it moves along the guild tube, returning to the reactor bottom. The guide tube is cone-shaped, where the upper part is far from the reactor wall while the lower part is close to the reactor wall. This design could effectively prevent the accumulation of the catalyst in the lower part. Semicircular head or oval head or butterfly head, etc., could be adapted in the reactor according to the practical needs.

This device has a distinct advantage, that is, without external circulation, the catalyst could be separated from the reaction solution inside the reactor and be recycled, improving the catalytic effect and productivity. The reactor is divided into two parts by the guild tube: the internal reaction zone and the settling zone between the exterior area and the reactor wall. Thus the settling and separation facilities as well as the circulation pump for catalyst recirculation are unnecessary, and can save equipment investment and reduce energy consumption as well as facilities' maintenance/management costs. Multilayer blades are adapted for the agitator. Different blades display different functions. The axial-flow propeller blades are applied to the lower blades to promote the catalyst and the material flow upward from the lower part of the reactor, while runoff-type disk turbine blades are adopted for the middle and upper blades, which could promote gas dispersion and strengthen the mixing of gas and liquid to facilitate the reaction. The cone-type guide tube is also provided, and the upper liquid moves along the reactor in the circumferential direction to increase the settlement time of the catalyst. After the upper liquid moves for almost one lap of the reactor, the aqueous solution containing the catalyst settles down to the lower part of the guild tube, and the reaction fluid is discharged from the discharging outlet through the overflow weir. The lower part of the guide tube is close to the reactor wall to prevent the accumulation of the catalyst. The separation of the catalyst and reaction fluid is carried out inside the reactor owing to the guild tube. By the combination of different types of agitator blades, the guild tube, and the baffle, the material flow could be controlled. On the one hand, the reaction fluid moves upward from the reactor bottom, and after it moves along the guild tube for one lap, it reaches the baffle, and finally is discharged via the overflow weir. On the other hand, the aqueous phase containing the catalyst moves to the upper part of the reactor from the bottom of the reactor, and then moves through the guide tube to settle down at the bottom of the guild tube to achieve internal circulation.

10.2.5 A Method to Produce Cyclohexene Using High-Purity Benzene [7]

A method to produce cyclohexene using high-purity benzene as raw material includes hydrogenation of high-purity benzene to cyclohexene and subsequent separation/purification. For hydrogenation, a ruthenium-containing catalyst with highly dispersed zirconium oxide is used. High-purity benzene (>99.9%, the content of sulfur is less than 5 ppm) is supplied as raw material. The concentration of product cyclohexene is higher than 99.5%, and the content of methyl cyclopentene is lower than 1000 ppm. Using high-purity benzene as raw material and controlling the sulfur content in hydrogen/benzene could avoid catalyst poisoning and extend the time scale of the catalyst to achieve long-term operation of the device. The reaction solution is dehydrated prior to the rectification to avoid the hydrolysis of the extraction agent, ensuring the extraction ability. The energy-saving design of the new partition wall distillation tower is used for rectification, which changes the traditional two-time extraction of benzene, cyclohexane, and cyclohexene. After pre-distillation, the separation of three components is achieved by once distillation, and the energy consumption is greatly reduced.

Figure 10.6 is a schematic diagram of cyclohexene production using high-purity benzene as raw material.

As shown in Fig. 10.6, R1 is a hydrogenation reactor, which is equipped with an agitator (one, or two, or more agitators could be used serially). High-purity benzene, hydrogen, and reaction slurry containing the catalyst enter the reactor. Under the reaction condition of 133–150 °C, 4.7–5.5 MPa, and optimum residence time of benzene, the desired benzene conversion could be reached, producing the maximum concentration of cyclohexene and by-product cyclohexane. The materials are then

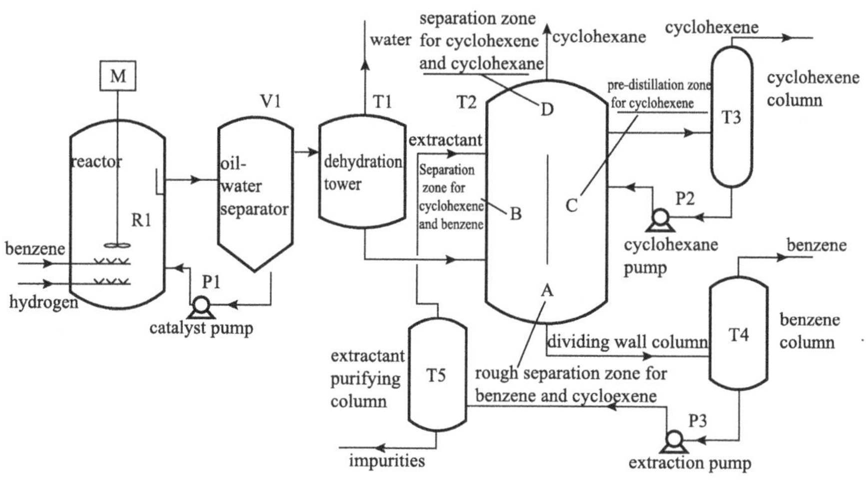

Fig. 10.6 Schematic diagram of cyclohexene production using high-purity benzene as raw material

fed into oil-water separator Vl to separate the oil phase containing cyclohexene, cyclohexane, and benzene from the aqueous phase containing the catalyst. After that, the oil phase enters the dehydration tower **Tl**, and the water phase returns to reactor **Rl** by pump **Pl**. The process is as follows:

(1) Partial hydrogenation of high-purity benzene to cyclohexene: in the presence of hydrogenation catalyst, high-purity benzene reacts with hydrogen to produce cyclohexene (133–150 °C, 4.7–5.5 MPa). To avoid catalyst poisoning, keep the catalyst activity/selectivity going, prolong the catalyst service life, and realize the long-time, stable operation of the device, benzene purity should be no less than 99.9% (mass fraction), and its total sulfur content should be no more than 10 ppm (for hydrogen, its purity is no less than 95% (volume fraction) and the sulfur content is no more than 0.1 ppm).
(2) Separation and purification of cyclohexene: first, the oil phase is separated from the water phase, then the cyclohexene in the oil phase is separated from the unreacted benzene and by-product cyclohexane. An appropriate settlement section is provided within the reactor to achieve oil and water separation by gravity settling. Separation can also be accomplished by concatenating a separated oil-water separator after the reactor, or by using an appropriate filter. The catalyst content of the oil phase is no more than 10 ppm. Prior to extractive distillation, the oil phase is dehydrated by a stripping water tower to make the water content less than 500 ppm, avoiding hydrolysis of extraction agent and keeping long-term extraction ability.

Single column, 4-column, and 3-column distillation units could be used for extractive distillation. Figure 10.6 shows a new single distillation tower with a partition wall, which is energy-saving. Prior to entering the rectification column **T2** with a partition wall, the oil phase enters dehydration column **T1** to dehydrate. After dehydration, the moisture content of the oil phase is lower than 500 ppm, preferably lower than 50 ppm (and lower than 5 ppm is most preferable). The dehydrated oil phase enters the rectification column **T2** for extraction. It is firstly pre-distilled in zone **A** to achieve rough separation of benzene and cyclohexane. The extractant is added from the middle of zone **D** and the heat for separation is provided by the reboiler in the bottom of zone **B**. Pre-distillation of cyclohexene is performed in zone **C**, where the top is rich in cyclohexane and the bottom is rich in benzene, and the middle part is rich in cyclohexene. Cyclohexene and the extractant are withdrawn from the middle of zone C, and then they are sent into the cyclohexene column **T3**, where cyclohexene is obtained at the top of the column. Zone **B** is used for the separation of cyclohexene and benzene. Benzene and the extractant in the bottom of zone **B** enter the benzene separation tower **T4**, where the purified benzene is obtained at the top and then it is recycled to the hydrogenation reactor. The purified extractant is obtained at the bottom of tower **T4**, and then it is recycled to zone **D** for reuse. The separation of cyclohexene and cyclohexane is achieved in zone **D**, where pure cyclohexane is obtained in the top and it could be used as a by-product after hydrorefining. Part of the extractant from the bottom of tower **T4** enters the refining tower **T5** to remove the impurities to maintain the extraction capacity of the extractant.

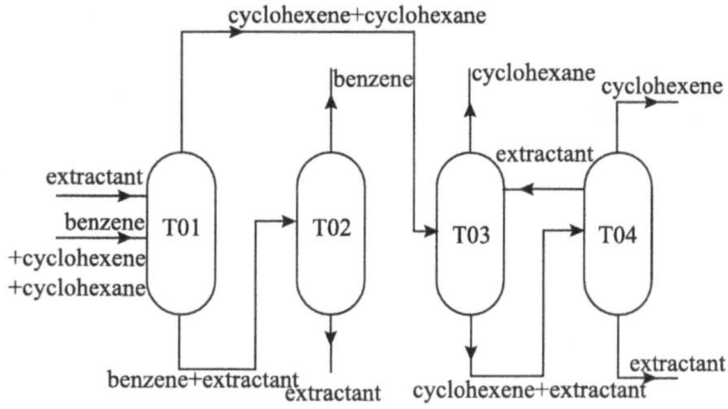

Fig. 10.7 Schematic illustration of a conventional 4-column extractive distillation system

When the production scale is small or the energy cost is low, the traditional secondary extraction method could be used to separate benzene, cyclohexene, and cyclohexane.

Figure 10.7 is a schematic illustration of a conventional 4-column extractive distillation system.

As shown in Fig. 10.7, benzene, cyclohexene, and cyclohexane could be separated by extractive distillation in column **T01**. Benzene and the extractant could be separated in tower **T02** by the conventional distillation, and cyclohexane and cyclohexene could be separated by extractive distillation in column **T03**. Using conventional distillation, cyclohexene and the extractant could be separated in column **T04**.

When the production scale is large, to facilitate operation, the four zones of **A**, **B**, **C**, and **D** in the single distillation tower with a partition wall could be redesigned as a combination of four distillation columns (**T2A**, **T2B**, **T2C**, and **T2D**), which could improve the operation stability further.

Figure 10.8 is a schematic illustration of the extractive distillation system after the extractive distillation column with a partition wall was split into four distillation columns.

As shown in Fig. 10.8, benzene, cyclohexene, and cyclohexane could be separated by extractive distillation in column **T2A**, and benzene and the extractant could

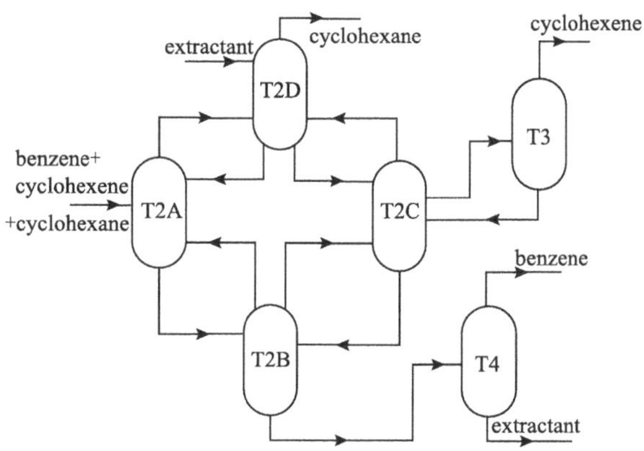

Fig. 10.8 Schematic illustration of the extractive distillation system after the extractive distillation column with a partition wall was split into four distillation columns

be separated by conventional distillation in column **T2B**. Cyclohexane and cyclohexene could be separated by extractive distillation in column **T2C**. Cyclohexene and the extractant enter column **T3** and they are separated by conventional distillation. Cyclohexane could be obtained at the top of column **T2D**. Because cyclohexene is extracted twice by distillation in this system, the energy consumption is high. To solve this problem, column T2B could be split into two parts and then combined with T2A and T2C, respectively. Hence, the twice extraction could be turned into a single extraction, only with a reboiler and a condenser needed, reducing the energy consumption greatly.

Figure 10.9 is a schematic diagram of splitting zone **T2B** into two parts and combining them with **T2A** and **T2C** towers.

In Fig. 10.9, **T2A (T2B)** column separates benzene, cyclohexene, and cyclohexane by extractive distillation. The mixture of cyclohexene and cyclohexane is obtained at the top of column **T2A (T2B)**, and benzene and the extractant are mainly at the bottom. The mixture of cyclohexene and cyclohexane enters **T2D**. After the extractant is added to **T2D**, cyclohexane is obtained at the top, then it enters **T3** tower and residual cyclohexene could be obtained at the top of tower **T3**. The main components in the bottom of column **T2D** are cyclohexene and the extractant, which subsequently enter tower **T2C (T2B)**, and a small amount of benzene containing the extractant from the bottom of **T2D** returns to tower **T2A (T2B)**. Cyclohexene is obtained at the top of **T2C (T2B)**, and the extractant is obtained at the bottom of the column, which will be combined with benzene and the extractant from **T2A (T2B)**, and then it enters column **T4**. In column **T4**, pure benzene is obtained at the top of the column, and the extractant is obtained at the bottom, which will be reused further.

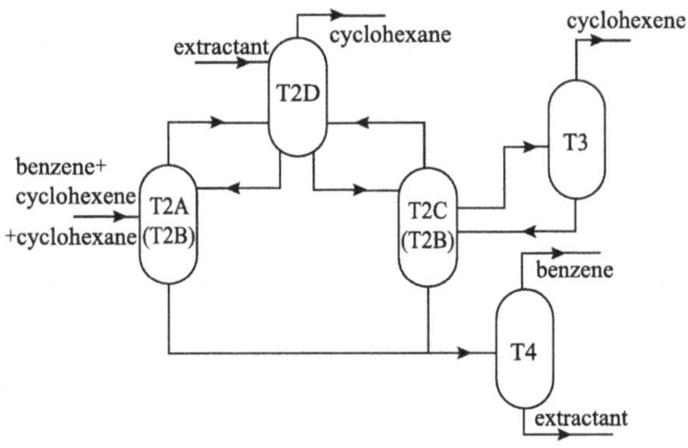

Fig. 10.9 Schematic illustration of splitting zone **T2B** into two parts and combining with **T2A** and **T2C** towers

10.2.6 A Method for Continuous Production of Cyclohexene [8]

The entire process for the continuous production of cyclohexene includes four steps: hydrogenation of benzene, separation of reaction products and catalyst, separation and purification of cyclohexene, and regeneration of the catalyst. In the step of regeneration of the catalyst, the proportion of the regenerated catalyst is controlled to be 1–5%. This method overcomes the shortcomings of former technologies in which the reaction is maintained only by controlling the operating conditions and the catalyst is not regenerated. While in this a new method, the catalyst could be regenerated continuously according to the proportion. The adsorbed oil, hydrogen, and other poisoning impurities could be removed through the regeneration of the catalyst, and the activity of the catalyst could be recovered. All the catalysts were regenerated within the time when catalyst performance is not affected, so that the catalysts are always of high reactivity and high selectivity, achieving long-time, continuous production of cyclohexene.

Figure 10.10 shows a schematic diagram of the continuous production of cyclohexene.

As shown in Fig. 10.10, the whole process includes benzene hydrogenation reaction, separation of reaction products and catalyst, separation and purification of cyclohexene, and regeneration of the catalyst. The amount of regenerated catalyst for each batch is around 1–5%, and all the catalysts are regenerated in a certain period of time to maintain the high activity and selectivity of the hydrogenation reaction.

(1) Benzene hydrogenation reaction: Benzene and hydrogen are selectively hydrogenated to cyclohexene and cyclohexane at the presence of a ruthenium catalyst. The conversion of benzene is no less than 40% and the selectivity of cyclohexene is no less than 75%.

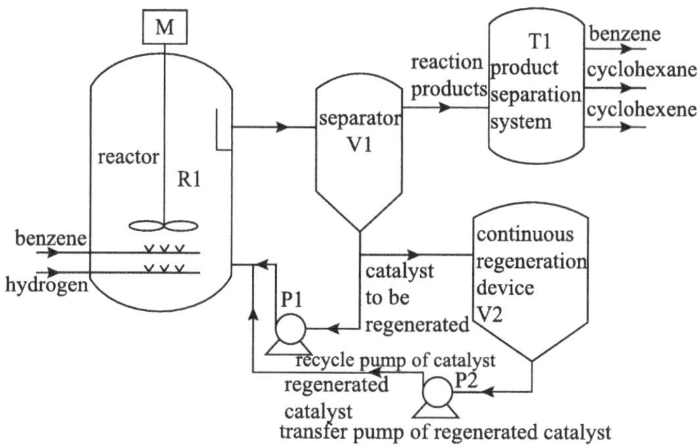

Fig. 10.10 Schematic diagram of continuous production of cyclohexene

(2) Separation of reaction products and catalyst: The reaction products and the catalyst are separated by sedimentation or centrifugation into an aqueous phase containing the catalyst and oil phase. The oil phase enters the cyclohexene separation parts for purification. Most of the water phase is returned directly to the reactor and a small part is sent to the catalyst regeneration part.
(3) Separation and purification of cyclohexene: The reaction products include cyclohexene, cyclohexane and reactive benzene, and they are separated by extractive distillation, and benzene is recirculated to reactors for further use while cyclohexene and cyclohexane are treated as products and by-products.
(4) Catalyst Regeneration: A small part of the catalyst from the sedimentation separation section is subjected firstly to gas stripping, oxidation, and hydrothermal stabilization to remove entrained oil, hydrogen, and other impurities from the catalyst to restore the activity of the catalyst, and then is pumped back to the reactor for further use. The catalyst slurry separated from the reaction mixture firstly enters the air stripper and then is stripped with nitrogen to remove the dissolved/entrained oil in the catalyst slurry. Gas stripping could be carried out directly under the pressure of 0.1–0.6 MPa and temperature of 60–110 °C after the separation of the catalyst slurry. The volume ratio of inert gas to catalyst slurry is about 0.5–1.5. After gas stripping, most of the oil and hydrogen are removed and then the slurry is sent to the oxidation tank to further oxidize the residual oil and excessively adsorbed hydrogen on the catalyst surface with oxygen-depleted air or hydrogen peroxide. In order to prevent the catalyst from being oxidized, the reaction conditions and the oxygen concentration must be strictly controlled. The oxygen content in the oxygen-depleted air is 3–6% (volume fraction), and the volume ratio of the oxygen-depleted air to the catalyst slurry is about 10–100, or employing the hydrogen peroxide (concentration of 15–30%) to oxidize the slurry under the pressure of 0–0.2 MPa and temperature of 50–110 °C. After oxidation, hydrogen is removed in the form of water, and the small amount of remaining organic compounds are also oxidized. In order to stabilize the catalyst, hydrothermal treatment is required. Hydrothermal treatment includes two steps: gas stripping and hydrothermal stabilization. Nitrogen is still employed in the gas stripping process and the volume ratio of nitrogen and catalyst slurry is 0.5–1.5. The hydrothermal stabilization process is carried out at a temperature close to the temperature of hydrogenation, preferably 120–160 °C, to stabilize the catalyst and prolong the catalyst regeneration cycle. The activity of the catalyst after the above treatment is maximally recovered and could be returned to the hydrogenation reaction system directly by the pump.

10.2.7 A Method of Producing Caprolactam Using High-Purity Benzene [9]

Caprolactam, one of the important downstream products of cyclohexene, is an important monomer for the production of polyamide 6 and nylon 6 engineering plastics. It has excellent thermal stability, processability, mechanical properties, and chemical resistance and is widely used in the fields like automobiles, ships, medical products, daily necessities, electrical and electronic components, etc. Production of caprolactam by selective hydrogenation of benzene to cyclohexene is a resource-saving and environment-friendly method.

Figure 10.11 shows a process diagram of the production of caprolactam using high-purity benzene as raw material.

As shown in Fig. 10.11, the whole process includes several sections: selective hydrogenation of benzene to cyclohexene (**A**); separation and purification of cyclohexene (**B**); hydration of cyclohexene to cyclohexanol (**C**); separation and purification of cyclohexanol (**D**); cyclohexanol dehydrogenation to cyclohexanone (**E**); cyclohexanone refining (**F**); cyclohexanone ammoximation to cyclohexanone oxime (**G**); refining of cyclohexanone oxime (**H**); rearrangement of cyclohexanone oxime to caprolactam (**I**); and caprolactam refining (**J**).

A. Selective hydrogenation of benzene to cyclohexene. Benzene was selectively hydrogenated to cyclohexene in a heterogeneous catalytic system containing Ru–Zn catalyst, additives, and dispersants under 135–150 °C, 4.8–5.5 MPa, and a suitable stirring speed. The conversion of benzene was controlled higher than 40%, and the selectivity of cyclohexene was higher than 80%, and the by-product was cyclohexane. The purity of benzene was greater than 99.95% (mass fraction), and the content of sulfur was less than 5 ppm and the content of toluene was not greater than 100 ppm. The purity of hydrogen was greater than 99% (volume fraction) with the content of sulfur less than 50 ppb. After flash evaporation, a small amount of catalyst and additives entrained in the oil phase was recycled.
B. Separation and purification of cyclohexene. This process included 6 columns, which were columns for dehydration of oil phase, benzene removal, benzene recovery, cyclohexene separation, cyclohexene refining, and solvent refining. After the oil phase containing cyclohexene, cyclohexane, and benzene was dehydrated, benzene removal, cyclohexene separation, refining of cyclohexene and cyclohexane, and recovery of benzene and the extractant were achieved by extractive distillation to obtain the product of cyclohexene with purity higher than 99.8% (mass fraction), a solvent content of less than 0.005% (mass fraction), and a by-product of cyclohexane with purity of more than 99.9% (mass fraction).
C. Hydration of cyclohexene to cyclohexanol. Hydration of cyclohexene to produce cyclohexanol was carried out under 110–130 °C and 0.5–0.8 MPa employing 1–3 levels of the reactor, preferably level 2 and a molecular sieve-based catalyst, preferably ZSM-5.
D. Separation and purification of cyclohexanol. This section included three columns: cyclohexanol separation, cyclohexene removal, and cyclohexanol refining. After the separation and refining of cyclohexanol, the purity of the reaction liquid of hydration was greater than 99.5% (mass fraction) with the light fraction less than 0.3% (mass fraction).

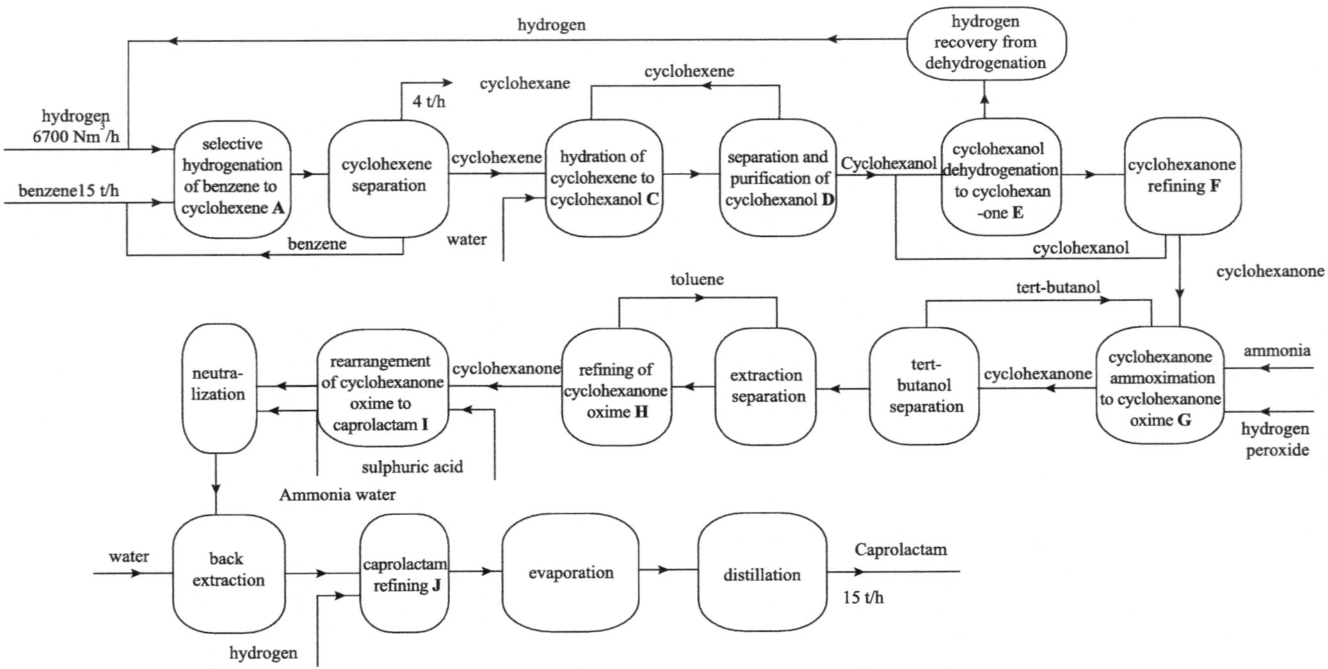

Fig. 10.11 Schematic diagram of caprolactam production using high purity benzene as raw material

E. Cyclohexanol dehydrogenation to cyclohexanone. The reaction was carried out in a fixed-bed tube reactor where the catalyst was packed in the tube and the heat conducive oil was introduced outside. At the presence of dehydration catalyst, dehydrogenation of cyclohexanol to cyclohexanone and hydrogen was performed at 210–240 °C and 0.01–0.1 MPa. Hydrogen was separated from the reaction liquid through a gas-liquid separator and then was recycled for reuse.
F. Cyclohexanone refining. This section included drying tower, light component separation tower, alcohol tower, and ketone tower. After light component separation, cyclohexanone separation, and cyclohexanol separation, the purity of cyclohexanone in the dehydrogenation reaction liquid was more than 99.95% (mass fraction), and the heptanone content was less than 200 ppm. The hydrogen recovery rate could reach more than 80% by employing a pressure swing adsorption device.
G. Cyclohexanone ammoximation to cyclohexanone oxime. The reaction solution was mixed with tert-butanol solution, hydrogen peroxide, and ammonia under the temperature of 75–83 °C and pressure of 0.35–0.45 MPa. At the presence of the titanium-silicon molecular sieve catalyst, cyclohexanone, ammonia, and hydrogen peroxide reacted to produce cyclohexanone oxime. After the reaction, the liquid was separated by filtration to separate the catalyst-free supernatant and the catalyst-containing solution. The catalyst-free solution was first separated into the gas phase and liquid phase by flash evaporation, then it entered the tert-butanol rectification column, where tert-butanol and water were obtained at the top of the column, and the cyclohexanone oxime-water solution with tert-butanol content less than 100 ppm was obtained at the bottom of the column.
H. Refining of cyclohexanone oxime. Cyclohexanone oxime and wastewater were separated using toluene as extractant. The solution containing cyclohexanone oxime and toluene entered the first distillation column firstly, where toluene was separated at the top of the tower, and the solution containing toluene and oxime (toluene fraction of 20–40%) was obtained at the bottom of the column. Then the mixture at the bottom entered the second distillation column and the cyclohexanone oxime was obtained at the bottom with purity greater than 99.9%. The reaction tail gas entered the exhaust emission cooler to condensate and recover the tert-butanol and ammonia, and the non-condensable gas entered the exhaust gas recovery tower for washing to absorb ammonia. The non-condensable gas at the top was de-composited or incinerated to reach the discharge standards and then emitted. The wastewater obtained by extraction and separation and wastewater obtained by distillation were stripped using steam in a wastewater stripping tower, where toluene in the wastewater was recovered at the top, and the content of toluene in wastewater after stripping was not more than 1%.
I. Rearrangement of cyclohexanone oxime to caprolactam. Cyclohexanone oxime entered the rearrangement reactor according to a certain proportion, and caprolactam was produced at the presence of fuming sulfuric acid in the rearrangement reactor (preferably secondary rearrangement reactor). Cyclohexanone oxime entered the first and second rearrangement reaction tank at the ratio of 4:1. The

catalyst was 102% fuming sulfuric acid with 8% of free SO_3. The ratio of fuming sulfuric acid to cyclohexanone oxime was 1.2:1–1.5:1. The temperature of the first-order rearrangement reaction was 100–110 °C, and the temperature of the second-order rearrangement reaction was 100–130 °C under atmospheric pressure. The product of crude caprolactam and oleum was neutralized with 20% aqueous ammonia, forming crude caprolactam and ammonium sulfate aqueous solution. The crude caprolactam ammonium sulfate solution was delaminated in the separator. The 70% (mass fraction) of crude caprolactam aqueous solution was obtained at the upper layer then was overflowed into the buffer tank, and was sent to the extraction system subsequently. The 39% (mass fraction) of the aqueous solution of ammonium sulfate was obtained at the lower layer, and then was sent to the extraction column to recover the caprolactam. In the ammonium sulfate extraction column, caprolactam was extracted using an extraction agent (preferably benzene, but not limited to benzene). Ammonium sulfate solution entered from the top of the column, and benzene entered from the bottom of the column which was separated at the top of the column. The resulting benzene-caprolactam solution phase entered the caprolactam extraction column via the overflow. The ammonium sulfate solution from the bottom of the column contained a trace amount of benzene, which was pumped to an ammonium sulfate stripper to remove traces of benzene. The operating temperature on top of the tower was 45 °C and the operating pressure was 10 kPa. An aqueous solution of 70% (mass fraction) of crude caprolactam was obtained from the top of the aqueous-ammonium sulfate separator and was pumped to the top of the caprolactam extraction column. The bottom liquid of the caprolactam extraction column was discharged to the condensate stripper. At the top of the caprolactam extraction column, the rising benzene extracted caprolactam from the aqueous solution and then overflowed into the pump tank for caprolactam-benzene solution and was pumped into the stripping column. The operating temperature of the top of the tower was 45 °C and the operating pressure was 10 kPa. The solution of benzene-caprolactam entered the bottom of the stripping column and the condensate from the evaporation process entered from the top of the column to extract the caprolactam from the rising droplets of benzene-caprolactam. The caprolactam with water was separated from benzene on the top of the stripping column and the caprolactam aqueous solution on the bottom entered the top of the benzene stripper with an operating temperature of 96 °C and operating pressure of 10 kPa. In order to avoid the accumulation of impurities, the extraction agent needed double-effect distillation. The extractant was pumped to the bottom of the distillation column. The heating source on the bottom was the low-pressure steam. The steam on the top was used as a heat source for the second distillation tower. The condensed extractant was recycled. The steam on the top of the second distillation column was recycled with the condensation of the heat exchanger. The operating temperature on the top of the first distillation column was 118 °C and the operating pressure was 0.2 MPa. The operating temperature on the top of the second distillation column was 80 °C and the operating pressure was atmospheric pressure.

J. Caprolactam refining. There were still trace impurities after extraction of caprolactam. The impurities would be removed through the refining unit. The refining unit at least included, but was not limited to, two steps of ion exchange and hydrogenation. The extracted caprolactam passed through an ion exchange resin column and was pumped to the hydrogenation reactor. The ion-exchanged caprolactam entered the first hydrogenation reactor and the reactants entered the second hydrogenation reactor by means of overflow. The hydrogenation temperature was 90 °C with 0.7 MPa pressure. The catalyst was Raney nickel, platinum, and so on. The hydrogenated caprolactam aqueous solution was concentrated in two steps and concentrated from 30 to 99.9%. In the first step, the caprolactam aqueous solution was concentrated to 90% in the three-effect evaporation tower. In order to reduce steam consumption, the steam distilled from the previous stage was used for the reboiler of next stage. In order to obtain a sufficient temperature difference for the reboiler, the evaporation was performed in a stepwise manner. The pressure at the top of the first-stage evaporator was 250–350 kPa, and the concentration of caprolactam at the bottom was 35–45% (mass fraction), which was discharged to the second-effect reboiler. The caprolactam solution was further concentrated to 52–60% from the steam of the primary evaporation tower and discharged to the three-stage reboiler. The pressure at the top of the second-stage evaporator was 130–240 kPa. In the three-effect evaporation tower, the caprolactam solution was further concentrated to 90% by the water vapor from the two-effect evaporation tower. The top pressure of the three-stage evaporator was 110–15 kPa. The concentrated caprolactam solution and the caprolactam flow from the bottom entered the flash tank to get caprolactam with 99.9% purity, which was sent to the distillation process, and the high boiling point impurities and traces of water were removed. In order to recover caprolactam from the bottom product of the main distillation column as much as possible, the product was subjected to two-stage distillation again. The 99.9% caprolactam was pumped to a caprolactam distillation evaporator, and 70% of the feed after evaporation at 127 °C was transferred into the caprolactam distillation separator for vapor-liquid separation. The pressure at the top of the separator was 0.5 kPa. The caprolactam vapor on the top was condensed by hot water and sent to the final tank. The bottom product was discharged to the distillation separation evaporator for residue. 70% of the feed after evaporation under the conditions of 127 °C entered the distillation separator of residue for vapor-liquid separation with the top pressure of 0.5 kPa.

10.2.8 A High Efficient Cyclohexanone Production Methods [10]

Cyclohexanone is the raw material for the preparation of caprolactam and a good organic solvent. Similar to caprolactam, it is one of the important downstream products of cyclohexene. Cyclohexanone production methods are mainly phenol method and benzene method; due to the higher price and poor availability of phenol, the current mainstream is the benzene method, accounting for more than 95% of the total capacity of the world. Using benzene as a raw material, the current methods are the traditional cyclohexane oxidation method and the cyclohexene method developed in recent years.

Figure 10.12 shows the flow chart for the production of cyclohexanone by the traditional cyclohexane oxidation process.

In Fig. 10.12, cyclohexane oxidation is a radical reaction, which is difficult to control and requires the addition of inhibitors to control benzene conversion rate to 3–5%. It causes large energy consumption resulting in a large number of emissions, thus, there are serious security risks. During saponification decomposition, a large amount of wastewater is produced. It has multiple steps, long process, benzene utilization rate of about 85%, waste of resources, and environmental pollution.

Figure 10.13 shows the flow chart for the production of cyclohexanone by benzene selective hydrogenation to cyclohexene.

In Fig. 10.13, selective hydrogenation of benzene fundamentally eliminates the potential safety hazard. The by-product cyclohexane can be fully utilized. The utilization rate of carbon atoms is nearly 100%. There is no wastewater, waste gas emission, and it is resource saving and environment friendly. Compared with the cyclohexane method, the costs can be reduced by 1000–1500 RMB per ton of cyclohexanone.

A. Benzene selective hydrogenation to cyclohexene: Cyclohexene and cyclohexane are formed from pure benzene and hydrogen at 135–145 °C and 4.5–5.5 MPa with hydrogenation catalyst. The requirements are benzene purity of greater than 99.9%, sulfur content of less than 10 ppm, non-aromatic hydrocarbons content of less than 100 ppm, hydrogen purity of greater than 99% (volume fraction), and sulfur content of less than 50 ppb.
B. Separation of cyclohexene, cyclohexane, and benzene: After the mixing of reaction liquids of hydrogenation and cyclohexane dehydrogenation, cyclohexene purity is greater than 99.8% (mass fraction) and the solvent content is less than 0.005% (mass fraction) through processes including water separation, benzene separation, benzene recovery, cyclohexene separation, cyclohexene purification, and solvent purification.
C. Cyclohexane dehydrogenation to prepare cyclohexene: The cyclohexane from Step **B** is heated and then enters a fixed-bed reactor containing an alum-based catalyst at a temperature of 420–500 °C and a pressure of 0.01–0.1 MPa. With oxidative dehydrogenation to produce cyclohexene, cyclohexene and unreacted cyclohexane are returned to Step **B** for separation after cooling and condensation of the reactants. Air is used as an oxidant to control cyclohexane conversion and reduce side reactions.
D. Cyclohexene hydration: Using a two-stage series reactor, ZSM-5 molecular sieve is used as a catalyst, under 110–130 °C and 0.5–0.8 MPa conditions. The cyclohexanol can be formed from cyclohexene and water. The conversion rate is controlled to be less than 15% to reduce side reactions.

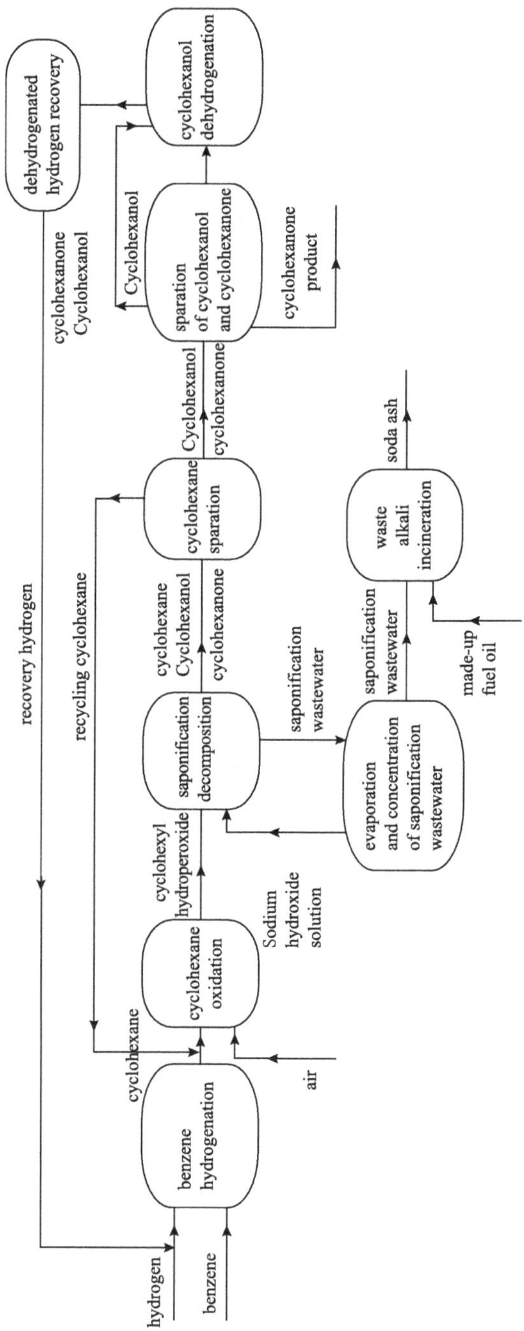

Fig. 10.12 Traditional cyclohexane oxidation process for the production of cyclohexanone

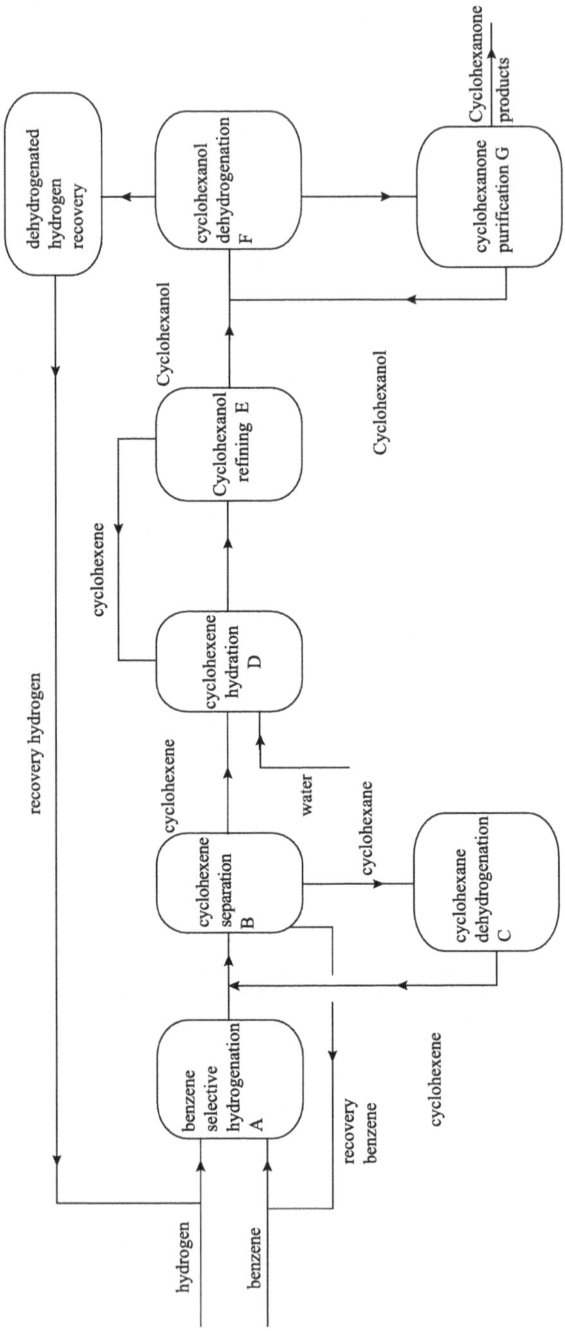

Fig. 10.13 Flow chart of benzene selective hydrogenation process to produce cyclohexanone

E. Cyclohexanol refining: It includes three columns for cyclohexanol separation, cyclohexene separation, and cyclohexanol refining. After the hydration reaction solution is separated through the three columns, the purity of the cyclohexanol is more than 99.5% (mass fraction). The content of light components is less than 0.3% (mass fraction) and the content of methylcyclopentanol is less than 0.15%. The cyclohexene is returned to Step **D**.
F. Cyclohexanol dehydrogenation to cyclohexanone: In a tubular reactor, cyclohexanone and hydrogen are produced using copper-silicon catalyst at 210–240 °C, 0.01–0.1 MPa. The conversion rate is controlled to be less than 60% and the temperature is lower than 240 °C to reduce the side reactions. The hydrogen produced by dehydrogenation is condensed, degreased, purified by adsorption, and returned to Step **A**, and the hydrogen recovery rate can reach over 90%.
G. Cyclohexanone purification: The light components, cyclohexanone, and cyclohexanol are separated from the dehydrogenation reaction liquid, and the cyclohexanone product with purity greater than 99.95% (mass fraction) is obtained through vacuum distillation. The light component is less than 100 ppm and the cyclohexanol is less than 100 ppm.

References

1. Ren, P.B., Ju, H.B., Wang, E.Q., et al.: Preparation of cyclohexanone from coking benzene. Chinese Patent, CN 100751298 (2005)
2. Yang, K.J., Yuan, X.M., Sun, S.Q., et al.: Apparatus for selective hydrogenation of benzene. Chinese Patent, CN 10210982.9 (2010)
3. Meng, Q.G., Liu, X.A., Chen, F.H., et al.: Apparatus and process for selective hydrogenation of benzene into cyclohexene. Chinese Patent, CN 10127661 (2011)
4. Meng, Q.G., Liu, X.A., Chen, F.H., et al.: Apparatus for selective hydrogenation of benzene into cyclohexene. Chinese Patent, CN 20157770.9 (2011)
5. Li, Y., Yuan, X.M., Yang, K.J., et al.: Recycling method of catalyst for partial hydrogenation of benzene. Chinese Patent, CN 10121986.9 (2012)
6. Yang, K.J., Yuan, X.M., Liu, X.W., et al.: Apparatus for gas-liquid-liquid reaction. Chinese Patent, CN 10074362 (2012)
7. Liu, X.W., Yang, K.J., Yuan, X.M., et al.: Preparation of cyclohexene from high pure benzene. Chinese Patent, CN 10176338.3 (2012)
8. Liu, X.W., Yuan, X.M., Wang, G., et al.: The method for continuous production of cyclohexene. Chinese Patent, CN 10377451.8 (2012)
9. Yang, K.J., Yuan, X.M., Liu, X.W., et al.: Preparation of caprolactam from high pure benzene. Chinese Patent, CN 10176320 (2012)
10. Liu, X.W., Li, Y., Xu, H., et al.: Production of cyclohexanone with high yield. Chinese Patent, CN 10377478.7 (2012)

The manufacturer's authorised representative in the EU is Springer Nature Customer Service Centre GmbH, Europaplatz 3, 69115 Heidelberg, Germany. If you have any concerns regarding our products, please contact ProductSafety@springernature.com

Printed and bound by CPI Group (UK) Ltd, Croydon, CR0 4YY

30/09/2026

02225504-0014